Annals of Mathematics Studies

Number 100

RECENT DEVELOPMENTS IN SEVERAL COMPLEX VARIABLES

EDITED BY

JOHN E. FORNAESS

PRINCETON UNIVERSITY PRESS

AND

UNIVERSITY OF TOKYO PRESS

PRINCETON, NEW JERSEY

1981

Published in Japan exclusively by
University of Tokyo Press;
In other parts of the world by
Princeton University Press

Printed in the United States of America
by Princeton University Press, Princeton, New Jersey

◆

Library of Congress Cataloging in Publication data will
be found on the last printed page of this book

THE FOUNDING OF THE ANNALS STUDIES

With the publication of Number 100 in the Annals of Mathematics Studies it seems appropriate to put on record an account of its beginnings. The Studies as a series had its origin during 1933-1939, when the Department of Mathematics of Princeton University and the School of Mathematics of the Institute for Advanced Study were housed together in Fine Hall. A. W. Tucker had charge of the production and distribution of mathematics lecture notes from both University and Institute. Students did the mimeographing and collating, and Miss Gwen Blake, the Institute mathematics secretary, handled the distribution. But demand grew as word of the Princeton lecture notes spread in the mathematical world. In 1936 the notes were shifted from hand production to planographing by Edwards Brothers, Ann Arbor, and in 1938, when Joseph Brandt became director of Princeton University Press, the Press agreed to handle storage and distribution. Over thirty volumes of these Princeton mathematical notes were "published" between 1933 and 1939.

With the experience and profits gained from the Princeton Mathematical Notes, the Annals of Mathematics Studies series was launched in 1940. It was, to quote from the 1939-40 annual report to the University president (H. W. Dodds) by the mathematics department chairman (L. P. Eisenhart), "a new type of mathematical publication, designed to meet the need in America for an inexpensive means of publishing material which lies between a full-fledged book and a journal-paper." As its name implied the new series was under the supervision of the editors (S. Lefschetz, J. von Neumann, H. F. Bohnenblust) of the journal *Annals of Mathematics* and aimed for the high quality represented by that journal. As with the notes, A. W. Tucker supervised production and finances; the Press handled storage, distribution, and advertising. The first three volumes appeared in 1940, by Weyl, Tukey, and Godel, and the second three in 1941, by Murray, Post, and Church.

In 1946 the Press, with D. C. Smith, Jr., director, and H. S. Bailey, Jr., science editor, accepted responsibility for the production as well as distribution of the Studies. A. W. Tucker continued as anonymous editor or "manager" until 1949, when Emil Artin and Marston Morse were named as editors.

The Studies, and preceding notes, were very much a team effort. Some of the people who helped in the early years were: A. H. Taub, J. F. Randolph, J. H. Giese, J. H. Lewis, Gwen Blake, Agnes Henry, Ellen Weber, and Ginny Dzurkoc.

Somewhat ironically, as the Studies became more established and widely known, particularly because some of the early volumes have become modern classics, some authors became reluctant to publish anything less than their most finished work in the series, in spite of its modest informal format. Thus in 1965 it became desirable to initiate a new series of notes, somewhat like the lecture notes from which the Studies grew. This was done, and the Princeton Mathematical Notes was started as an adjunct to the Annals of Mathematics Studies, with the same editors in charge of both series.

FOREWORD

This volume contains the proceedings of a conference on several complex variables, held at Princeton University, April 16-20, 1979.

The meeting was sponsored in part by the National Science Foundation.

Well over a hundred mathematicians participated and made the conference a great success. There were three one-hour lectures daily and, most days, eight half-hour seminar talks.

J. E. FORNAESS

TABLE OF CONTENTS

Recent Developments in
Several Complex Variables

PROJECTIVE CAPACITY

H. Alexander*

Introduction

We shall study a notion of capacity for compact subsets of complex projective space. Our principal motivation for this comes from a theorem of Sibony and Wong [8]. Let K be a compact circled subset of the unit sphere in C^2. One can view K as a subset of complex projective 1-space, P^1 and then, by a choice of an inhomogeneous coordinate in P^1, one can imbed C^1 into P^1 and suppose that K lies in the complex plane. If then K has positive logarithmic capacity as a subset of the complex plane, Sibony and Wong showed that the polynomial convex hull of K (back in C^2) contains, about the origin, a full neighborhood whose size depends only on the logarithmic capacity of K and its distance from the origin in C^1. Moreover, by employing Ronkin's [6] notion of the Γ-capacity of a compact subset of C^N, they extend this result to the N-dimensional case. For some interesting applications of the theorem we refer the reader to the original paper [8]—the general theme is that the growth of a global hypersurface is determined by the growth of its intersections with complex lines for a set of complex lines of positive capacity.

We shall prove a version of the Sibony-Wong theorem using projective capacity. This will obviate the choice of an imbedding map of affine

*Supported in part by a grant from the National Science Foundation.

Recent Developments in Several Complex Variables
0-691-08285-5/81/000003-25$01.25/0 (cloth)
0-691-08281-2/81/000003-25$01.25/0 (paperback)
For copying information, see copyright page.

space into projective space and will also replace Γ-capacity, a notion with a rather complicated inductive definition which, by an example in Sadullaev ([7], p. 513) which originated with C. O. Kiselman, results in the unfortunate property that sets of zero Γ-capacity need not be polar sets.

The logarithmic capacity of a compact set in the complex plane has many equivalent definitions; one can arrive at it from the logarithmic potential of a mass distribution, from the transfinite diameter, and from Tchebycheff polynomials, to name three. Working on the Riemann sphere, P^1, Tsuji ([11], [12]) has shown that these three approaches lead to a single notion of capacity, elliptic capacity in his terminology, which has the property of being invariant under isometries of P^1 for the chordal metric, a property not possessed by logarithmic capacity. In defining capacity for compact subsets of P^N we shall utilize Tchebycheff polynomials; as we are working on projective space, our polynomials will be homogeneous. The fact that homogeneous polynomials in two variables always factor into linear factors, while this is not true for more than two variables, seems to account for the fact that Tchebycheff polynomials are useful and that the transfinite diameter, while easily definable, appears to be not so closely linked to the higher dimensional analytic structure.

Using the natural projection $C^N \setminus \{0\} \to P^{N-1}$, we shall identify compact subsets of P^{N-1} with compact circled subsets of the unit sphere in C^N. It is for the latter sets that we shall define a capacity by using homogeneous polynomials in N variables. In the complex plane one can consider monic polynomials; in our case, as a substitute, we shall introduce a convenient way to normalize homogeneous polynomials. In doing so, we shall consider the extremal problem of minimizing the integral $\int \log|f|d\sigma$ (where σ is unit "surface area" on the unit sphere) for all homogeneous polynomials of degree n with sup norm one on the unit sphere.

For subsets of C^N, definitions of capacity have been given by Zaharjuta [13], Siciak [9], and Bedford [1]. Their points of view are somewhat different from ours.

Here is a summary of what follows. In §1, we define normalized polynomials and projective capacity. In §2, the extremal problem on homogeneous polynomials is studied, and in §3 we derive bounds on capacity. Our version of the Sibony-Wong theorem and its converse are obtained in §4. Projective capacity in P^1 is shown to coincide with Tsuji's elliptic capacity in §5. Finally in §6 we derive some further properties of projective capacity. In particular, we show that a compact subset of P^N has zero capacity if and only if it is a locally polar set. The proof of this requires a rather subtle proposition of Josefson [5].

1. *Normalized polynomials and projective capacity*

We begin with some notation. Let $\Pi: C^N \setminus \{0\} \to P^{N-1}$ be the natural projection, B the open unit ball in C^N and σ normalized "surface" measure on ∂B. It follows from Jensen's inequality that $\int \log |f| d\sigma > -\infty$ for every polynomial $f(\neq 0)$ on C^N. If K is a compact circled subset of ∂B, then $\Pi(K)$ is a subset of P^{N-1}. This obviously establishes a 1-to-1 correspondence between the compact circled subsets of ∂B and arbitrary compact subsets of P^{N-1}. Usually we shall make this identification without explicit mention. If f is a continuous complex valued function on a set K, we write $\|f\|_K = \sup\{|f(x)| : x \in K\}$.

Now consider a homogeneous polynomial $f(\neq 0)$ of degree n, first for $N = 2$ variables. We can factor as

$$f = C \prod_{k=1}^{n} (\alpha_k z_1 + \beta_k z_2) \tag{1.1}$$

with $C > 0$ and $|\alpha_k|^2 + |\beta_k|^2 = 1$ and where C is unique. We shall say that f is normalized iff $C = 1$. It is obvious that any homogeneous polynomial can be "normalized" by multiplying it by an appropriate constant. We want to make the same definition for $N > 2$, but homogeneous polynomials of more than two variables need not split into factors. We can obviate this difficulty in the following way. Observe that because the measure σ on ∂B is invariant under unitary transformations we have

$$\int \log |z_2| d\sigma = \int \log |\alpha z_1 + \beta z_2| d\sigma$$

provided $|\alpha|^2 + |\beta|^2 = 1$. Thus taking the logarithm in (1.1) and integrating gives

$$\int \log |f| d\sigma = \log C + n \int \log |z_2| d\sigma \,. \tag{1.2}$$

Hence f is normalized iff the two integrals in (1.2) are equal. Now, in general, for a polynomial f of degree n in C^N we say that f is *normalized* iff

$$\int \log |f| d\sigma = n \int \log |z_N| d\sigma \,. \tag{1.3}$$

It is obvious that (a) z_j^n is normalized, (b) if f is normalized of degree n and g is normalized of degree m then fg is normalized of degree $n+m$, (c) if $f(\neq 0)$ is an arbitrary homogeneous polynomial, then there is a unique positive constant C such that Cf is normalized.

We shall use normalized polynomials in place of monic polynomials to define the Tchebycheff constants in the classical manner. Namely, let K be a compact circled subset of ∂B (i.e., an arbitrary compact subset of P^{N-1}), define for $j = 1, 2, \cdots$,

$$m_j = m_j(K) = \inf \{\|f\|_K : f \text{ a normalized polynomial of degree } j\}. \tag{1.4}$$

From (a) z_1^j is normalized and so

$$0 \leq m_j \leq \|z_1^j\|_K \leq \|z_1^j\|_{\partial B} = 1 \,.$$

It follows easily from (b) that $m_{j+k} \leq m_j \cdot m_k$. Now we quote a well-known classical fact.

LEMMA 1.1. *If* $\{a_j\}_1^\infty$ *is a sequence of real numbers and* $a_{j+k} \leq a_j + a_k$, *then* $\lim_{j\to\infty} \frac{a_j}{j}$ *exists and equals* $\inf \frac{a_j}{j} \geq -\infty$.

Now if none of the m_j are zero, set $a_j = \log m_j$ and apply the lemma to conclude that we can define the projective capacity of K, denoted cap (K), by

(1.5) $$\operatorname{cap} K = \lim_{j\to\infty} m_j^{\frac{1}{j}} = \inf_j m_j^{\frac{1}{j}} .$$

When some $m_s = 0$, it follows (see the proof of Proposition 2.4) that K, as a subset of P^{N-1}, lies in a hypersurface of degree s, and therefore $m_j = 0$ for all $j \geq s$. Thus, putting $\operatorname{cap} K = 0$, we see that (1.5) is equally valid in this case. From $0 \leq m_j \leq 1$, we get $0 \leq \operatorname{cap} K \leq 1$.

Also note that (1.5) entails the fact that $\operatorname{cap} K \leq m_j^{\frac{1}{j}}$ for each j. Clearly, cap is a monotone set function and is invariant under unitary transformations.

In the remaining sections cap K will denote the projective capacity of K.

2. *Main estimate for homogeneous polynomials*

The following upper bound for normalized polynomials will be quite useful.

THEOREM 2.1. *Let f be a normalized polynomial of degree n in C^N. Then*

(2.1) $$\|f\|_{\partial B} \leq 1 .$$

Moreover, equality holds in (2.1) iff $f = L^n$ with $L = a_1 z_1 + a_2 z_2 + \cdots + a_N z_N$ where $\Sigma |a_k|^2 = 1$.

For a convenient proof, we shall reformulate this as an entirely equivalent statement.

THEOREM 2.2. *Let f be a homogeneous polynomial of degree n in C^N. Suppose that $\|f\|_{\partial B} = 1$. Then*

$$\int \log |f| d\sigma \geq n \int \log |z_N| d\sigma . \tag{2.2}$$

Moreover, equality holds in (2.2) iff $f = L^n$ *where* L *is as above.*

COROLLARY 2.3. *Let* h *be a homogeneous polynomial of degree* n *in* C^N. *Then*

$$\int \log |h| d\sigma \geq \log \|h\|_{\partial B} + n \int \log |z_N| d\sigma . \tag{2.3}$$

Proof. Apply Theorem 2.2 to $f = h/\|h\|_{\partial B}$.

We shall indicate two proofs for Theorem 2.2. The first works only for $N = 2$; the second, in general. I do not know how to deduce the general case directly from the 2-dimensional case.

Proof 1: $(N = 2)$. Factor f as in (1.1). Then $\|f\|_{\partial B} = 1$ implies $C \geq 1$. Hence $\int \log |f| d\sigma = \log C + n \int \log |z_N| d\sigma \geq n \int \log |z_N| d\sigma$, as claimed in (2.2). Moreover, if equality holds, then $C = 1$ and each of the linear factors attains its maximum over ∂B on the same circle and so these factors coincide except for a possible constant factor. This proves the theorem.

Proof 2. Consider the projection map $p : \partial B \to C$ given by $p(z_1, \cdots, z_N) = z_1$. For $|\lambda| < 1$, $p^{-1}(\lambda)$ is a sphere Γ_λ in C^{N-1} (= the hyperplane in C^N obtained by fixing $z_1 = \lambda$). Let σ_λ be normalized "surface" measure on Γ_λ. It is easy to see that there is a positive continuous function $A(z_1)$ on the open unit disc D such that

$$\int g d\sigma = \int\!\!\int_D \left\{ \int_{\Gamma_\lambda} g \mid \Gamma_\lambda \, d\sigma_\lambda \right\} A(\lambda) \, du dv \tag{2.4}$$

for every σ-integrable function g on ∂B, where $\lambda = u + iv$. This follows from introducing the appropriate coordinates on ∂B.

Now suppose f is a homogeneous polynomial of degree n such that $\|f\|_{\partial B} = 1$. Our conclusion is clearly invariant under a unitary transformation and so we may assume that $f(1; 0,\cdots,0) = 1$. This means that we can write $f(z_1, z_2,\cdots, z_N) = z_1^n + q(z_1,\cdots,z_N)$ where q is a linear combination of monomials of degree n other than z_1^n. Thus $q(z_1; 0,\cdots,0) \equiv 0$. Now fix $\lambda \in D$ and apply Jensen's inequality to the function of $N-1$ variables $f(\lambda, z_2,\cdots, z_N)$ on the $(N-1)$-ball whose boundary is Γ_λ and whose center is $(\lambda, 0,\cdots,0)$. We get

$$\int_{\Gamma_\lambda} \log |f| d\sigma_\lambda \geq \log |f(\lambda, 0, \cdots, 0)| . \tag{2.5}$$

But $f(\lambda, 0, \cdots, 0) = \lambda^n + q(\lambda, 0, \cdots, 0) = \lambda^n$ and so $\log |f(\lambda, 0, \cdots, 0)| = n \log |\lambda|$. Applying (2.4) to $g = \log |f|$ yields

$$\int \log |f| d\sigma \geq n \int_D \log |\lambda| A(\lambda) dudv . \tag{2.6}$$

On the other hand, applying (2.4) to $\log |z_1^n| = n \log |z_1|$ gives

$$n \int \log |z_1| d\sigma = n \int_D \log |\lambda| A(\lambda) dudv . \tag{2.7}$$

Together, (2.6) and (2.7) yield (2.2).

Now suppose we have equality in (2.2). Then (2.5) must be an equality for almost all λ in D. This implies, for almost all λ in D, that $f(\lambda, z_2, \cdots, z_n)$ has no zeros in B. Now if for some $\lambda \neq 0$ in D, $f(\lambda, z_2, \cdots, z_n)$ had a zero in B, then the same would be true on a neighborhood of λ because the zero set of a homogeneous polynomial is a union of complex lines through the origin. The only other possibility is that all of the zeros of $f(\lambda, z_2, \cdots, z_N)$ lie on $\lambda = 0$. From this it easily follows that $f \equiv z_1^n$. This completes the proof of Theorem 2.2.

We shall now define Tchebycheff polynomials. Their existence is not quite as evident as in the classical case.

PROPOSITION 2.4. *Let* K *be a compact circled subset of* ∂B. *For each positive integer* n *there exists a normalized polynomial* T_n *such that* $\|T_n\|_K = m_n(K)$.

REMARK. We say that T_n is an nth *Tchebycheff polynomial* for K.

We cannot expect, in general, that T_n will be unique. For example, if K is invariant under some unitary map, then any Tchebycheff polynomial will be transformed by this map into another one.

Proof. Let $\{f_j\}$ be a sequence of normalized polynomials of degree n such that $\|f_j\|_K \to m_n(K)$. Since the vector space of homogeneous polynomials of degree n is finite dimensional and $\|f_j\|_B \le 1$, a standard compactness argument implies that a subsequence of $\{f_j\}$ converges uniformly on B to a homogeneous polynomial f of degree n. Without loss of generality, we suppose $f_j \to f$ uniformly on B. Then $\|f\|_K = m_n$. By Fatou's lemma,

$$\int \log |f| d\sigma \ge \overline{\lim_{j\to\infty}} \int \log |f_j| d\sigma = n \int \log |z_N| d\sigma . \tag{2.8}$$

In particular, $f \not\equiv 0$, even if $m_n = 0$. We consider two cases. (i) $m_n = 0$. Then $f \equiv 0$ on K and so K (in P^{N-1}) lies in a hypersurface of degree n in P^{N-1}. Now to get T_n we only have to normalize f; i.e., set $T_n = Cf$ where $C > 0$ is chosen so that T_n is normalized. (ii) $m_n > 0$. We claim that f is already normalized in this case (and so $T_n = f$ works). Suppose not; then by (2.8), $\int \log |f| d\sigma > n \int \log |z_N| d\sigma$. Hence, there is a constant C with $0 < C < 1$ and that Cf is normalized. But then, as $m_n > 0$, $m_n(K) \le \|Cf\|_K = C\|f\|_K = Cm_n(K) < m_n(K)$, a contradiction. This proves Proposition 2.4.

3. *A lower bound on capacity*

We now show that sets of positive measure have positive capacity. This shows that the theory is not trivial and also serves to justify the definition of normalized polynomial.

PROPOSITION 3.1. *Let* K *be a compact circled subset of* ∂B *such that* $\sigma(K) > 0$. *Then*

$$\text{cap}\,(K) \geq \exp\left\{\frac{1}{\sigma(K)}\int \log |z_N|\,d\sigma\right\}. \tag{3.1}$$

In particular, $\text{cap}\,(K) > 0$.

REMARK. The converse of this last statement is false. The torus $\{z : |z_1| = |z_2| = \cdots = |z_N|\} \cap \partial B$ provides an example; cf. Theorem 4.4.

Proof. Let f be a normalized polynomial of degree k. Then

$$k\int \log |z_N|\,d\sigma = \int_{\partial B} \log |f|\,d\sigma \leq \int \log |f|\,d\sigma\,, \tag{3.2}$$

where the inequality follows from the fact that $\|f\|_{\partial B} \leq 1$ by Theorem 2.1. Hence

$$k\int \log |z_N|\,d\sigma \leq \sigma(K)\log \|f\|_K\,. \tag{3.3}$$

Taking the inf over all such f gives

$$k\int \log |z_N|\,d\sigma \leq \sigma(K)\log m_k(K)\,. \tag{3.4}$$

Hence

$$\frac{1}{\sigma(K)}\int \log |z_N|\,d\sigma \leq \log (m_k(K))^{\frac{1}{k}}\,.$$

Letting $k \to \infty$ gives (3.1).

COROLLARY 3.2. *We have*

$$\text{cap}\,(P^{N-1}) \geq \exp\left\{\int \log |z_N| d\sigma\right\}. \tag{3.5}$$

When $N = 2$, the right-hand side of (3.5) is seen to be $e^{-1/2}$ and so

$$\text{cap}\,(P^1) \geq e^{-1/2}. \tag{3.6}$$

By invoking Tsuji's work on elliptic capacity we shall show in Proposition 5.2 that equality holds in (3.6).

In the other direction, by considering the normalized polynomial $f = (z_1 z_2 \cdots z_N)^s$ of degree Ns with $\|f\|_{\partial B} = N^{-sN/2}$, one sees that $m_{Ns}(P^{N-1}) \leq N^{-Ns/2}$ and hence that

$$\text{cap}\,(P^{N-1}) \leq N^{-1/2}. \tag{3.7}$$

Using $\int \log |z_N| d\sigma = -\frac{1}{2} \sum_1^{N-1} 1/k$, a computation pointed out to me by W. Stoll and the classical fact that $\sum_1^{N-1} 1/k = \log N + \gamma + \varepsilon_N$ where γ is Euler's constant and $\varepsilon_N \to 0$ as $N \to \infty$, we get

$$e^{-1/2(\gamma+\varepsilon_N)} \leq \frac{\text{cap}\,(P^{N-1})}{N^{-1/2}} \leq 1. \tag{3.8}$$

4. *The Sibony-Wong Theorem*

We can now prove our version of the Sibony-Wong theorem. For a compact set K in C^N, $\hat{K}$ will denote its polynomially convex hull:

$$\hat{K} = \{z \in C^N : |F(z)| \leq \|F\|_K \text{ for every polynomial } F \text{ on } C^N\}. \tag{4.1}$$

THEOREM 4.1. *Let* K *be a compact circled subset of* ∂B *with* $\text{cap}\,(K) > 0$. *Then* $\hat{K}$ *contains the ball* $\{z : \|z\| \leq \text{cap}\,(K)\}$.

LEMMA 4.2. *Let* F_k *be a homogeneous polynomial of degree* k. *Then*

$$\|F_k\|_{\partial B} \le \frac{1}{(\text{cap } K)^k} \|F_k\|_K . \tag{4.2}$$

Proof. Let $m_k = m_k(K)$. Since (4.2) is homogeneous in F_k, we may assume that F_k is normalized. Thus $(\text{cap } K)^k \le m_k \le \|F_k\|_K$, by the definition of capacity and m_k. From Theorem 2.1, we get

$$\|F_k\|_{\partial B} \le 1 \le \frac{\|F_k\|_K}{m_k} \le \frac{\|F_k\|_K}{(\text{cap } K)^k} . \tag{4.3}$$

LEMMA 4.3. *Let* $F = \sum_{k=0}^{d} F_k$ *be a polynomial of degree* d *written as a sum of its homogeneous parts,* F_k *homogeneous of degree* k. *Then if* K *is a compact circled subset of* ∂B, *we have for* $0 \le k \le d$,

$$\|F_k\|_K \le \|F\|_K . \tag{4.4}$$

Proof. Let $a \in K$. For a complex number λ with $|\lambda| = 1$, $\lambda a \in K$ because K is circled. Write

$$F(\lambda a) = \sum_{k=0}^{d} \lambda^k F_k(a) .$$

Hence, by the Cauchy integral formula,

$$F_k(a) = \frac{1}{2\pi i} \int_{|\lambda|=1} \frac{F(\lambda a)}{\lambda^{k+1}} \, d\lambda . \tag{4.5}$$

The obvious estimate gives (4.4).

Proof of Theorem 4.1. Let $F = \sum_{0}^{d} F_k$ be a polynomial and fix z such that $0 < \|z\| < \text{cap}\,(K)$. Since $F_k(z) = \|z\|^k F_k\left(\frac{z}{\|z\|}\right)$ we have

$$(4.6)\qquad |F_k(z)| \le \|z\|^k \, \|F_k\|_{\partial B} \le \|z\|^k \, \frac{\|F_k\|_K}{(\mathrm{cap}\,(K))^k}$$

by Lemma 4.2. Applying Lemma 4.3 gives $|F_k(z)| \le (\|z\|/\mathrm{cap}\,(K))^k \|F\|_K$. Hence

$$(4.7)\qquad |F(z)| \le \sum_0^d |F_k(z)| \le \left(\sum_{k=0}^d \left(\frac{\|z\|}{\mathrm{cap}\,(K)}\right)^k\right) \|F\|_K$$

$$< \frac{1}{1 - \dfrac{\|z\|}{\mathrm{cap}\,(K)}} \cdot \|F\|_K .$$

Now (4.7) implies that $z \in \hat{K}$ and this proves Theorem 4.1. Our next result is the converse of Theorem 4.1.

THEOREM 4.4. *Let* K *be a compact circled subset of* ∂B. *Suppose that* $\hat{K}$ *contains a ball about the origin,* $B_\delta = \{z : \|z\| < \delta\}$ *for some* δ *with* $0 < \delta < 1$. *Then* $\mathrm{cap}\,(K) > 0$ *and, in fact,*

$$(4.8)\qquad \mathrm{cap}\,(K) \ge \delta \cdot \mathrm{cap}\,(P^{N-1}) .$$

Proof. Let f be a homogeneous polynomial of degree n. Then

$$(4.9)\qquad \|f\|_{B_\delta} = \delta^n \|f\|_{\partial B} .$$

Since $\hat{K} \ge B_\delta$ we have

$$(4.10)\qquad \|f\|_{B_\delta} \le \|f\|_K .$$

Thus

$$(4.11)\qquad \|f\|_{\partial B}^{\frac{1}{n}} \le \frac{1}{\delta} \|f\|_K^{\frac{1}{n}}$$

Taking f in (4.11) to be a normalized polynomial we have

$$(4.12)\qquad (m_n(P^{N-1}))^{\frac{1}{n}} \le \|f\|_{\partial B}^{\frac{1}{n}} \le \frac{1}{\delta} \|f\|_K^{\frac{1}{n}} .$$

Taking the inf over all such f yields

$$(m_n(P^{N-1}))^{\frac{1}{n}} \leq \frac{1}{\delta}(m_n(K))^{\frac{1}{n}} . \tag{4.13}$$

Now let $n \to \infty$ to get (4.8).

Putting the previous two theorems together gives

$$\text{cap } K \leq \sup \{\nu : B_\nu \subseteq \hat{K}\} \leq \frac{\text{cap } K}{\text{cap } P^{N-1}} , \tag{4.14}$$

where we have seen that $0 < \text{cap } P^{N-1} < 1$.

5. *Elliptic capacity*

Tsuji's elliptic capacity ([12]), for a compact set E of the Riemann sphere $\overline{C}(= C \cup \{\infty\})$ can be defined as follows. Let

$$[a, b] = |a-b| \Big/ \sqrt{(1+|a|^2)(1+|b|^2)}$$

be the chordal distance for two points a and b of $\overline{C}$. Let $\mathcal{P}_n$ be the set of all functions P on $\overline{C}$ of the form

$$P(z) = \prod_{k=1}^{n} [z, z_k] \tag{5.1}$$

where $z_1, z_2, \cdots, z_n$ are in $\overline{C}$. Although P is not a holomorphic polynomial of z, it plays the role of the modulus of a monic polynomial. Now define a Tchebycheff constant

$$\tilde{m}_n(E) = \inf_{P \in \mathcal{P}_n} \|P\|_E . \tag{5.2}$$

One shows in the usual way that $\lim \tilde{m}_n^{\frac{1}{n}}(E)$ exists. This limit is taken as the definition of *elliptic capacity*.

PROPOSITION 5.1. *Under the natural identification of* $\overline{C}$ *and* P^1, *the elliptic and projective capacities of a set coincide.*

Proof. Let K be a compact circled subset of ∂B, $B \subseteq C^2$. Let $E = \{\lambda \in C : \lambda = z_1/z_2 \text{ for } (z_1, z_2) \in K\}$. We show that the elliptic capacity of E equals the projective capacity of K. Let f be an arbitrary normalized polynomial in z_1, z_2 of degree n. Then we can write

$$f(z_1, z_2) = \prod_{k=1}^{n} (\alpha_k z_1 + \beta_k z_2) \tag{5.3}$$

where $|\alpha_k|^2 + |\beta_k|^2 = 1$. Now let

$$P(\lambda) = \prod_{k=1}^{n} \left[\lambda, \frac{-\beta_k}{\alpha_k}\right], \tag{5.4}$$

then $P \in \mathcal{P}_n$ and an easy computation gives

$$P(\lambda) = (1 + |\lambda|^2)^{-\frac{n}{2}} |f(\lambda, 1)| . \tag{5.5}$$

Now putting $\lambda = z_1/z_2$ in (5.5) for $(z_1, z_2) \in K$ and using $|z_1|^2 + |z_2|^2 = 1$, we get $P(\lambda) = |f(z_1, z_2)|$. It follows that we have a correspondence between normalized polynomials f of degree n and $P \in \mathcal{P}_n$ such that $\|f\|_K = \|P\|_E$. This implies that $m_n(K) = \tilde{m}_n(E)$ and hence the equality of the two capacities.

Tsuji shows that the elliptic capacity of E can also be obtained from potentials of the form

$$u(z) = \int \log \frac{1}{[z, \zeta]} \, d\mu(\zeta) \tag{5.6}$$

where μ is a probability measure on E. We shall apply this when $E = \overline{C}$. It is clear then by symmetry that the equilibrium distribution μ becomes the normalized invariant measure on $\overline{C} = P^1$; i.e.,

$$(5.7) \qquad \mu(\zeta) = \frac{1}{\pi}\frac{dudv}{(1+|\zeta|^2)^2}, \quad \zeta = u+iv\ .$$

Then since the elliptic capacity of E equals e^{-V} where $V = u(a)$ for every interior point $a \in E$ (see [12], p. 90), we compute in (5.6) using $z = 0$ and μ as in (5.7) to get $V = u(0) = \frac{1}{2}$. This proves

PROPOSITION 5.2.

$$(5.8) \qquad \text{cap}\,(P^1) = e^{-\frac{1}{2}}\ .$$

6. *Further properties of projective capacity*

Finally we shall examine the behavior of capacity under set theoretical operations and shall characterize sets of zero capacity.

PROPOSITION 6.1. *Let* K, K_n, $n = 1, 2, \cdots$, *be compact circled subsets of* ∂B *and suppose that* $K_n \downarrow K$. *Then* $\text{cap}\,K_n \downarrow \text{cap}\,K$.

Proof. By monotonicity, $\text{cap}\,K_n \downarrow \delta$ for some $\delta \geq \text{cap}\,K$. Arguing by contradiction, we suppose that $\delta > \text{cap}\,K$. Choose n so that $(m_n(K))^{\frac{1}{n}} < \delta$. Let T_n be an nth Tchebycheff polynomial for K. Thus $\|T_n\|_K < \delta^n$. By continuity we can choose k such that $\|T_n\|_{K_k} < \delta^n$. Since T_n is normalized, this implies that

$$\text{cap}\,(K_k) \leq (m_n(K_k))^{\frac{1}{n}} \leq \|T_n\|_{K_k}^{\frac{1}{n}} < \delta\ .$$

This contradicts the definition of δ and proves the proposition.

Our next result asserts that a countable union of sets of zero capacity is also of zero capacity. This allows us to localize the notion of zero capacity. We shall employ an extension of a lemma of Bishop [2] on Jensen measure. We refer to Stout [10] for terminology on uniform algebras.

LEMMA 6.2. *Let* $\mathfrak{A}$ *be a uniform algebra on the compact space* X, *and* $\Gamma \subseteq X$ *its Shilov boundary. Let* μ *be a probability measure on* X. *Then there exists a probability measure* ν *on* Γ *such that*

$$\int_X \log |f| d\mu \leq \int_\Gamma \log |f| d\nu \tag{6.1}$$

for all $f \in \mathfrak{A}$.

REMARK. Bishop proved this in the case when μ is a point mass δ_x for some $x \in X$; then ν is a representing measure for x, a so-called Jensen representing measure. Examination of Bishop's argument shows that it works in the generality of Lemma 6.2 with only the obvious notational changes. We shall say that ν is a *Jensen measure* for μ.

THEOREM 6.3. *Let* K, K_n, $n = 1, 2, 3, \cdots$, *be compact circled subsets of* ∂B *such that*

$$K = \bigcup_{n=1}^{\infty} K_n . \tag{6.2}$$

If $\operatorname{cap} K_n = 0$ *for all* $n = 1, 2, 3, \cdots$, *then* $\operatorname{cap} K = 0$.

Proof. We assume, arguing by contradiction, that $\operatorname{cap} K \equiv \delta > 0$. Then, by Theorem 4.1, $\hat{K}$ contains $\overline{B}_\delta$. Let μ be normalized "surface area" on ∂B_δ; i.e., μ is defined by

$$\int_{\partial B_\delta} g d\mu = \int_{\partial B} g(\delta z) d\sigma(z) . \tag{6.3}$$

In Lemma 6.2, let X be $\hat{K}$ and $\mathfrak{A}$ be $P(X)$, the closure of the polynomials in the supremum norm over X. Since $\operatorname{spt}\mu = \partial B_\delta \subseteq X$, we get a Jensen measure ν for μ; ν is supported on K. Choose k such that $\nu(K_k) \equiv \theta > 0$. We shall show that $\operatorname{cap}(K_k) > 0$, giving us the contradiction which we seek.

Let h be a homogeneous polynomial of degree n. Applying (6.3) to $\log |h|$ gives

$$(6.4)\qquad \int \log|h|d\mu = \int \log|h(\delta z)|d\sigma(z) = \int \log|h|d\sigma + n\log\delta\,.$$

By (6.1) we have

$$(6.5)\qquad \int \log|h|d\mu \le \int \log|h|d\nu = \int_{K_k} + \int_{K\setminus K_k}$$

$$\le \theta\log\|h\|_{K_k} + (1-\theta)\log\|h\|_{\partial B}\,.$$

From (6.4) and (6.5) we get

$$(6.6)\qquad \int_{\partial B} \log|h|d\sigma \le -n\log\delta + \theta\log\|h\|_{K_k} + (1-\theta)\log\|h\|_{\partial B}\,.$$

Now recall (2.3),

$$(6.7)\qquad \log\|h\|_{\partial B} + n\int \log|z_N|d\sigma \le \int \log|h|d\sigma\,.$$

Together (6.6) and (6.7) yield

$$(6.8)\qquad \|h\|_{\partial B} \le Q^n\|h\|_{K_k}\,,$$

where $Q = \exp((-\log\delta - \int\log|z_N|d\sigma)/\theta)$. Arguing as in the proof of Theorem 4.4, one sees that (6.8) implies that $\operatorname{cap}(K_k) \ge Q^{-1}\operatorname{cap}(P^{N-1}) > 0$, a contradiction.

We recall that a subset E of a complex manifold is a *locally polar* set if for each $p \in E$ there exists an open neighborhood Ω and a plurisubharmonic (psh) function u defined on Ω such that (i) $u = -\infty$ on $\Omega \cap E$ and (ii) $u \not\equiv -\infty$ on Ω. We shall show that a compact subset of projective space has zero capacity if and only if it is a locally polar set. Note that since projective space is compact, it possesses no globally polar sets. It will be useful to maintain the distinction between subsets of projective space and the corresponding circled sets in the unit sphere. Thus, for a compact set $E \subseteq P^{N-1}$, we let $K(E)$ be $\Pi^{-1}(E) \cap \partial B$ where $\Pi : C^N \setminus \{0\} \to P^{N-1}$ is the natural projection. We first prove the easier half of the equivalence.

THEOREM 6.4. *If a compact subset of complex projective space has zero projective capacity, then it is a locally polar set.*

Proof. It clearly suffices to prove the following assertion. Let E be a compact subset of P^{N-1} which is contained in $\Omega \equiv \Pi(\{z_N \neq 0\})$ (where $\Pi : C^N \setminus \{0\} \to P^{N-1}$ is the natural projection) and such that $\text{cap}(E) = 0$, then there exists a psh function u defined on Ω such that (i) $u \equiv -\infty$ on E and (ii) $u \not\equiv -\infty$ on Ω.

Let $K = K(E)$ be the corresponding compact circled subset of ∂B. Let T_n be an nth Tchebycheff polynomial for K. Put $\Omega' = \{z \in C^N : \frac{1}{2} < \|z\| < 2$ and $z_N \neq 0\}$. Then Ω' is open in C^N, $\Omega' \supseteq K$, and $\Pi(\Omega') = \Omega$. Since $\|T_n\|_{\partial B} \leq 1$, we have $\|T_n\|_{\Omega'} \leq 2^n$. Define functions w_n on Ω' by

$$w_n = \frac{1}{n} \log |T_n| . \tag{6.9}$$

Then w_n is psh on Ω' and satisfies $-\infty \leq w_n \leq \log 2$. Put $w = \limsup_{n\to\infty} w_n$ on Ω'. Since T_n is normalized, we have

$$\int_{\partial B \cap \Omega'} w_n d\sigma = \int_{\partial B \cap \Omega'} \log |z_N| d\sigma . \tag{6.10}$$

As the $\{w_n\}$ are uniformly bounded above we can apply Fatou's lemma to get

$$\int_{\partial B \cap \Omega'} w d\sigma \geq \limsup_{n\to\infty} \int_{\partial B \cap \Omega'} w_n d\sigma = \int_{\partial B \cap \Omega'} \log |z_N| d\sigma > -\infty . \tag{6.11}$$

We conclude that $w > -\infty$ σ a.e. on $\partial B \cap \Omega'$. Fix $p \in \partial B \cap \Omega'$ such that

$$\limsup_{n\to\infty} w_n(p) \equiv w(p) > -\infty . \tag{6.12}$$

Since $\operatorname{cap}(E) = \operatorname{cap}(K) = 0$, we have $\|T_n\|_K^{\frac{1}{n}} = m_n(K)^{\frac{1}{n}} \to 0$; i.e.,

$$\sup_K w_n - \frac{1}{n}\log\|T_n\|_K = \frac{1}{n}\log m_n \to -\infty \tag{6.13}$$

as $n \to \infty$. Because of (6.12) and (6.13) we can choose a sequence $\{t_n\}$ of non-negative real numbers $t_n \geq 0$ such that

$$\text{(a)} \quad \sum_{n=1}^{\infty} t_n \frac{\log m_n}{n} = -\infty ,$$

$$\text{(b)} \quad \sum_{n=1}^{\infty} t_n w_n(p) > -\infty , \tag{6.14}$$

$$\text{(c)} \quad \sum_{n=1}^{\infty} t_n \equiv T < \infty .$$

Now define a function

$$W = \sum_{n=1}^{\infty} t_n w_n \tag{6.15}$$

on Ω'. Writing W as

$$W = \sum_{1}^{\infty} t_n(w_n - \log 2) + (\log 2)\cdot T \tag{6.16}$$

and noting that the infinite sum on the right of (6.16) is the limit of a decreasing sequence of psh functions on Ω' because $w_n - \log 2 \leq 0$, we conclude that W is psh on Ω'. By (6.14a), $W \equiv -\infty$ on K and by (6.14b), $W(p) \not\equiv -\infty$.

Since $w_n(\lambda z) = \log|\lambda| + w_n(z)$ for $z \in \Omega'$, λ a complex number such that $\lambda z \in \Omega'$, we have $W(\lambda z) = T\cdot\log|\lambda| + W(z)$. Now define a function

W_1 on Ω' by

$$W_1(z) = W(z) - T \cdot \log |z_N| . \tag{6.17}$$

As $z_N \neq 0$ on Ω', W_1 is psh on Ω' and for $z \in \Omega'$, $W(z) = -\infty$ if and only if $W_1(z) = -\infty$. Moreover $W_1(\lambda z) \equiv W_1(z)$. Therefore W_1 induces a unique psh function u on Ω such that $u \circ \Pi = W_1$. Then $u \not\equiv -\infty$ on Ω because $u(\Pi(p)) = W_1(p) \neq -\infty$ and $u = -\infty$ on E as $W_1 = -\infty$ on K and E is $\Pi(K)$. This proves the theorem.

REMARK. Ω is naturally identified with $C^{N-1} \subseteq P^{N-1}$. The argument shows that in $C^{N-1} \subseteq P^{N-1}$, a compact set of zero capacity is "globally polar"; i.e., u above is defined on all of C^{N-1}.

In order to prove the converse of the last theorem we need to relate capacity to polynomials in affine space. Suppose, as above, that the compact set $E \subseteq P^{N-1}$ is contained in $\Pi(\{z_N \neq 0\}) \equiv \Omega$. We identify Ω with C^{N-1} via the global coordinates $\zeta_k = z_k/z_N$, $k = 1, 2, \cdots, N-1$, where $[z_1, z_2, \cdots, z_N]$ are the homogeneous coordinates induced on P^{N-1} from the natural projection $\Pi : C^N \setminus \{0\} \to P^{N-1}$.

LEMMA 6.5. *Suppose* $\operatorname{cap} E > 0$. *Then for every (not necessarily homogeneous) polynomial* $g(\zeta_1, \cdots, \zeta_{N-1})$ *of degree* n *and every point* $t = (t_1, \cdots, t_{N-1}) \in C^{N-1}$, *we have*

$$|g(t)| \leq Q^n (1 + \|t\|^2)^{\frac{n}{2}} \|g\|_E \tag{6.18}$$

with $Q = 1/\operatorname{cap}(E)$ *and* $\|t\|^2 = \Sigma |t_k|^2$, *where* E *is viewed as a subset of* C^{N-1}.

Proof. Let $f(z_1, \cdots, z_N)$ be the associated homogeneous polynomial for g; i.e.,

$$f(z_1, z_2, \cdots, z_N) = z_N^n \, g\left(\frac{z_1}{z_N}, \frac{z_2}{z_N}, \cdots, \frac{z_{N-1}}{z_N}\right). \tag{6.19}$$

Applying Lemma 4.2 to f gives

$$\|f\|_{\partial B} \leq Q^n \|f\|_K \tag{6.20}$$

where $K = K(E) = \Pi^{-1}(E) \cap \partial B$. By (6.19), $f(z) = z_N^n\, g(\zeta)$. As z varies over K, ζ varies over E and $|z_N| \leq 1$. Hence $\|f\|_K \leq \|g\|_E$. Also for $t = (t_1, \cdots, t_N)$ in C^N, $|f(t)| \leq \|t\|^n \|f\|_{\partial B}$. We thus get

$$|f(t)| \leq Q^n \|t\|^n \|g\|_E \tag{6.21}$$

for t in C^N. Now putting $t_N = 1$ in (6.21) gives (6.18).

REMARK. This lemma has a converse; namely, if $E \subseteq C^{N-1} \subseteq P^{N-1}$ and (6.18) holds for every polynomial g, then $\operatorname{cap} E > 0$. In fact, for a homogeneous polynomial f of degree n we have, for $z \in \partial B$, $|f(z_1, \cdots, z_N)|$ $= |z_N^n\, g(\zeta_1, \cdots, \zeta_{N-1})| \leq |z_N|^n \cdot Q^n (1 + \|\zeta\|^2)^{\frac{n}{2}} \|g\|_E$. One checks that $|z_N|^n \cdot (1 + \|\zeta\|^2)^{\frac{n}{2}} = 1$ since $z \in \partial B$. Also as E is compact there is a $C > 0$ such that $|z_N| \geq C$ on K. This implies that $\|g\|_E \leq C^{-n} \|f\|_K$. We then have that

$$\|f\|_{\partial B} \leq A^n \|f\|_K \tag{6.22}$$

with $A = Q/C$, for every homogeneous polynomial of degree n. We have seen that (6.22) implies that $\operatorname{cap}(K) > 0$. Moreover it is easy to check that if (6.18) holds for every g and some E, then it continues to hold for the image of E under an affine motion with a possibly different value of Q. This means that the property of having positive capacity is invariant under translation.

LEMMA 6.6. *Let f_j, $j = 1, 2, \cdots$, be a sequence of holomorphic functions on an open set Ω with $|f_j| \leq M$. Define*

$$u(z) = \overline{\lim_{z' \to z}}\ \overline{\lim_{j \to \infty}}\ \frac{1}{j} \log |f_j(z')| . \tag{6.23}$$

If $u(z_0) = -\infty$ *for some* z_0 *in* Ω, *then*

(6.24) $$u(z_0) = -\infty = \lim_{j\to\infty} \frac{1}{j} \log |f_j(z_0)| .$$

Proof. Put

(6.25) $$\Psi(z) = \overline{\lim_{j\to\infty}} \frac{1}{j} \log |f_j(z)| .$$

Then $u(z) = \overline{\lim_{z'\to z}} \Psi(z')$. Hence $-\infty = \overline{\lim_{z\to z_0}} \Psi(z)$. Take $A > 0$ arbitrary. Then there exists a $\delta > 0$ such that $\Psi(z) < -A$ for z in the set $\Omega_1 = \{z : 0 < |z-z_0| < \delta\} \subseteq \Omega$. Now take $0 < \delta_1 < \delta$ and apply Hartogs' lemma ([4], p. 21) to the uniformly bounded above sequence $\frac{1}{j} \log |f_j|$ on the set Ω_1 We conclude that for the compact set $K = \{z : |z-z_0| = \delta_1\} \subset \Omega_1$, there is a j_0 such that

(6.26) $$\frac{1}{j} \log |f_j(z)| < -A$$

if $z \in K$ and $j \geq j_0$. By the maximum principle, (6.26) then holds at $z = z_0$ for $j \geq j_0$. This proves (6.24).

THEOREM 6.7. *If a compact subset of complex projective space is locally polar, then it has zero projective capacity.*

Proof. Our proof is patterned after an argument of Josefson [5]. Let E be a compact subset of P^{N-1} which is locally polar. Since "locally polar" and "zero capacity" are both local concepts, invariant under affine motions, we may, without loss of generality, suppose (following Josefson) that $E \subseteq B_{1/2} \subseteq C^{N-1} \subseteq P^{N-1}$ and that there is a psh function $u(\neq -\infty)$ defined on B_4 such that $u = -\infty$ on E and $u < 0$ on $\overline{B}_2$. It suffices to show that, under these assumptions, $\operatorname{cap} E = 0$. Arguing by contradictions, we suppose otherwise; i.e., that $\operatorname{cap} E > 0$.

As Josefson observes, it follows from an argument of Bremermann [3] that there exist holomorphic functions f_j on B_4 such that $|f_j| \leq 1$ on B_2 and

(6.27) $$u(z) = \overline{\lim_{z'\to z}}\ \overline{\lim_{j\to\infty}} \frac{1}{j} \log|f_j(z')| .$$

Let Ψ be given by formula (6.25). Then Ψ is not identically $-\infty$ on a neighborhood of 0 (otherwise it would follow that $u \equiv -\infty$ on a neighborhood of 0). Therefore we can choose a z_0 near 0 such that $\Psi(z_0) > -\infty$ and such that E is contained in the ball centered at z_0 of radius 1/2. Now if $q > 0$ is such that $\Psi(z_0) > -q > -\infty$, there is an infinite set of j's such that

(6.28) $$\frac{1}{j} \log|f_j(z_0)| > -q ;$$

i.e.,

(6.29) $$|f_j(z_0)| > e^{-jq} .$$

Let S be the infinite set of j's for which (6.28) and (6.29) hold. Replacing u by a possibly smaller function we may assume that

(6.30) $$u(z) = \overline{\lim_{z'\to z}}\ \overline{\lim_{\substack{j\to\infty \\ j\in S}}} \frac{1}{j} \log|f_j(z')| .$$

Furthermore, by a translation, we may suppose that $z_0 = 0$ and still have f_j holomorphic on B_3, $|f_j(0)| > \exp(-jq)$, and $|f_j| \leq 1$ on B_1 for each $j \in S$.

Now we claim that the sequence of functions $\left\{\frac{1}{j} \log|f_j|\right\}$ converges uniformly to $-\infty$ on E as $j \in S \to \infty$. Take $M > 0$. Define

(6.31) $$\Psi(z) = \overline{\lim_{\substack{j\in S \\ j\to\infty}}} \frac{1}{j} \log|f_j(z)| .$$

By Lemma 6.6, for $z \in E$, $u(z) = -\infty = \Psi(z)$. Thus for each $z \in E$ there is a neighborhood V_z contained in $B_{1/2}$ such that $\Psi(\zeta) < -M$ for $\zeta \in V_z$ (since $\overline{\lim_{z'\to z}} \Psi(z') = u(z) = -\infty$). Taking the union of the sets V_z yields an open set Ω in C^{N-1} such that $E \subseteq \Omega \subseteq B_{1/2}$ and $\Psi < -M$ on Ω.

Now applying Hartogs' lemma to the uniformly bounded above sequence $\left\{\frac{1}{j}\log|f_j|\right\}_{j\in S}$ on Ω we conclude that there is a j_0 such that if $j \geq j_0$ and $j \in S$, then for $z \in E$,

$$\frac{1}{j}\log|f_j(z)| \leq -M\,. \tag{6.32}$$

This establishes our claim.

Now define, for $j \in S$, $\phi_j = -(1/(j\cdot q))\log\|f_j\|_E$; $\phi_j \to \infty$ and $|f_j(z)| \leq \exp(-jq\phi_j)$ for $z \in E$. Now we appeal to a basic result of Josefson ([5], Proposition) according to which there exists a polynomial g_j for each $j \in S$ (j assumed sufficiently large) of (some) degree d_j such that

$$1 \leq \|g_j\|_B \leq 2^{d_j} \tag{6.33}$$

and

$$|g_j(z)| < \exp\left(-C\,d_j\,\phi_j^{\frac{1}{N-1}}\right) \tag{6.34}$$

for $z \in E$, where C is a universal constant. This is possible because we have arranged that $f_j \in N(qj)$, in the terminology of Josefson.

Now we use the assumption that $\operatorname{cap} E > 0$ and apply Lemma 6.5 to g_j. We get, for $t \in C^{N-1}$ with $\|t\| \leq 1$ that

$$|g_j(t)| \leq Q^{d_j}2^{d_j/2}\|g_j\|_E\,. \tag{6.35}$$

Taking a d_j^{th} root in (6.35) and using (6.34) yields

$$\|g_j\|_B^{\frac{1}{d_j}} \leq Q\sqrt{2}\exp\left(-C\phi_j^{\frac{1}{N-1}}\right). \tag{6.36}$$

Letting $j \to \infty$, $j \in S$, we conclude that $\|g_j\|_B^{\frac{1}{d_j}} \to 0$. But this contradicts (6.33) and completes the proof.

REMARK. If E is a σ-compact locally polar subset of C^N, then our arguments show that E is globally polar in C^N; namely, by Theorem 6.7,

E is a countable union of compact sets of zero capacity, by the remark after Theorem 6.4, these sets are globally polar and finally, it is well known that a countable union of globally polar sets is itself globally polar. This is a special case of Josefson's result [5] where σ-compactness is not required of E. Of course we have used the main proposition of [5] –our arguments serve to replace the second part of Josefson's proof for this special case.

REFERENCES

[1] E. Bedford, Extremal plurisubharmonic functions and pluripolar sets in C^2, preprint.

[2] E. Bishop, Holomorphic completions, analytic continuations and the interpolations of seminorms, Ann. Math. 78 (1963), 468-500.

[3] H. J. Bremermann, On the conjecture of the equivalence of the plurisubharmonic functions and the Hartogs functions, Math. Ann. 131 (1956), 76-86.

[4] L. Hörmander, An Introduction to Complex Analysis in Several Variables, Van Nostrand, 1966.

[5] B. Josefson, On the equivalence between locally polar and globally polar sets for plurisubharmonic functions on C^n, Arkiv for Math. 16 (1978), 109-115.

[6] L. Ronkin, Introduction to the Theory of Entire Functions of Several Variables, English trans., Transl. Math. Monographs, vol. 44, Amer. Math. Soc., Providence, R.I., 1974.

[7] A. Sadullaev, A boundary uniqueness theorem in C^n, Math. USSR Sbornik 30 (1976), 501-535.

[8] N. Sibony and P.-M. Wong, Some results on global analytic sets, preprint.

[9] J. Siciak, Extremal plurisubharmonic functions in C^n, Proc. First Finnish-Polish Summer School in Complex Analysis at Podlesice, 1977, 115-152.

[10] E. L. Stout, The Theory of Uniform Algebras, Bogden & Quigley, Tarrytown-on-Hudson, N. Y., 1971.

[11] M. Tsuji, Some metrical theorems on Fuchsian groups, Jap. J. Math. 19 (1947), 483-516.

[12] ———, Potential Theory in Modern Function Theory, Maruzen, Tokyo, 1959.

[13] V. Zaharjuta, Transfinite diameter, Čebyšev constants, and capacity for compacta in C^n, Math USSR Sbornik 25 (1975), 350-364.

ANOTHER PROOF OF THE LEMMA OF THE LOGARITHMIC DERIVATIVE IN SEVERAL COMPLEX VARIABLES

Aldo Biancofiore* and Wilhelm Stoll**

1. *Introduction*

Al Vitter [9] extended Nevanlinna's well-known Lemma of the Logarithmic Derivative to meromorphic functions of several variables. His proof uses the method of non-negative curvature introduced by Carlson and Griffiths [1], Griffiths and King [3], Cowen and Griffiths [2] and Wong [10]. Here we shall give a direct and elementary proof, which reduces the several variable case to the one variable one by fiber integration. As Vitter [9] has shown, the defect relation for meromorphic maps $f: \mathbb{C}^n \to \mathbb{P}_m$ can be easily derived from the Lemma of the Logarithmic Derivative. Also see Griffiths and King [3] Proposition 9.3 and Stoll [8] Theorem 21.16.

For the formulation of the results some notations have to be introduced. If N is an ordered set, if $M \subseteq N$, if $a \in N$ and $b \in N$ with $a \leq b$, denote

$$M(a, b) = \{x \in M \mid a < x < b\} \qquad M[a, b] = \{x \in M \mid a \leq x \leq b\}$$

$$M[a, b) = \{x \in M \mid a \leq x < b\} \qquad M(a, b] = \{x \in M \mid a < x \leq b\} .$$

If $z = (z_1, \cdots, z_n) \in \mathbb{C}^n$, define $|z| = (|z_1|^2 + \cdots + |z_n|^2)^{1/2}$ and $\tau(z) = |z|^2$. If $r > 0$ define

*Supported in part by a grant from the Consiglio Nazionale delle Ricerche.

**Supported in part by National Science Foundation Grant No. MCS 78-02099.

Recent Developments in Several Complex Variables
0-691-08285-5/81/000029-17 $00.85/0 (cloth)
0-691-08281-2/81/000029-17 $00.85/0 (paperback)

$$\mathbb{C}^n(r) = \{z \in \mathbb{C}^n \mid |z| < r\} \qquad \mathbb{C}^n[r] = \{z \in \mathbb{C}^n \mid |z| \leq r\}$$

$$\mathbb{C}^n\langle r\rangle = \{z \in \mathbb{C}^n \mid |z| = r\}.$$

If $n = 1$, omit the upper index 1. The sphere $\mathbb{C}^n\langle r\rangle$ is considered to be a real analytic manifold oriented to the exterior of $\mathbb{C}^n(r)$. Define $d^c = (i/4\pi)(\bar{\partial} - \partial)$. The pullback of the form

$$\sigma_n = d^c \log \tau \wedge (dd^c \log \tau)^{n-1} \quad \text{on} \quad \mathbb{C}^n - \{0\}$$

to $\mathbb{C}^n\langle r\rangle$ defines a positive measure on $\mathbb{C}^n\langle r\rangle$ with total measure 1. On $\mathbb{C}^n$ define

$$\upsilon_n = dd^c \tau > 0 \qquad \rho_n = \upsilon_n^n.$$

Then ρ_n is the Lebesque measure on $\mathbb{C}^n$ normalized such that $\mathbb{C}^n(r)$ has measure r^{2n}. The complex projective line $\mathbb{P}_1$ is identified with $\mathbb{C} \cup \{\infty\}$. For a and b in $\mathbb{P}_1$ the chordal distance from a to b is denoted by $\|a, b\|$ and defined by

$$\|a, b\| = \frac{|a - b|}{\sqrt{1 + |a|^2}\,\sqrt{1 + |b|^2}} \qquad \text{if } a \in \mathbb{C} \text{ and } b \in \mathbb{C}$$

$$\|a, \infty\| = \frac{1}{\sqrt{1 + |a|^2}} \qquad \text{if } a \in \mathbb{C}$$

where $\|a, a\| = 0$ and $0 \leq \|a, b\| = \|b, a\| \leq 1$. There exists one and only one positive form ω of degree 2 on $\mathbb{P}_1$ such that

$$\omega(z) = dd^c \log(1/\|a, z\|^2) \qquad \forall z \in \mathbb{P}_1 - \{a\} \quad \forall a \in \mathbb{P}_1$$

On $\mathbb{C}$ we have $(1 + \tau)^2 \omega = \upsilon_1$.

Let f be a meromorphic function on $\mathbb{C}^n$. Then f is identified with a meromorphic map $f: \mathbb{C}^n \to \mathbb{P}_1$ such that $f^{-1}(\infty) \neq \mathbb{C}^n$. For all $0 < s < r$ the *characteristic* of f is defined by

$$T_f(r, s) = \int_s^r t^{1-2n} \int_{\mathbb{C}^n(t)} f^*(\omega) \wedge \upsilon_n^{n-1}\, dt$$

If $j \in \mathbb{N}[1, n]$, let f_{z_j} be the jth partial derivative of f. Take $s \in \mathbb{R}$. Let g and h real valued functions on $\mathbb{R}(s, \infty)$. We write $g(r) \underset{\cdot}{\leq} h(r)$ if there exists a subset E of $\mathbb{R}(s, \infty)$ with finite Lebesgue measure such that $g(r) \leq h(r)$ for all $r \in \mathbb{R}[s, \infty) - E$. Now we are ready to formulate the *Lemma on the Logarithmic Derivative.*

THEOREM. *For every meromorphic function* f *on* $\mathbb{C}^n$ *and every* $j \in \mathbb{N}[1, n]$ *and* $0 < s \in \mathbb{R}$, *we have*

$$\int_{\mathbb{C}^n\langle r\rangle} \log^+ \frac{|f_{z_j}|}{|f|} \sigma_n \underset{\cdot}{\leq} 17 \log^+ (r T_f(r, s)) .$$

Let us recall the standard value distribution functions. Compare Stoll [7] as reference. Let ν be a divisor on $\mathbb{C}^n$. Since $\mathbb{C}^n$ is a manifold, we can identify ν with its multiplicity function.

Abbreviate

$$S(r) = \mathbb{C}^n[r] \cap \operatorname{supp} \nu \qquad \text{if } 0 < r \in \mathbb{R} .$$

The *counting function* of ν is defined by

$$n_\nu(r) = r^{2-2n} \int_{S(r)} \nu \, v_n^{n-1} \qquad \text{if } n > 1$$

$$n_\nu(r) = \sum_{z \in S(r)} \nu(z) \qquad \text{if } n = 1 .$$

For all $0 < s < r$ the *valence function* of ν is defined by

$$N_\nu(r, s) = \int_s^r n_\nu(t) \frac{dt}{t} . \tag{1.1}$$

If $\nu \geq 0$, then $n_\nu \geq 0$ increases and if $\theta > 1$, then

$$(1.2) \qquad n_\nu(r) \leq \frac{\theta}{\theta-1} N_\nu(\theta r, s) \qquad \text{if } 0 < s < r .$$

Let f be a meromorphic function on $\mathbb{C}^n$. If $a \in \mathbb{P}_1$ and if $f^{-1}(a) \neq \mathbb{C}$, the a-divisor $\nu_f^a \geq 0$ is defined. Its counting function and valence function are denoted by $n_f(r, a)$ and $N_f(r, s; a)$ respectively. If $f^{-1}(0) \neq \mathbb{C}$, then $\nu_f = \nu_f^0 - \nu_f^\infty$ is the divisor of f and its counting function and valence function are denoted by $n_f(r)$ and $N_f(r, s)$ respectively.

Jensen's formula asserts

$$N_f(r, s) = \int_{\mathbb{C}^n\langle r\rangle} \log|f|\sigma_n - \int_{\mathbb{C}^n\langle s\rangle} \log|f|\sigma_n$$

if $0 < s < r$. Let $g \not\equiv 0$ and $h \not\equiv 0$ be holomorphic functions on $\mathbb{C}^n$ such that $hf = g$. Let $\mu \geq 0$ be the greatest common divisor of (g, h). Then

$$T_f(r, s) + N_\mu(r, s)$$

$$= \int_{\mathbb{C}^n\langle r\rangle} \log \sqrt{|g|^2 + |h|^2}\, \sigma_n - \int_{\mathbb{C}^n\langle s\rangle} \log \sqrt{|g|^2 + |h|^2}\, \sigma_n$$

if $0 < s < r$. If $a \in \mathbb{P}_1$ and $f^{-1}(a) \neq \mathbb{C}^n$, the *compensation function* of f is defined by

$$m_f(r, a) = \int_{\mathbb{C}^n\langle r\rangle} \log \frac{1}{\|f, a\|}\, \sigma_n \geq 0 \qquad \forall r > 0 .$$

For $0 < s < r$ the *First Main Theorem* states

$$(1.3) \qquad T_f(r, s) = N_f(r, s; a) + m_f(r; a) - m_f(s; a)$$

2. *A one variable estimate*

The reduction to the one variable case requires the explicit evaluation of the constants. Therefore we recalculate some parts of the one variable proof. If $a_j > 0$ for $j = 1, \cdots, p$, we recall

$$\log^+(a_1 \cdots a_p) \leq \log^+ a_1 + \cdots + \log^+ a_p$$
$$\log^+(a_1 + \cdots + a_p) \leq \log^+ a_1 + \cdots + \log^+ a_p + \log p\,.$$

LEMMA 2.1. *Let* $f \not\equiv 0$ *be a meromorphic function on* $\mathbb{C}$. *If* $0 < r < R$, *then*

$$\int_{\mathbb{C}\langle r\rangle} \log^+ |f'/f| \sigma_1 \leq \log^+ m_f(R, 0) + \log^+ m_f(R, \infty)$$
$$+ 3 \log^+ n_f(R, 0) + 3 \log^+ n_f(R, \infty) + \log^+ \frac{R}{(R-r)^2}$$
$$+ \log^+ \frac{1}{r} + \log^+ \frac{r}{R-r} + 8 \log 2\,.$$

(Compare Hayman [5], p. 37 and Gross [4] Lemma 3.6.)

Proof. Differentiation of the Jensen-Poisson formula yields (Hayman [5] Theorem 1.1, p. 1 and (2.2), p. 36)

$$\frac{f'(z)}{f(z)} = \int_{\mathbb{C}\langle R\rangle} \log|f(\zeta)| \frac{2\zeta}{(\zeta - z)^2} \sigma_1(\zeta) + \sum_{u \in \mathbb{C}[R]} \nu(u) \left[\frac{1}{z-u} + \frac{\bar{u}}{R^2 - \bar{u}z}\right]$$

for all $z \in \mathbb{C}[R] - \operatorname{supp} \nu$. If $\mathbb{C}[R] \cap \operatorname{supp} \nu \neq \emptyset$, define η on $\mathbb{C}[R] - \operatorname{supp} \nu$ by

$$\eta(z) = \operatorname{Max} \left\{ \frac{1}{|z-u|} \,\middle|\, u \in \mathbb{C}[R] \cap \operatorname{supp} \gamma \right\}$$

if $\mathbb{C}[R] \cap \operatorname{supp} \nu = \emptyset$, define $\eta = 0$. If $z \in \mathbb{C}[r] - \operatorname{supp} \nu$, then

$$\left|\frac{f'(z)}{f(z)}\right| \leq \frac{2R}{(R-r)^2} \int_{\mathbb{C}\langle R\rangle} |\log|f|| \sigma_1$$
$$+ (1/r)(n_f(R, 0) + n_f(R, \infty)) \left(r\, \eta(z) + \frac{r}{R-r}\right).$$

We have

$$\int_{C\langle R\rangle} |\log|f||\sigma_1 \leq \int_{C\langle R\rangle} \log\sqrt{1+|f|^2}\,\sigma_1 + \int_{C\langle R\rangle} \log\frac{\sqrt{1+|f|^2}}{|f|}\,\sigma_1$$

$$= m_f(R,0) + m_f(R,\infty) .$$

Therefore

$$\log^+\left|\frac{f'(z)}{f(z)}\right| \leq \log^+ m_f(R,0) + \log^+ m_f(R,\infty)$$

$$+ \log^+ n_f(R,0) + \log^+ n_f(R,\infty)$$

$$+ \log^+\frac{1}{r} + \log^+(r\,\eta(z)) + \log^+\frac{r}{R-r} + \log^+\frac{R}{(R-r)^2} + 5\log 2$$

for all $z \in C[r] - \operatorname{supp}\nu$. By Hayman [5], Lemma 2.2, we have

$$\int_{C\langle r\rangle} \log^+(r\,\eta)\,\sigma_1 \leq 2\log^+(n_f(R,0)+n_f(R,\infty)) + \frac{1}{2}$$

$$\leq 2\log^+ n_f(R,0) + 2\log^+ n_f(R,\infty) + 2\log 2 + \frac{1}{2} .$$

Since $1/2 < \log 2$, we obtain

$$\int_{C\langle r\rangle} \log^+|f'/f|\,\sigma_1 \leq \log^+ m_f(R,0) + \log^+ m_f(R,\infty)$$

$$+ 3\log^+ n_f(R,0) + 3\log^+ n_f(R,\infty)$$

$$+ \log^+\frac{1}{r} + \log^+\frac{r}{R-r} + \log^+\frac{R}{(R-r)^2} + 8\log 2 . \qquad \text{q.e.d.}$$

Now, the lemma on the logarithmic derivative in one variable can be derived along classical lines, which shall not concern us here if $n = 1$.

3. *The case of several variables*

Assume that $n > 1$. As before define $\tau(z) = |z|^2$ if $z \in \mathbb{C}^n$. Define $\tilde{\tau}$ on $\mathbb{C}^{n-1}$ by $\tilde{\tau}(w) = |w|^2$ for all $w \in \mathbb{C}^{n-1}$. Take $r > 0$. Define p on $\mathbb{C}^{n-1}[r]$ by

$$p = \sqrt{r^2 - \tilde{\tau}}$$

LEMMA 3.1. *Take* $r > 0$. *Let* h *be a function on* $\mathbb{C}^n\langle r\rangle$ *such that* $h\sigma_n$ *is integrable over* $\mathbb{C}^n\langle r\rangle$, *then*

$$\int_{\mathbb{C}^n\langle r\rangle} h\sigma_n = r^{2-2n} \int_{\mathbb{C}^{n-1}[r]} \left(\int_{\mathbb{C}\langle p(w)\rangle} h(w, \zeta)\sigma_1(\zeta) \right) \rho_{n-1}(w) .$$

(Compare Stoll [6], (3.16), p. 141.)

Proof. Define $E = \{(z_1, \cdots, z_n) \in \mathbb{C}^n\langle r\rangle \mid 0 \leq z_n \in \mathbb{R}\}$ and $S = \mathbb{C}^n\langle r\rangle - E$. Then E is a closed subset of $\mathbb{C}^n\langle r\rangle$ with zero (2n–1)-dimensional Hausdorff measure. A bijective map

$$g: \mathbb{R}(0, 2\pi) \times \mathbb{C}^{n-1}(r) \to S$$

of class C^∞ is defined by $g(\phi, w) = (w, p(w)e^{i\phi})$ for all $\phi \in \mathbb{R}(0, 2\pi)$ and all $w \in \mathbb{C}^{n-1}(r)$. Let ϕ be the variable in $\mathbb{R}(0, 2\pi)$ and let $w_1, \cdots, w_{n-1}$ be the complex variables on $\mathbb{C}^{n-1}$ with x_j the real part of w_j and y_j be the imaginary part of w_j. The partial derivatives $g_\phi, g_{x_1}, g_{y_1}, \cdots, g_{x_{n-1}}, g_{y_{n-1}}$ are pointwise linearly independent over $\mathbb{R}$ and g is perpendicular to these derivatives. Here $g(\phi, w)$ points in the direction of the exterior normal of $\mathbb{C}^{n-1}\langle r\rangle$ at $g(\phi, w)$. Since

$$\det(g, g_\phi, g_{x_1}, g_{y_1}, \cdots, g_{x_{n-1}}, g_{y_{n-1}}) = r^2 > 0$$

the map g is an orientation preserving diffeomorphism. We shall compute $g^*(\sigma_n)$. We have

$$g^*(d^c\tau) = d^c\tilde{\tau} + (1/2\pi)(r^2-\tilde{\tau})\,d\phi$$

$$g^*(dd^c\tau) = dd^c\tilde{\tau} - (1/2\pi)\,d\tilde{\tau}\wedge d\phi$$

$$g^*(dd^c\tau)^{n-1} = (dd^c\tilde{\tau})^{n-1} - \frac{n-1}{2\pi}\,d\tilde{\tau}\wedge d\phi\wedge(dd^c\tilde{\tau})^{n-2}$$

$$g^*(d^c\tau\wedge(dd^c\tau)^{n-1}) = (1/2\pi)(r^2-\tilde{\tau})\,d\phi\wedge(dd^c\tilde{\tau})^{n-1}$$

$$+\frac{n-1}{2\pi}\,d\phi\wedge d\tilde{\tau}\wedge d^c\tilde{\tau}\wedge(dd^c\tilde{\tau})^{n-2}$$

Since $\tilde{\tau}$ is a parabolic exhaustion of $\mathbb{C}^{n-1}$ (Stoll [8] (10.10), pp. 74 and 79) we have

$$\tilde{\tau}(dd^c\tilde{\tau})^{n-1} = (n-1)\,d\tilde{\tau}\wedge d^c\tilde{\tau}\wedge(dd^c\tilde{\tau})^{n-2}$$

Hence

$$g^*(d^c\tau\wedge(dd^c\tau)^{n-1}) = (1/2\pi)\,r^2 d\phi\wedge(dd^c\tilde{\tau})^{n-1}$$

Since $\tau\circ g = r^2$ and $(dd^c\tilde{\tau})^{n-1} = \rho_{n-1}$, we obtain

$$g^*(\sigma_n) = g^*(d^c\log\tau\wedge(dd^c\log\tau)^{n-1}) = g^*(\tau^{-n}d^c\tau\wedge(dd^c\tau)^{n-1})$$

$$g^*((\sigma_n) = (1/2\pi)\,r^{2-2n}\,d\phi\wedge\rho_{n-1}\,.$$

Fubini's theorem implies

$$\int_{\mathbb{C}^n\langle r\rangle} h\sigma_n = r^{2-2n}\int_{\mathbb{C}^{n-1}[r]}\frac{1}{2\pi}\int_0^{2\pi} h(w, p(w)e^{i\phi})\,d\phi\wedge\rho_{n-1}(w)\,.$$

q.e.d.

Let f be a meromorphic function on $\mathbb{C}^n$. Take $w\in\mathbb{C}^{n-1}$ and define a holomorphic map $j_w:\mathbb{C}\to\mathbb{C}^n$ by $j_w(z) = (w, z)$. If $j_w(\mathbb{C})$ is not contained in supp ν_f^∞, then f pulls back to a meromorphic function $f[w] = j_w^*(f)$. If $(w, z)\in\mathbb{C}^n - \operatorname{supp}\nu_f^\infty$, then $f[w](z) = f(w, z)$. Take $a\in\mathbb{P}_1$. If also $j_w(\mathbb{C})$ is not contained in supp ν_f^a, the a-divisor of $f[w]$ is defined: $\nu_{f[w]}^a$.

LEMMA 3.2 (Stoll [6], Hilfssatz 7). *Take* $a \in \mathbb{P}_1$. *Let* f *be a meromorphic function on* $\mathbb{C}^n$ *with* $f \neq a$. *If* $r > 0$, *then*

$$r^{2-2n} \int_{\mathbb{C}^{n-1}[r]} n_{f[w]}(p(w), a)\, \rho_{n-1}(w) \leq n_f(r, a) .$$

Proof. For each $j \in \mathbb{N}[1, n]$ define $\pi_j : \mathbb{C}^n \to \mathbb{C}^{n-1}$ by

$$\pi_j(z_1, \cdots, z_n) = (z_1, \cdots, z_{j-1}, z_{j+1}, \cdots, z_n) .$$

Then

$$\upsilon_n^{n-1} = \sum_{j=1}^{n} \pi_j^*(\upsilon_{n-1}^{n-1}) \geq \pi_n^*(\rho_{n-1}) .$$

Abbreviate $\nu = \nu_f^a$ and $\nu_w = \nu_{f[w]}^a$ if $w \in \mathbb{C}^{n-1}$. Define $N = \mathbb{C}^n(r) \cap \operatorname{supp} \nu$. If $N = \emptyset$, the lemma is trivial. Assume that $N \neq \emptyset$. Then N is a pure (n–1)-dimensional analytic subset of $\mathbb{C}^n(r)$. Define $\pi = \pi_n | N$. Then $\pi(N) \subseteq \mathbb{C}^{n-1}(r)$. The set $E = \{z \in N | \operatorname{rank}_z \pi < n-1\}$ is analytic in $\mathbb{C}^n(r)$. Let E_0 be the union of all (n–1)-dimensional branches of E and let E_1 be the union of all other branches of E. Then E_0 and E_1 are analytic in $\mathbb{C}^n(r)$ with $\dim E_1 \leq n-2$ and $\dim E_0 = n-1$ if $E_0 \neq \emptyset$. Also $E'_0 = \pi(E_0)$ is analytic in $\mathbb{C}^{n-1}(r)$ with $E_0 = \pi_n^{-1}(E'_0) \cap \mathbb{C}^n(r)$. Therefore $\pi^*(\rho_{n-1}) = 0$ on E_0. The complement $N_0 = N - E_0$ is open in N and $N_0 \cap E_1$ is thin analytic in N_0 if $N_0 \neq \emptyset$. Then $N_1 = N_0 - E_1 = N - E$ is open in N. A thin analytic subset E_2 of N_1 exists such that π is locally biholomorphic on $N_2 = N_1 - E_2$. Then $D = \pi(E_0 \cup E_1 \cup E_2)$ has measure zero in $\mathbb{C}^{n-1}$. Also $M_2 = \pi(N_2)$ is open in $\mathbb{C}^{n-1}$. Here $M = \pi(N) = M_2 \cup D$ and $M_3 = M_2 - D$ differ by sets of measure zero from M_2.

The intersection $\mathbb{C}^n\langle r\rangle \cap \operatorname{supp} \nu$ has (2n–2)-dimensional Hausdorff measure zero. If $w \in \mathbb{C}^{n-1}(r)$, then

$$\pi^{-1}(w) = (\{w\}\times \mathbb{C}(p(w))) \cap N$$

and for almost all $w \in \mathbb{C}^{n-1}(r)$ we have

$$\pi^{-1}(w) = (\{w\}\times \mathbb{C}[p(w)]) \cap \operatorname{supp} \nu .$$

For all $w \in \mathbb{C}^{n-1}(r) - M$, we have $\pi^{-1}(w) = \emptyset$, hence $n_{f[w]}(p(w), a) = 0$ for almost all $w \in \mathbb{C}^{n-1}(r) - M$. If $w \in M_3$, then

$$\pi^{-1}(w) = (\{w\}\times \mathbb{C}(p(w))) \cap N .$$

Take $a = (b, c) \in N_2$ with $b = \pi(a) \in \mathbb{C}^{n-1}(r)$ and $c \in \mathbb{C}$. There exist open, connected neighborhoods V of b in $\mathbb{C}^{n-1}(r)$ and W of c and $U = V\times W$ of a in $\mathbb{C}^n(r)$ with $U\cap N_2 = U\cap N$ such that $\pi : U\cap N \to V$ in biholomorphic. Let λ be the inverse map. Then a holomorphic function $h : V \to W$ exists such that $\lambda(w) = (w, h(w))$ for all $w \in V$. The multiplicity $\nu_f^a(w, z) = q$ is constant for all $(w, z) \in U\cap N$. A holomorphic function $F : U \to \mathbb{C}-\{0\}$ exists such that

$$f(w, z) = a+(z-h(w))^q F(w, z)$$

for all $(w, z) \in U$. Therefore $\nu(w, z) = \nu_f^a(w, z) = q = \nu_{f[w]}^a(z) = \nu_w(z)$ for all $(w, z) \in U\cap N$. Altogether we obtain

$$r^{2n-2} n_f(r, a) = \int_N \nu\, \upsilon_n^{n-1} \geq \int_N \nu\, \pi^*(\rho_{n-1}) = \int_{N_0} \nu\, \pi^*(\rho_{n-1})$$

$$= \int_{N_2} \nu\, \pi^*(\rho_{n-1}) = \int_{M_2} \sum_{(w,z)\in N_2} \nu(w, z)\, \rho_{n-1}(w)$$

$$= \int_{M_3} \Big(\sum_{z\in \mathbb{C}(p(w))} \nu_w(z) \Big) \rho_{n-1}(w) = \int_{M_3} n_{f[w]}(p(w), a)\, \rho_{n-1}(w)$$

$$= \int_M n_{f[w]}(p(w), a)\, \rho_{n-1}(w) = \int_{\mathbb{C}^{n-1}(r)} n_{f[w]}(p(w), a)\, \rho_{n-1}(w). \quad \text{q.e.d.}$$

LEMMA 3.3. *Take* $a \in \mathbb{P}_1$. *Let* f *be a meromorphic function on* $\mathbb{C}^n$ *with* $f \not\equiv a$. *Take* $0 < r \in \mathbb{R}$. *Then*

$$m_f(r, a) = r^{2-2n} \int_{\mathbb{C}^{n-1}(r)} m_{f[w]}(p(w), a)\, \rho_{n-1}(w)\,.$$

Proof. By Lemma 3.1 we have

$$\begin{aligned} m_f(r, a) &= -\int_{\mathbb{C}^n\langle r\rangle} \log \|a, f\|\, \sigma_n \\ &= -r^{2-2n} \int_{\mathbb{C}^{n-1}(r)} \int_{\mathbb{C}\langle p(w)\rangle} \log \|a, f(w, z)\|\, \sigma_1(z)\, \rho_{n-1}(w) \\ &= r^{2-2n} \int_{\mathbb{C}^{n-1}(r)} m_{f[w]}(p(w), a)\, \rho_{n-1}(w)\,. \end{aligned}$$

q.e.d.

LEMMA 3.4. *If* $1 \leq r \in \mathbb{R}$, *then*

$$\frac{1}{r^{2n}} \int_{\mathbb{C}^n[r]} \log^+ \frac{1}{\sqrt{r^2 - \tau}}\, \rho_n \leq \frac{1}{2} \sum_{q=1}^{n} \frac{1}{q}\,.$$

Proof. We have

$$\frac{1}{r^{2n}} \int_{\mathbb{C}^n[r]} \log^+ \frac{1}{\sqrt{r^2 - \tau}}\, \rho_n = \int_{\mathbb{C}^n[1]} \log^+ \frac{1}{r\sqrt{1-\tau}}\, \rho_n \leq \int_{\mathbb{C}^n[1]} \log \frac{1}{\sqrt{1-\tau}}\, \rho_n$$

where

$$\begin{aligned} \int_{\mathbb{C}^n[1]} \log \frac{1}{\sqrt{1-\tau}}\, \rho_n &= \frac{n}{2} \int_{\mathbb{C}^n(1)} \log \frac{1}{1-\tau}\, \tau^{n-1}\, d\tau \wedge \sigma_n \\ &= \frac{n}{2} \int_0^1 \log \frac{1}{1-t}\, t^{n-1}\, dt = \frac{1}{2} \sum_{q=1}^{n} \frac{1}{q}\,. \end{aligned}$$

q.e.d.

LEMMA 3.5. *Let* $h \geq 0$ *be a non-negative function on* $\mathbb{C}^n(r)$ *such that* $\log^+ h$ *is integrable over* $\mathbb{C}^n(r)$. *Then*

$$\frac{1}{r^{2n}} \int_{\mathbb{C}^n(r)} \log^+ h \, \rho_n \leq \log^+\left(\frac{1}{r^{2n}} \int_{\mathbb{C}^n(r)} h \, \rho_n\right) + \log 2 .$$

Proof. We have

$$\frac{1}{r^{2n}} \int_{\mathbb{C}^n(r)} \log^+ h \, \rho_n \leq \frac{1}{r^{2n}} \int_{\mathbb{C}^n(r)} \log(1+h)\, \rho_n \leq \log\left(\frac{1}{r^{2n}} \int_{\mathbb{C}^n(r)} (1+h)\, \rho_n\right)$$

$$= \log\left(1+\frac{1}{r^{2n}} \int_{\mathbb{C}^n(r)} h \, \rho_n\right) \leq \log^+\left(\frac{1}{r^{2n}} \int_{\mathbb{C}^n(r)} h \, \rho_n\right) + \log 2 .$$

q.e.d.

LEMMA 3.6. *Let* $f \not\equiv 0$ *be a meromorphic function on* $\mathbb{C}^n$ *with* $n > 1$. *Let* r *and* R *be real numbers with* $1 \leq r < R$. *Then*

$$\int_{\mathbb{C}^n\langle r\rangle} \log^+ \frac{|f_{z_n}|}{|f|} \sigma_n \leq \log^+ m_f(R,0) + \log^+ m_f(R,\infty)$$

$$+ 3 \log^+ n_f(R,0) + 3 \log^+ n_f(R,\infty)$$

$$+ \log^+ \frac{1}{\sqrt{R^2 - r^2}} + \log^+ \frac{R^2}{(R-r)^2} + \log^+ \frac{r}{R-r}$$

$$+ 16(n-1) \log \frac{R}{r} + \frac{1}{2} \sum_{q=1}^{n-1} \frac{1}{q} + 16 \log 2 .$$

Proof. As before define $p(w) = \sqrt{r^2 - |w|^2}$ if $w \in \mathbb{C}^{n-1}[r]$. Also define $P(w) = \sqrt{R^2 - |w|^2}$ for all $w \in \mathbb{C}^{n-1}[R]$. Lemma 3.1 implies

$$\int_{\mathbb{C}^n\langle r\rangle} \log^+ \frac{|f_{z_n}|}{|f|}\,\sigma_n = r^{2-2n} \int_{\mathbb{C}^{n-1}(r)} \int_{\mathbb{C}\langle p(w)\rangle} \log^+ \frac{|f'[w](z)|}{|f[w](z)|}\,\sigma_1(z)\,\rho_{n-1}(w).$$

By Lemma 2.1 we have

$$\begin{aligned}\int_{\mathbb{C}^n\langle r\rangle} \log^+ \frac{|f_{z_n}|}{|f|}\,\sigma_n \le r^{2-2n} \Bigg[&\int_{\mathbb{C}^{n-1}(r)} \log^+ m_{f[w]}(P(w),0)\,\rho_{n-1}(w) \\ &+ \int_{\mathbb{C}^{n-1}(r)} \log^+ m_{f[w]}(P(w),\infty)\,\rho_{n-1}(w) \\ &+ 3 \int_{\mathbb{C}^{n-1}(r)} \log^+ n_{f[w]}(P(w),0)\,\rho_{n-1}(w) \\ &+ 3 \int_{\mathbb{C}^{n-1}(r)} \log^+ n_{f[w]}(P(w),\infty)\,\rho_{n-1}(w) \\ &+ \int_{\mathbb{C}^{n-1}(r)} \log^+ \frac{P(w)}{(P(w)-p(w))^2}\,\rho_{n-1}(w) \\ &+ \int_{\mathbb{C}^{n-1}(r)} \log^+ \frac{1}{p(w)}\,\rho_{n-1}(w) \\ &+ \int_{\mathbb{C}^{n-1}(r)} \log^+ \frac{p(w)}{P(w)-p(w)}\,\rho_{n-1}(w)\Bigg] + 8\log 2\,.\end{aligned}$$

If $a = 0$ or $a = \infty$, Lemma 3.5 and Lemma 3.3 imply

$$r^{2-2n} \int_{\mathbb{C}^{n-1}(r)} \log^+ m_{f[w]}(P(w), a)\, \rho_{n-1}(w)$$

$$\leq \log^+ \Big(r^{2-2n} \int_{\mathbb{C}^{n-1}(r)} m_{f[w]}(P(w), a)\, \rho_{n-1}(w)\Big) + \log 2$$

$$\leq \log^+ \Big(R^{2-2n} \int_{\mathbb{C}^{n-1}(R)} m_{f[w]}(P(w), a)\, \rho_{n-1}(w)\Big) + (2n-2) \log \frac{R}{r} + \log 2$$

$$= \log^+ m_f(R, a) + (2n-2) \log \frac{R}{r} + \log 2 .$$

If $a = 0$ or $a = \infty$, Lemma 3.5 and Lemma 3.2 imply

$$r^{2-2n} \int_{\mathbb{C}^{n-1}(r)} \log^+ n_{f[w]}(P(w), a)\, \rho_{n-1}(w)$$

$$\leq \log^+ \Big(r^{2-2n} \int_{\mathbb{C}^{n-1}(r)} \log^+ n_{f[w]}(P(w), a)\, \rho_{n-1}(w)\Big) + \log 2$$

$$\leq \log^+ \Big(R^{2-2n} \int_{\mathbb{C}^{n-1}(R)} n_{f[w]}(P(w), a)\, \rho_{n-1}(w)\Big) + (2n-2) \log \frac{R}{r} + \log 2$$

$$\leq \log^+ n_f(R, a) + (2n-2) \log \frac{R}{r} + \log 2 .$$

If $w \in \mathbb{C}^{n-1}[r]$, then $p(w) \leq (r/R)\, P(w)$ and $P(w) \geq \sqrt{R^2 - r^2}$. Hence

$$\frac{P(w)}{(P(w)-p(w))^2} = \frac{1}{P(w)} \frac{1}{\left(1-\frac{p(w)}{P(w)}\right)^2} \leq \frac{1}{\sqrt{R^2-r^2}} \frac{1}{\left(1-\frac{r}{R}\right)^2} .$$

Therefore

$$r^{2-2n} \int_{C^{n-1}[r]} \log^+ \frac{P(w)}{(P(w)-p(w))^2} \rho_{n-1}(w) \leq \log^+ \frac{1}{\sqrt{R^2-r^2}} + \log^+ \frac{R^2}{(R-r)^2} .$$

By Lemma 3.4 we have

$$r^{2-2n} \int_{C^{n-1}(r)} \log^+ \frac{1}{p} \rho_{n-1} \leq \frac{1}{2} \sum_{q=1}^{n-1} \frac{1}{q} .$$

If $w \in C^{n-1}[r]$, then $p(w)/(P(w)-p(w)) \leq r/(R-r)$. Hence

$$r^{2-2n} \int_{C^{n-1}(r)} \log^+ \frac{p}{P-p} \rho_{n-1}(w) \leq \log^+ \frac{r}{R-r} .$$

Putting these estimates together we obtain Lemma 3.6. q.e.d.

The next result can be interpreted as a Lemma of the Logarithmic Derivative without exceptional intervals.

PROPOSITION 3.7. *Let* f *be a meromorphic function on* $\mathbb{C}^n$ *with* $n > 1$. *Take* $r > 1$, $s > 0$ *and* $\theta > 1$ *with* $r > s$. *Then*

$$\int_{\mathbb{C}^n\langle r\rangle} \log^+ \frac{|f_{z_n}|}{|f|} \sigma_n \leq 8 \log^+ T_f(\theta r, s) + 4 \log^+ m_f(s,0) + 4 \log^+ m_f(s,\infty)$$
$$+ 8(2n-1) \log \theta + 10 \log^+ \frac{1}{\theta-1} + 33 \log 2 + \frac{1}{2} \sum_{q=1}^{n-1} \frac{1}{q} .$$

Proof. Take $R = \frac{1+\theta}{2} r$. Then $r < R < \theta r$. Then $1 < \frac{\theta r}{R} = \frac{2\theta}{1+\theta}$. For $a = 0$ or $a = \infty$, the First Main Theorem (1.3) and (1.2) imply

$$\log^+ m_f(R, a) \leq \log^+ T_f(\theta r, s) + \log^+ m_f(s, a) + \log 2$$

$$\log^+ n_f(R, a) \leq \log^+ \frac{2\theta}{\theta - 1} N_f(\theta r, s; a) \leq \log^+ T_f(\theta r, s) + \log^+ m_f(s, a)$$
$$+ \log \theta + \log^+ \frac{1}{\theta - 1} + 2 \log 2$$

$$\log^+ \frac{R^2}{(R-r)^2} = 2 \log \frac{\theta+1}{\theta-1} \leq 2 \log \theta + 2 \log^+ \frac{1}{\theta - 1} + 2 \log 2$$

$$\log^+ \frac{1}{\sqrt{R^2 - r^2}} = \frac{1}{2} \log^+ \left(\frac{1}{r^2} \frac{4}{(\theta+1)^2 - 4} \right) \leq \frac{1}{2} \log^+ \frac{1}{\theta - 1}$$

$$\log^+ \frac{r}{R-r} = \log^+ \frac{2}{\theta - 1} \leq \log^+ \frac{1}{\theta - 1} + \log 2$$

$$\log^+ \frac{R}{r} = \log \frac{\theta+1}{2} \leq \log \theta \; .$$

These estimates and Lemma 3.6 complete the proof. q.e.d.

THEOREM 3.8 (The Lemma on the Logarithmic Derivative). *Let* f *be a meromorphic function on* $\mathbb{C}^n$ *with* $n > 1$. *Take* $j \in \mathbb{N}[1, n]$ *and* $0 < s < r$. *Then*

$$\int_{\mathbb{C}^n\langle r\rangle} \log^+ \frac{|f_{z_j}|}{|f|} \sigma_n \;\dot{\leq}\; 17 \log^+ (r\, T_f(r, s)) \; .$$

Proof. W.l.o.g. we can assume that $f \not\equiv 0$ and that $j = n$. Take $r_0 > s + 1$ such that $T_f(r, s) \geq e$ for all $r \geq r_0$. By Gross [4] Lemma 3.7, p. 57 and (3.41), p. 59 we have

$$T_f\left(r + \frac{1}{\log T_f(r, s)}, s\right) \;\dot{\leq}\; T_f(r, s)^2 \; .$$

Take $r > r_0$ and $\theta = 1 + 1/(r \log T_f(r, s)) < 2$. Proposition 3.7 implies

$$\int_{\mathbb{C}^n\langle r\rangle} \log^+ \frac{|f_{z_n}|}{|f|} \sigma_n \;\dot{\leq}\; 16 \log T_f(r, s) + 4 \log^+ m_f(s, 0) + 4 \log^+ m_f(s, \infty)$$
$$+ (8(2n-1) + 33) \log 2 + 10 \log (r \log T_f(r, s))$$
$$+ \frac{1}{2} \sum_{q=1}^{n-1} \frac{1}{q}$$
$$\dot{\leq}\; 17 \log (r\, T_f(r, s)) \; .$$

q.e.d.

Following Vitter [9] the defect relation can be derived easily. Some care has to be taken to interpret the proceedings on page 103 properly and to define D correctly.

DEPARTMENT OF MATHEMATICS
UNIVERSITY OF NOTRE DAME
NOTRE DAME, INDIANA 46556
U.S.A.

REFERENCES

[1] Carlson, J. and Ph. Griffiths: A defect relation for equidimensional holomorphic mappings between algebraic varieties. Ann. of Math. *95*(1972), 557-584.

[2] Cowen, M. and Ph. Griffiths: Holomorphic curves and metrics of negative curvature. J. Analyse Math. *29*(1976), 93-152.

[3] Griffiths, Ph. and J. King: Nevanlinna theory and holomorphic mappings between algebraic varieties. Acta Math. *130*(1973), 145-220.

[4] Gross, F.: Factorization of meromorphic functions. Math. Research Center. Naval Research Lab. Washington, D.C. (1972), pp. 258.

[5] Hayman, W.: Meromorphic functions. Oxford University Press (1964), pp. 191.

[6] Stoll, W.: Mehrfache Integrale auf komplexen Mannigfaltigkeiten, Math. Zeitschr. *57*(1952), 116-152.

[7] ________: Holomorphic functions of finite order. Conf. board math. sciences. Reg. conf. series *21*(1974), pp. 83.

[8] ________: Value distribution on parabolic spaces. Lecture Notes in Math. *600*(1977), pp. 216. Springer-Verlag.

[9] Vitter, A.: The lemma of the logarithmic derivative in several complex variables. Duke Math. J. *44*(1977), 89-104.

[10] Wong, P.: Defect relation for meromorphic maps on parabolic manifolds. (To appear in Duke J.)

GRAPH THEORETIC TECHNIQUES IN ALGEBRAIC GEOMETRY I: THE EXTENDED DYNKIN DIAGRAM $\bar{E}_8$ AND MINIMAL SINGULAR COMPACTIFICATIONS OF $\mathbb{C}^2$

Lawrence Brenton*, Daniel Drucker, and Geert C. E. Prins

Introduction

This is the first of a series of papers exemplifying the use of graph theory in the construction of singular two-(complex-) dimensional complex analytic spaces with interesting topological and analytic properties. In this paper we are concerned with complex surfaces which compactify affine 2-space $\mathbb{C}^2$, and certain other affine varieties. The main result (Theorem 6 below) is the classification of all compactifications of $\mathbb{C}^2$ which are Gorenstein surfaces with vanishing geometric genus and which satisfy a natural minimality condition. In particular they are all rational projective surfaces whose singular points are rational double points of type E_k, $0 \leq k \leq 8$, with an appropriate convention for the meaning of "E_k" when $k \neq 6, 7, 8$.

1. DEFINITION. A connected normal two-dimensional compact complex space $(X, \mathcal{O}_X)$ is an *analytic compactification of* $\mathbb{C}^2$ if X contains a subvariety C (the "curve at infinity") such that $X - C$ is biholomorphic to $\mathbb{C}^2$.

In case X is non-singular, compactifications of $\mathbb{C}^2$ have been objects of general interest at least since 1954 when Hirzebruch listed the

*Supported in part by National Science Foundation Grant No. MCS 77-03540.

Recent Developments in Several Complex Variables
0-691-08285-5/81/000047/17 $00.85/0 (cloth)
0-691-08281-2/81/000047/17 $00.85/0 (paperback)
For copying information, see copyright page.

problem of finding all such compactifications as one of the 34 noteworthy problems in complex and differentiable manifolds ([10], Problem 26). In particular it was conjectured that every non-singular analytic compactification is rational (bimeromorphically equivalent to $P^2(C)$). This conjecture was proved independently by Kodaira [13] and Morrow [16], after partial results had been provided by Remmert and Van de Ven [19], Van de Ven [20], and Ramanujam [18]. Indeed, [16] gives the complete classification of all non-singular analytic compactifications of C^2 (modulo blowing up extraneous points at infinity) and thus lays the non-singular case to rest.

In the meantime, very important advances were being made in understanding the topology of singular points of complex surfaces. This led to greatly accelerated activity in the topic of compact two-dimensional spaces with singularities, especially with respect to relations between their global topological and analytic properties and the local geometry of their singular points. In this spirit, the following result appeared in [3] as an application of the classification of Gorenstein analytic surfaces with negative canonical bundle. (A normal complex surface $(X, \mathcal{O}_X)$ is *Gorenstein* if each stalk $\mathcal{O}_{X,x}$ is a Gorenstein ring. Such surfaces are characterized by the property that the holomorphic line bundle K_{X_0} of differential 2-forms on the regular points X_0 of X extends across the singular points to provide a "canonical line bundle" K_X on X.)

2. THEOREM ([3], Proposition 2). *Let* X *be a Gorenstein surface which is an analytic compactification of* C^2 *by an irreducible curve* $C = X - C^2$. *Then the following are equivalent:*

(a) *the canonical bundle* K_X *is negative (that is,* K_X *admits a fibre metric* ρ *whose level sets are strictly pseudo-convex),*

(b) *the geometric genus* $p_g = \dim_C H^2(X, \mathcal{O}_X)$ *of* X *vanishes, and*

(c) *each singular point (if any) of* X *is a rational double point.*

Furthermore, if any of these conditions holds, then either

(i) X *is biholomorphic to the complex projective plane* $P^2 = P^2(C)$ ($\Longleftrightarrow$ X *is non-singular; Remmert and Van de Ven* [19]),

(ii) X *is biholomorphic to the singular quadric cone* $Q_0^2 = \{x^2+y^2+z^2 = 0\} \subset \overline{C^3} = P^3$, *or*

(iii) X *is a rational projective algebraic surface obtained from* P^2 *by the application of* k *successive monoidal point-transformations,* $3 \leq k \leq 8$, *followed by the collapsing of precisely* k *non-singular rational curves, each with self-intersection* -2, *to one or more singular points.*

For $k = 8$ this procedure provides examples of singular complex spaces whose rational cohomology rings are isomorphic to that of P^2 ([2]). In part II of this series we will apply graph theory to find and characterize all cohomology and homotopy P^2's that arise in this way. In the present paper, similar techniques are used to solve the corresponding problem for compactifications of C^2. For a discussion of Dynkin diagrams and their relation to root systems of semisimple complex Lie algebras and to exceptional sets of resolutions of rational double points, the reader is referred to [12], [1], and [9].

The construction

Let P be a point of the complex projective plane P^2 and let D be an analytic curve through P, non-singular at P and with a flex point there. Denote by L_0 the projective line tangent to D at P, and consider the spaces $\tilde{E}_0 = P^2$, $\tilde{E}_1, \cdots, \tilde{E}_8$ derived from P^2 by blowing up points as pictured. Here the number accompanying each line is the self-intersection. D, of course, might meet L_0 in other points besides P, but this is not relevant to the construction. (Indeed, the curve D serves only to identify, at each stage, the point to be blown up next. If, however, D is taken to be an irreducible cubic, then for $k \geq 3$ its proper transform gives rise to an anti-canonical divisor K^* on the resulting compactification, which generates the cohomology ring H^*. See [2], example 2, page 430.)

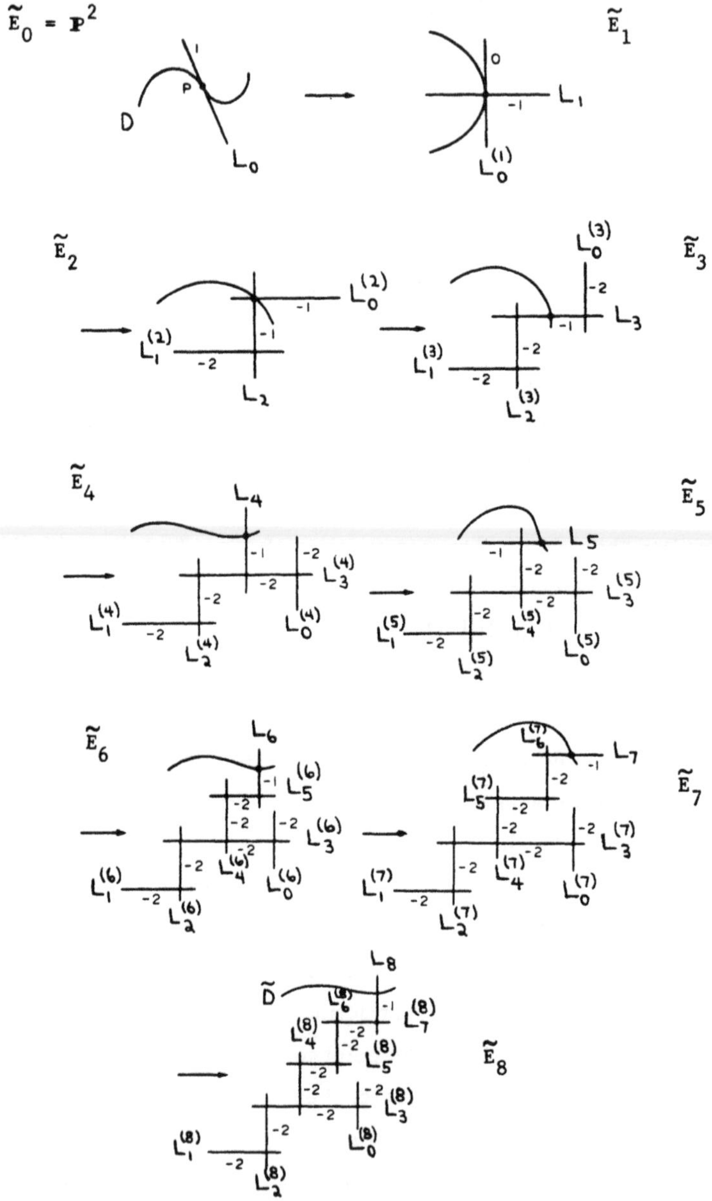

$\tilde{E}_0 = \mathbb{P}^2$
$\tilde{E}_1$
$\tilde{E}_2$
$\tilde{E}_3$
$\tilde{E}_4$
$\tilde{E}_5$
$\tilde{E}_6$
$\tilde{E}_7$
$\tilde{E}_8$

Now define singular surfaces $E_k(=E_k(P,D),\ 0 \le k \le 8)$ as follows:

$E_0 = \tilde{E}_0$,

E_1 = the surface obtained from $\tilde{E}_1$ by performing the σ-exchange on $L_0^{(1)}$ —that is, by blowing up a point on $L_0^{(1)}$ (not the point of intersection), then blowing down the proper transform of $L_0^{(1)}$ to a regular point,

E_2 = the surface obtained from $\tilde{E}_2$ by blowing down $L_0^{(2)}$ to a regular point and $L_1^{(2)}$ to a rational double point of type A_1, and

E_k $(3 \le k \le 8)$ = the space obtained from $\tilde{E}_k$ by collapsing, to one or more singular points, the k lines $L_j^{(k)}$, $j = 0, \cdots, k-1$, with self-intersection -2.

In particular, $E_0 = P^2$, $E_1 \cong$ the non-singular complex quadric surface Q^2 (isomorphic to $P^1 \times P^1$), and $E_2 \cong Q_0^2$. Since for $k > 2$ each surface E_k is obtained from a smooth variety by blowing down non-singular rational curves with self-intersection -2, the singularities of the E_k's are all rational double points ([1]). Furthermore, the spaces E_k are compactifications of C^2, for if C_k is the curve on E_k which is the image under the blowing down map $\pi_k : \tilde{E}_k \to E_k$ of the line $L_k = L_k^{(k)}$ (the line in the kth picture with self-intersection -1), then

$$E_k - C_k \cong \tilde{E}_k - \bigcup_{j=0}^{k} L_j^{(k)} \cong P^2 - L_0 \cong C^2 .$$

(For $k = 1$, C_k = the union of the base and the fibre after the σ-exchange. We note also that if b_2 denotes the second Betti number, then $b_2(E_k) = 1$ for $k \ne 1$, though $b_2(E_1) = 2$.) We want to show that under suitable hypotheses of minimality, etc., this construction yields *all* analytic compactifications of C^2. First we will systematize the construction in graph theoretic terms.

The "Dynkin diagrams" E_k, $k = 0, 1, \cdots, 8$

Consider the graph on 9 vertices (numbered as shown)

$\bar{E}_8$:
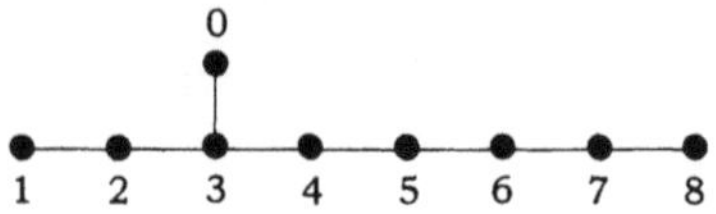

This is the extended Dynkin diagram associated to the affine root system of the exceptional simple complex Lie algebra $\mathfrak{e}_8$. Such extended Dynkin diagrams have recently come into the spotlight in finite group theory with the discovery by John McKay that the eigenvectors of their Cartan matrices are given by the columns of the character tables of the corresponding finite subgroups of $SL(2, \mathbb{C})$.

For $k = 8, 7$, and 6, the (non-extended) Dynkin diagrams E_k associated to the exceptional algebras $\mathfrak{e}_k$ are the graphs obtained from $\bar{E}_8$ by successively deleting vertices from the right:

E_k :
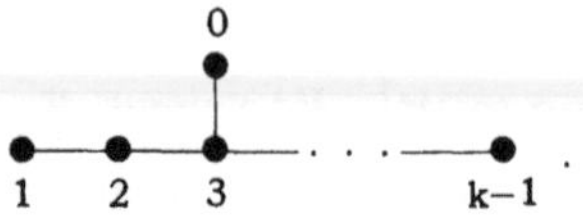

If we continue this process for $k = 5, 4$, and 3, we obtain the graphs D_5, A_4, and $A_2 + A_1$, where D_k is the graph  and A_k the graph . There is also a sequence of natural Lie algebra containments corresponding to this succession of deletions, namely $\mathfrak{e}_8 \supset \mathfrak{e}_7 \supset \mathfrak{e}_6 \supset \mathfrak{o}(10, \mathbb{C}) \supset \mathfrak{sl}(5, \mathbb{C}) \supset \mathfrak{sl}(3, \mathbb{C}) \oplus \mathfrak{sl}(2, \mathbb{C})$, where $\mathfrak{o}(2k, \mathbb{C})$ is the orthogonal Lie algebra associated to D_k and where $\mathfrak{sl}(k+1, \mathbb{C})$, the Lie algebra of the special linear group $SL(k+1, \mathbb{C})$, corresponds to the diagram A_k. If for the sake of uniformity of notation we define $E_5 = D_5$, $E_4 = A_4$, and $E_3 = A_2 + A_1$, then a comparison of these graphs with the configurations of the exceptional curves of the maps $\pi_k: \tilde{E}_k \to E_k$ described above shows that for $3 \leq k \leq 8$ the graphs E_k exactly describe the singularities of the surfaces E_k.

To complete the program, it will be convenient to have some sort of graph theoretic interpretation for the degenerate cases $k = 0, 1$, and 2. Let us use the notation A_0 for all of the following:

(1) the two-dimensional non-singular point

$$(0, (\mathcal{O}_{\mathbf{C}^3}/(x^1 + y^2 + z^2))_0) \subset (0, \mathcal{O}_{\mathbf{C}^3,0}),$$

(2) the trivial complex Lie algebra $\mathfrak{sl}(1, \mathbf{C})$, and (with all due apologies to graph theorists)

(3) the "quasi-graph"  .

Our purpose in introducing (1) and (2) is to adjoin the trivial case $k = 0$ to the family of singularities $\{x^{k+1} + y^2 + z^2 = 0\}$ and the family of simple Lie algebras $\mathfrak{sl}(k+1, \mathbf{C})$ associated to the linear Dynkin diagram A_k.

For $k = 0, 1, \cdots, 8$, define E'_k to be the subgraph of $\overline{E}_8$ consisting of the vertices $k+1, k+2, \cdots, 8$ (numbered as in the diagram above) and all the edges connecting them. E'_k, then, is just a copy of A_{8-k} embedded in $\overline{E}_8$. Define E_k to be the complementary (quasi-) graph $\overline{E}_8 - E'_k -$ $\{$the kth vertex and all edges adjoining it$\}$. That is,

$E_8 =$ 

$E_7 =$ 

$E_6 =$ 

$E_5 =$  (D_5)

$E_4 =$ 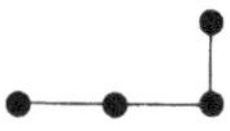 (A_4)

E_3 =  (A_2+A_1)

E_2 =  (A_1+A_0)

E_1 =  (A_0)

E_0 = the empty graph.

For future use we also record here the "determinants" of these graphs. For an undirected weighted graph Γ with vertices $v_0, \cdots, v_{k-1}$, the *determinant of* Γ is the determinant of the $k \times k$ symmetric matrix (a_{ij}) where a_{ii} is the weight of v_i and $a_{ij} = -1$ if $\{v_i v_j\}$ is an edge of Γ, 0 otherwise, for $i \neq j$. That is, (a_{ij}) is the Cartan matrix if Γ is a Dynkin diagram with all roots of equal length, and is the negative of the intersection matrix if Γ is the graph dual to a collection of curves meeting normally on a surface. In this section the weights are taken to be the negatives of the self-intersection numbers. This has the advantage that the determinants of the E_k's will all be positive. For the graphs defined above,

$$\det(E'_k) = \det(A_{8-k}) = 9-k\,, \quad \text{for all } k\,,$$

while for $k > 2$

$$\det(E_k) = \det(E'_k) = 9-k \quad \text{also}$$

(cf. [10]). We set det() = $\det(\emptyset) = 1$ so that

$$\det(E_1) = \det(E_0) = 1 \quad \text{and} \quad \det(E_2) = 2\,.$$

These determinants will turn out to have topological significance for the corresponding spaces E_k.

Again let Γ be a weighted graph on vertices $v_0, \cdots, v_{k-1}$, with weight n_i on v_i. Adjoin a new vertex v_k not in Γ, and for $i = 0, 1, \cdots, k-1$, let $\sigma_{v_i}(\Gamma)$ denote the graph $\Gamma \cup \{v_k\} \cup \{v_k v_i\}$ with weights 1 on v_k, $n_i + 1$ on v_i, and n_j on v_j for $j \neq i, k$. Similarly, for each edge $v_i v_j$ of Γ, let $\sigma_{v_i v_j}(\Gamma)$ denote the graph $(\Gamma - \{v_i v_j\}) \cup \mapsto$ $\{v_k\} \cup \{v_k v_i, v_k v_j\}$ with weights 1 on v_k, $n_i + 1$ on v_i, $n_j + 1$ on v_j, and n_ℓ on v_ℓ for $\ell \neq i, j, k$. The operations $\sigma_{v_i}, \sigma_{v_i v_j}$ are collectively called *σ-processes*, or *blow-ups*, and are dual (after changing the signs of the weights) to the Hopf monoidal (quadric) transform on a complex surface. (We shall not consider triple or tangential intersections of curves or curves with singular points in this paper.)

3. LEMMA. *Let Γ_0 be the graph on one vertex v_0 with weight -1 (no edges). Let $\tilde{\Gamma} = \sigma^k(\Gamma_0)$ be a graph on $k+1$ vertices obtained from Γ_0 by the successive application of k σ-processes. Suppose that $\tilde{\Gamma}$ contains k vertices of weight 2. Then $k \geq 3$ and $\tilde{\Gamma} \cong E_k \cup \{v_k\} \cup \{v_k v_{k-1}\}$, with $v_0, \cdots, v_{k-1}$ of weight 2 and v_k of weight 1. (If $k > 8$, E_k is the graph obtained from E_8 by adjoining $k-8$ vertices, in the obvious fashion, to the longest arm.) Indeed, the sequence $\sigma_1, \cdots, \sigma_k$ of blow-ups is exactly $\sigma_1 = \sigma_{v_0}$, $\sigma_2 = \sigma_{v_0 v_1}$, $\sigma_3 = \sigma_{v_0 v_2}$, $\sigma_j = \sigma_{v_{j-1}}$ for $j \geq 4$:*

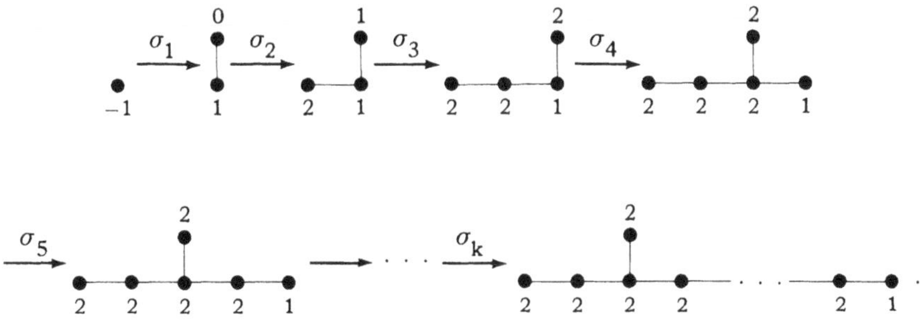

Proof. By definition of the σ-process, the last vertex v_k to be adjoined has weight $n_k = 1$, so the remaining vertices $v_0, \cdots, v_{k-1}$ must all have weight 2. The total weight of the graph, $\sum_{j=0}^{k} n_j$, is therefore $2k+1$.

Suppose that among the k σ-processes m edges were blown up, together with k–m vertices. Since blowing up an edge adds 3 to the total weight of a graph, while blowing up a vertex adds 2, we have $2k+1 = -1+3m+2(k-m)$, or $m = 2$.

We claim that the two edges blown up are of the form v_0v_j, $j > 0$. To prove this, suppose the contrary; i.e., suppose that for some index i, $\sigma_i = \sigma_{v_\ell v_j}$, $0 < \ell < j < i$. For each r, let Γ_r be the graph obtained by applying the first r σ-processes to Γ_0. Then the weight of v_ℓ is 1 in Γ_ℓ, is at least 2 in Γ_j, and therefore must be greater than 2 in Γ_i. It follows that v_ℓ has weight > 2 in $\Gamma_k = \tilde{\Gamma}$, contrary to hypothesis. We will denote the two σ-processes which blow up edges by σ_{ℓ_1} and σ_{ℓ_2}.

Now clearly $\sigma_1 = \sigma_{v_0}$. Thus v_0 takes part in the 3 blow-ups σ_1, σ_{ℓ_1}, and σ_{ℓ_2}. Each blow-up raises the weight of v_0 by 1, so v_0 cannot take part in any other blow-up. It follows that all the remaining σ-processes are of the form σ_{v_j}, where $j > 0$, and that v_0 belongs to only one edge of $\tilde{\Gamma}$. Consider now the first steps in the sequence:

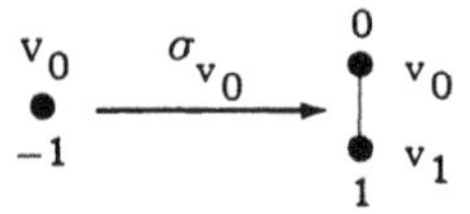

σ_2 cannot be σ_{v_1}, for then v_1 has weight 2 in Γ_2 and weight at least 3 in Γ_{ℓ_1}, so σ_2 must be $\sigma_{v_0v_1}$:

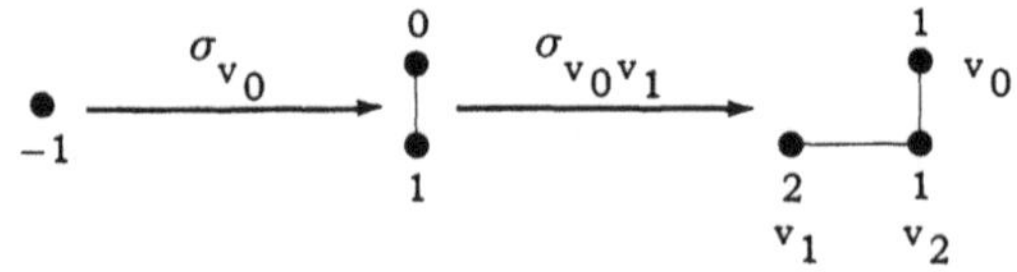

By the same argument σ_3 must be $\sigma_{v_0v_2}$, lest either v_1 have weight 3 in Γ_3 or v_2 have weight ≥ 3 in Γ_{ℓ_2}.

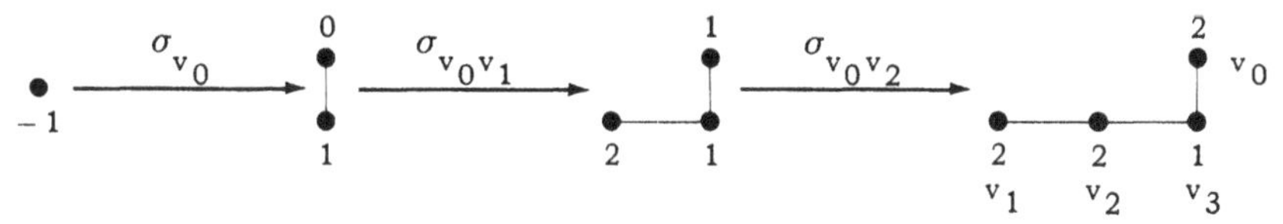

The only choice for the next blow-up σ_4 is σ_{v_3}, and now this pattern continues inductively.

The main theorem

Comparison of the derivations of the graphs of Lemma 3 with the constructions of the spaces $\tilde{E}_k$ in the preceding section shows that Lemma 3 provides a uniqueness theorem about certain compactifications of $\mathbb{C}^2$. The main result of this paper will make this precise. Now if X is any compactification of $\mathbb{C}^2$, additional compactifications can always be achieved by blowing up points at infinity. Thus a classification theory must concern itself with compactifications which are in some sense minimal (cf. [16], [18]). For the purposes of this paper, we define "minimal" as follows:

4. DEFINITION. A normal compact two-dimensional complex space X is *minimal* if it has smallest second Betti number among all spaces dominated by its non-singular model. That is, if $\pi: \tilde{X} \to X$ is the minimal resolution of singularities of X and if $\rho: \tilde{X} \to \overline{X}$ is a holomorphic bimeromorphic mapping of $\tilde{X}$ onto a normal surface $\overline{X}$, then $b_2(X) \leq b_2(\overline{X})$.

The point of the definition is that a minimal surface X is one obtained from a non-singular surface $\tilde{X}$ by blowing down negatively embedded curves in the most efficient way possible. In particular, X itself should have no exceptional curves with vanishing first homology group (cf. the notions of "minimal" and "relatively minimal" for non-singular surfaces, as, for instance, in [13], p. 789).

5. REMARK. If X is a minimal *rational* surface then either $X = \mathbf{P}^1 \times \mathbf{P}^1$ or else $b_2(X) = 1$. For let $\pi: \tilde{X} \to X$ resolve the singularities of X. Then $\tilde{X}$ is a non-singular rational surface. By the classification of rational surfaces, there exists a holomorphic mapping ρ, consisting of successive blowings down of exceptional curves of the first kind, of $\tilde{X}$ onto a relatively minimal non-singular rational surface S_n, the total space of a $\mathbf{P}^1$-bundle on $\mathbf{P}^1$ with Chern class $-n$, $n = 0, 2, 3, \cdots$, or else onto $\mathbf{P}^2$. But for $n \neq 0$ the base curve (zero section) of S_n is negatively embedded and can be collapsed to a normal analytic point. The resulting space has $b_2 = 1$ and is dominated by $\tilde{X}$. On the other hand, if $\rho: \tilde{X} \to S_0 = \mathbf{P}^1 \times \mathbf{P}^1$, but $\tilde{X} \neq S_0$, then $\tilde{X}$ is derived from S_0 by blowing up points. However, S_0 with a point blown up is the same as $\mathbf{P}^2$ with 2 points blown up, so $\tilde{X}$ dominates $\mathbf{P}^2$ as well. Hence either $\tilde{X} = \mathbf{P}^1 \times \mathbf{P}^1$ or $\tilde{X}$ maps onto a normal surface with $b_2 = 1$, so X minimal $\Longrightarrow$ either $b_2(X) = 1$ or $\tilde{X} = \mathbf{P}^1 \times \mathbf{P}^1$. If $\tilde{X} = \mathbf{P}^1 \times \mathbf{P}^1$ then $X = \tilde{X}$, for $\mathbf{P}^1 \times \mathbf{P}^1$ contains no negatively embedded curves.

6. THEOREM (A classification theorem for minimal singular compactifications of $\mathbf{C}^2$). *Let X be an analytic compactification of $\mathbf{C}^2$ which is a minimal Gorenstein space with vanishing geometric genus. Then X is a surface of type E_k for some $k \leq 8$. In particular, X has at most 2 singular points (at most 1 for $k \neq 3$); these are all rational double points and are of types A_1 $(k=2)$, $A_2 + A_1$ $(k=3)$, A_4 $(k=4)$, D_5 $(k=5)$, and E_k $(k=6,7,8)$. ($E_0 = \mathbf{P}^2$ and $E_1 = Q^2$ are non-singular). Moreover, if $k \neq 1$, X is an integral homology $\mathbf{P}^2$ with cohomology ring structure determined by $\det(E_k)$, while if $k \neq 2$ the Chern invariant c_1^2 is determined by $\det(E_k')$. That is,* $H_i(X, \mathbf{Z}) \cong \begin{cases} \mathbf{Z} & \text{for } i \text{ even} \\ 0 & \text{for } i \text{ odd} \end{cases}$, *except that $H_2(X, \mathbf{Z}) \cong \mathbf{Z} \oplus \mathbf{Z}$ if $k = 1$. Let g be the positive generator of $H^2(X, \mathbf{Z})$. (If $k = 1$, $g =$ the unique positive class with smallest square.*

Here a cohomology class x *is* positive *if some positive integral multiple* nx *is the Chern class of a very ample line bundle.) Denote by* c_1 *(the "first Chern class" of the Gorenstein space* X*) the Chern class of the dual of the canonical bundle* K_X. *Then*

(1) $g^2 = \det(E_k)$ $(k \neq 1)$, *and*

(2) $c_1^2 = \det(E'_k) = 9-k$ $(k \neq 2)$.

(Recall that for $k \geq 3$, $\det(E_k) = \det(E'_k)$, *reflecting the fact that* $c_1 = g$ *in that case.)*

Proof. Every compactification of $\mathbf{C}^2$ is rational, so Remark 5 implies that either $X = \mathbf{P}^1 \times \mathbf{P}^1$ $(= \mathbf{Q}^2$, the unique surface of type E_1) or else $b_2(X) = 1$. $b_2(X) = 1 \Longrightarrow$ the curve at infinity is irreducible. (In general for compactifications Y of $\mathbf{C}^n$, $H_{2n-2}(Y, \mathbf{Z})$ is generated freely by the analytic components of the hypersurface A at infinity. This is immediate from the sequence $\cdots \to H_{2n-1}(Y, A; \mathbf{Z}) \to H_{2n-2}(A; \mathbf{Z}) \to H_{2n-2}(Y, \mathbf{Z}) \to H_{2n-2}(Y, A; \mathbf{Z}) \to \cdots$. $H_i(Y, A; \mathbf{Z}) \cong \hat{H}_i(S^{2n}; \mathbf{Z}) = 0$ for $i \neq 2n$, since the sphere S^{2n} is the 1-point compactification of $\mathbf{C}^n \cong Y - A$ ([20], Propositions 2.1 and 2.2). In particular, if $Y = \mathbf{P}^n$, then A must be a linear hyperplane, for hyperplanes are the only analytic varieties that generate $H_{2n-2}(\mathbf{P}^n; \mathbf{Z}) \cong \mathbf{Z}$. This fact will be used shortly.) From Theorem 2, then, we conclude that either X is biholomorphic to $\mathbf{P}^2$, $\mathbf{Q}^2$, or $\mathbf{Q}_0^2$, or else X is derived from $\mathbf{P}^2$ by blowing up k points, $3 \leq k \leq 8$, then blowing down k non-singular rational curves, each with self-intersection -2.

In this last case, let $\pi: \tilde{X} \to X$ be the minimal resolution of singularities and let $\rho: \tilde{X} \to \mathbf{P}^2$ be inverse to the k blowings up. Let $\tilde{C} = \bigcup_{i=1}^{k} \tilde{C}_i$ be the exceptional curve of π, with $\tilde{C}_i$ irreducible. Let $C_0 = X - \mathbf{C}^2$. Then C_0 contains all the singular points of X (since $X - C_0 \cong \mathbf{C}^2$ is non-singular), and $\pi^{-1}(C_0) = \tilde{C}_0 \cup \tilde{C}$ for $\tilde{C}_0$ the proper transform of C_0 on $\tilde{X}$. Now $\rho \circ \pi^{-1}|_{X - C_0} : X - C_0 \to \mathbf{P}^2$ exhibits $\mathbf{P}^2$ as a compactification of C^2. Thus $\rho(\tilde{C}_0 \cup \tilde{C})$ is a projective line L_0 on P^2, since, as remarked parenthetically above, projective lines are the only analytic subvarieties of P^2 whose complements are homeomorphic to C^2. Therefore $L_0^2 = 1$, so $\tilde{X}$ is derived from P^2 in a manner dual to that described in the hypotheses of Lemma 3. By that lemma then, if $k \geq 3$, $\tilde{X}$ is one of the spaces $\tilde{E}_k$, whence X is a space of type E_k. (The requisite curve D in the construction can be taken to be any curve on P^2 with the necessary higher order tangent directions to L_0 at the point to be blown up.)

The topological properties of the spaces E_k are easy to check (cf. [2]). This completes the proof.

REMARK. The spaces of type E_8 are especially interesting, for in addition to being compactifications of C^2 they have the property of being homotopy equivalent to P^2 ([2], example 2). Note that topologically the singularity E_8 is the cone on an integral homology 3-sphere (cf. [7] for higher dimensional analogues).

Some other compactifications

Although it is compactifications of C^2 that have produced the most literature, there is some interest in compactifying other 2-dimensional affine varieties, and our technique provides many examples. For instance, if we start with two intersecting projective lines on P^2, blow up k points, then blow down k curves including the proper transforms of the two we started with, the resulting singular surface will be a compactification of P^2 –two lines $= C \times C^*$, where $C^* = C - \{0\}$. Here is one such derivation described by means of the associated graphs. A $\star$ indicates which edge or vertex is to be blown up next.

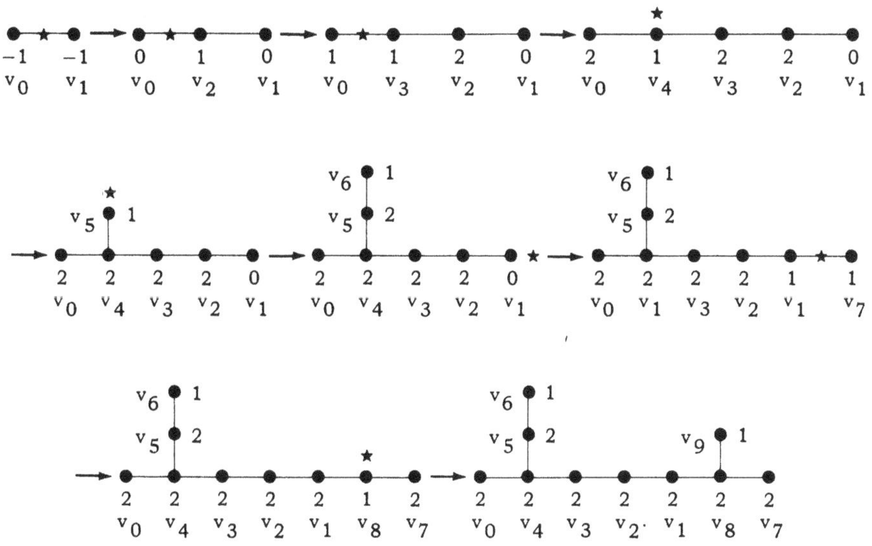

If the 8 curves corresponding to the vertices of weight 2 are now blown down, the resulting space X will be a singular rational surface with one singular point, which is a rational double point of type D_8. If C_6 and C_9 are the curves on X corresponding to the vertices v_6 and v_9, then $X - C_6 - C_9 \cong C \times C^*$. The surface X also has the rational cohomology type of P^2, with vanishing irregularity and geometric genus.

By blowing up different sequences of points, it is also possible to construct compactifications of $C \times C^*$ having the following combinations of singular points: $A_3 + A_1 + A_1$, $A_5 + A_1$, $A_2 + A_2 + A_2$, A_7, $D_6 + A_1$, $A_5 + A_2$, A_8, $E_7 + A_1$, and $E_6 + A_2$. These, together with D_8, comprise the complete list of singularity types of singular compactifications of $C \times C^*$ which can be constructed by the method of this paper. We have similar complete lists for compactifications of $(C^*)^2$, $C \times (C - \{0, 1\})$, $C^* \times (C - \{0, 1\}$, $(C - \{0, 1\})^2$, and $(C - \{0, 1\})^2 - \{(z, z) : z \in C\}$. Except for C^2, however, we do not have uniqueness results or nice characterizations of these compactifications among all possible singular compactifications of these affine spaces.

Added in proof

We have recently established that our spaces E_k, $k \neq 1$, are the only complex projective algebraic surfaces X satisfying

(1) X is Gorenstein and admits an effective anti-canonical divisor K^*, and

(2) $H^i(X, Z) \cong H^i(P^2, Z) \,\forall i$.

It follows that, among algebraic surfaces with property (1), the spaces E_8 are the only singular surfaces whose cohomology rings are isomorphic to that of P^2. In particular they are the only such spaces homotopy equivalent to P^2.

We also have the following improvement of a result of [3] in dimension 3 (cf. also [6], Corollary 2.2).

PROPOSITION. *Let* M *be a non-singular analytic compactification of* C^3 *by a normal subvariety* X. *Suppose that* $b_3(M) = 0$ *and* K_X *is not trivial. Then* X *is a surface of type* E_k *for some* $k \neq 1$.

Proofs will appear in part III of this series.

REFERENCES

[1] Artin, M., On isolated rational singularities of surfaces, Am. J. Math. *88* (1966), 129-136.

[2] Brenton, L., Some examples of singular compact analytic surfaces which are homotopy equivalent to the complex projective plane, Topology *16* (1977), 423-433.

[3] ________, On singular complex surfaces with negative canonical bundle, with applications to singular compactifications of C^2 and to 3-dimensional rational singularities, Math. Ann. 248 (1980), 117-124.

[4] Brenton, L., Drucker, D., and Prins, G., Graph theoretic techniques in algebraic geometry II: construction of singular complex surfaces of the rational cohomology type of CP^2, to appear in Comment Math. Helv.

[5] Brenton, L., and Morrow, J., Compactifying C^n, Proc. Symp. Pure Math. *30*: Several Complex Variables, Williams College, 1977, I, 241-246.

[6] ________, Compactifications of C^n, A.M.S. Transactions *246* (1978), 139-153.

[7] Brieskorn, E., Examples of singular normal complex spaces which are topological manifolds, Proc. Nat. Acad. Sci. U.S.A. *55* (1966), 1395-1397.

[8] Brieskorn, E., Rationale Singularitäten komplexer Flächen, Inv. math. *4*(1967/68), 336-358.

[9] ———, Singular elements of semi-simple algebraic groups, Actes Congrès intern. Math., 1970, Tome 2, 279-284.

[10] Drucker, D., and Goldschmidt, D., Graphical evaluation of sparse determinants, Proc. AMS *77*(1979), 35-39.

[11] Hirzebruch, F., Some problems on differentiable and complex manifolds, Ann. Math. *60*(1954), 212-236.

[12] Humphreys, J., Introduction to Lie Algebras and Representation Theory, Graduate Texts in Mathematics *9*, Springer-Verlag, New York, Heidelberg, Berlin, 1972.

[13] Kodaira, K., On the structure of compact complex analytic surfaces, I-IV, Am. J. Math. *86*(1964), 751-798; *88*(1966), 682-721; *90*(1968), 55-83, 1048-1065.

[14] ———, Holomorphic mappings of polydiscs into compact complex manifolds, J. Diff. Geom. *6*(1971), 33-46.

[15] Laufer, H., Normal two-dimensional singularities, Ann. Math. Studies No. 71, Princeton Univ. Press, Princeton, 1971.

[16] Morrow, J., Minimal normal compactifications of $\mathbb{C}^2$, Proc. of the Conf. on Complex Analysis, Rice Univ. Studies *59*(1973), 97-112.

[17] Mumford, D., The topology of normal singularities of an algebraic surface and a criterion for simplicity, Inst. des Hautes Etudes Scientifiques, Publ. math. no. *9*(1961), 5-22.

[18] Ramanujam, C. P., A topological characterization of the affine plane as an algebraic variety, Ann. Math. *94*(1971), 69-88.

[19] Remmert, R., and Van de Ven, T., Zwei Sätze über die komplex-projective Ebene, Nieuw Arch. Wisk. (3) *8*(1960), 147-157.

[20] Van de Ven, A., Analytic compactifications of complex homology cells, Math. Ann. *147*(1962), 189-204.

ON METRICS AND DISTORTION THEOREMS

Jacob Burbea

Abstract. A family of pseudo-distances, generalizing the Möbius and the Carathéodory pseudo-distances of a manifold M, is introduced and studied. The fact that these pseudo-distances are continuous and logarithmically plurisubharmonic is established. This extends a recent result of Vesentini. A similar result for the Bergman metric is proven. Other results related to the Bergman and the Carathéodory metrics are established.

§1. *Introduction*

As is well known, the classical Schwarz-Pick lemma may be stated by means of the Mobius distance function

$$\delta \equiv \delta(z,\zeta) = \left|\frac{z-\zeta}{1-z\bar{\zeta}}\right|$$

on the unit disk U of $\mathbf{C}$. The Poincaré distance on U is then $\rho = 2^{-1}\log(1+\delta)/(1-\delta)$. In this paper we introduce a generalization of these distances by setting $\delta_q = [1-(1-\delta^2)^q]^{1/2}$ and $\rho_q = 2^{-1}\log(1+\delta_q)/(1-\delta_q)$, where $q > 0$.

These distances induce intrinsic pseudo-distances $\delta_{M,q}$ and $\rho_{M,q}$ on a complex manifold M of $\mathbf{C}^n$ which are labeled as the "q-Möbius" and the "q-Carathéodory" pseudo-distances, respectfully. The fact that pseudo-distances have a built-in Schwarz lemma (Theorem 2) yields several distortion theorems. We shall show (Theorem 4) that for any $q > 0$ and any $\zeta \in M$, $\log \rho_{M,q}(\cdot,\zeta)$ is a continuous plurisubharmonic function on M.

Recent Developments in Several Complex Variables
0-691-08285-5/81/000065-28$01.40/0 (cloth)
0-691-08281-2/81/000065-28$01.40/0 (paperback)
For copying information, see copyright page.

This generalizes a recent result of Vesentini [9] for the case $q = 1$. A similar result (Theorem 3) holds for $\delta_{M,q}$ provided $0 < q \leq 1$.

We also study metrics which are induced by reproducing kernels of spaces of holomorphic functions or forms on the manifold M. For example, a result in this area is the fact that the classical Bergman metric $b_M(z:v)$, $(z, v) \in M \times C^n$, is a real analytic logarithmically plurisubharmonic function on $M \times C^n$ (see Theorem 8 and its corollary for details and a generalization).

In the remaining parts of this paper we shall study the interrelationships amongst the various metrics and pseudo-distances of M (see, for example, Theorem 10). This is done mainly by using a rather strong distortion theorem (Theorem 9) which generalizes a classical theorem of Pick (cf. Ahlfors [1, p. 3-4]). These investigations extend our earlier work [2, 3, 4].

§2. *Preliminaries*

Let $U = \{z : |z| < 1\}$ be the open unit disk in C and let $G = \mathrm{Aut}(U)$ denote the group of holomorphic automorphism of U. Any $\phi \in G$ is given by

$$\phi(z) = e^{i\theta} \frac{z-\zeta}{1-\bar{\zeta}z}, \qquad z \in U,$$

for some $\zeta \in U$ and some $\theta \in [0, 2\pi)$. On U one considers the "*Möbius distance*" function

$$\delta \equiv \delta(z, \zeta) = \left|\frac{z-\zeta}{1-\bar{\zeta}z}\right|; \quad z, \zeta \in U.$$

It is well known that this is indeed a distance on U and that it is invariant under the action of the group G on U.

The "*Poincaré metric*" $c(z:dz) = (1-|z|^2)^{-1}|dz|$ has a constant Gaussian curvature, equal to -4 and the "*Poincaré distance*" induced by this metric is given by

$$\rho \equiv \rho(z, \zeta) = \frac{1}{2} \log \frac{1+\delta}{1-\delta}; \quad \delta = \delta(z, \zeta), \quad z, \zeta \in U.$$

We note that

$$\delta = \tanh \rho, \quad 0 \leq \delta < 1,$$

and that

$$\rho \geq \delta .$$

We recall the following classical "*Schwarz-Pick lemma*"

PROPOSITION 1. *Let* $\phi: U \to U$ *be a holomorphic mapping of* U *into* U. *Then*

$$\delta(\phi(z), \phi(\zeta)) \leq \delta(z, \zeta)$$

and

$$\rho(\phi(z), \phi(\zeta)) \leq \rho(z, \zeta)$$

for any $z, \zeta \in U$. *Moreover, for a fixed pair* $(z, \zeta) \in U \times U$, $z \neq \zeta$, *equality in any of the above inequalities occurs if and only if* $\phi \in G$. *Furthermore,*

$$c(\phi(z) : d\phi(z)) \leq c(z : dz)$$

for all $z \in U$ *and equality, at one fixed point* $z \in U$, *occurs if and only if* $\phi \in G$.

We now introduce a generalization of the above distance functions δ and ρ. Let $q > 0$ and define

(2.1) $$\delta_q \equiv \delta_q(z, \zeta) = [1-(1-\delta^2)^q]^{1/2},$$

(2.2) $$\rho_q \equiv \rho_q(z, \zeta) = \frac{1}{2} \log \frac{1+\delta_q}{1-\delta_q},$$

where

$$\delta = \delta(z, \zeta); \qquad z, \zeta \in U .$$

Clearly, $\delta_1 = \delta$ and $\rho_1 = \rho$ and both δ_q and ρ_q are increasing functions of $q > 0$. We also note the additional relationships

$$\delta_q = \tanh \rho_q, \quad \operatorname{sech} \rho_q = \operatorname{sech}^q \rho; \quad 0 \leq \delta_q < 1 .$$

Since the functions

(2.3) $$f_q(r) = [1-(1-r^2)^q]^{1/2}$$

and

$$g(r) = \frac{1}{2}\log\frac{1+r}{1-r}, \quad g_q(r) = g[f_q(r)], \tag{2.4}$$

are increasing in $r \in [0, 1)$, Proposition 1 admits the following generalization:

LEMMA 1. *Let* $\phi: U \to U$ *be a holomorphic mapping of* U *into* U. *Then*

$$\delta_q(\phi(z), \phi(\zeta)) \le \delta_q(z, \zeta)$$

and

$$\rho_q(\phi(z), \phi(\zeta)) \le \rho_q(z, \zeta)$$

for any $z, \zeta \in U$. *Moreover, for a fixed pair* $(z, \zeta) \in U \times U$, $z \neq \zeta$, *equality in any of the above inequalities occurs if and only if* $\phi \in G$.

Obviously, the functions δ_q and ρ_q are invariant under the action of the group G on U. They also satisfy all the axioms of a distance function except perhaps the triangle inequality axiom. The function δ_q does indeed satisfy the triangle inequality and this property may be verified by direct computation. However, we shall see later (in Theorem 7) that this fact is a special case of a much more general result from the theory of reproducing kernels. Meanwhile we shall prove:

THEOREM 1. *The function* ρ_q *satisfies the triangle inequality if and only if* $q \in (0, 1]$.

Proof. Since the group G is transitive on U the triangle inequality for ρ_q is equivalent to

$$\rho_q(z_1, z_2) \le \rho_q(z_1, 0) + \rho_q(z_2, 0)$$

for any $z_1, z_2 \in U$. Writing

$$\lambda_1 = \delta_q(z_1, 0), \qquad \lambda_2 = \delta_q(z_2, 0), \qquad \lambda_{12} = \delta_q(z_1, z_2)$$

the above inequality is equivalent to

$$\frac{1+\lambda_{12}}{1-\lambda_{12}} \le \frac{1+\lambda_1}{1-\lambda_1}\,\frac{1+\lambda_2}{1-\lambda_2}\,.$$

This is equivalent to

$$\lambda_{12}^2 \le \left[\frac{\lambda_1+\lambda_2}{1+\lambda_1\lambda_2}\right]^2$$

which is equivalent to

$$\left[1-\left|\frac{z_1-z_2}{1-\overline{z}_2 z_1}\right|^2\right]^q \ge 1-\left[\frac{\lambda_1+\lambda_2}{1+\lambda_1\lambda_2}\right]^2 = \frac{(1-\lambda_1^2)(1-\lambda_2^2)}{(1+\lambda_1\lambda_2)^2}$$

or to

$$\frac{(1-|z_1|^2)(1-|z_2|^2)}{|1-\overline{z}_2 z_1|^2} \ge \frac{(1-\lambda_1^2)^{1/q}(1-\lambda_2^2)^{1/q}}{(1+\lambda_1\lambda_2)^{2/q}} = \frac{(1-|z_1|^2)(1-|z_2|^2)}{(1+\lambda_1\lambda_2)^{2/q}}\,.$$

The last inequality is equivalent to

$$1+\lambda_1\lambda_2 \ge |1-z_1\overline{z}_2|^q\,. \tag{2.5}$$

Inequality (2.5) is obviously false, in general, when $q > 1$. For $0<q\le 1$, however, we have

$$\begin{aligned}(1+\lambda_1\lambda_2)^{1/q} &\ge 1+\frac{1}{q}\lambda_1\lambda_2 = 1+\frac{1}{q}[1-(1-|z_1|^2)^q]^{1/2}[1-(1-|z_2|^2)^q]^{1/2}\\ &\ge 1+\frac{1}{q}[1-(1-q|z_1|^2)]^{1/2}[1-(1-q|z_2|^2)]^{1/2}\\ &= 1+|z_1|\,|z_2| \ge |1-z_1\overline{z}_2|\,.\end{aligned}$$

This concludes the proof.

Let $f = f(z,\overline{z})$ be a C^2-function. Its Laplacian is given by

$$\Delta f = 4\partial_z\partial_{\overline{z}} f$$

and we note that if f is also positive then

$$\Delta f = f\Delta \log f + 4f^{-1}|\partial_z f|^2\,. \tag{2.6}$$

For $q > 0$ we consider the function

$$f_q \equiv f_q(|z|) = [1-(1-|z|^2)^q]^{1/2}, \qquad z \in U .$$

A direct computation shows that

$$\Delta \log f_q = \frac{2q(1-|z|^2)^{q-2}}{f_q^4} \cdot (f_q^2 - q|z|^2) . \tag{2.7}$$

We also consider the function

$$g_q \equiv g_q(|z|) = g[f_q(|z|)] = \frac{1}{2} \log \frac{1+f_q}{1-f_q} .$$

Here we have

$$\Delta \log g_q = \frac{1}{2} \frac{q^2}{g_q^2} \frac{1}{h(q)} H(q) , \tag{2.8}$$

where

$$h(q) = q[1-(1-s^2)^{1/q}], \qquad s = f_q(|z|) , \tag{2.9}$$

and

$$H(q) = 2s^2 \log \frac{1+s}{1-s} - \left[2s + (1-s^2) \log \frac{1+s}{1-s}\right] h(q) . \tag{2.10}$$

LEMMA 2. *Let* $0 < q \leq 1$. *The function* $\log f_q(|z|)$ *is subharmonic in* $z \in U$.

Proof. We use (2.7). Then

$$f_q^2 - q|z|^2 = 1 - q|z|^2 - (1-|z|^2)^q \geq 0$$

when $0 < q \leq 1$ and $z \in U$. Hence $\Delta \log f_q \geq 0$ which concludes the proof.

LEMMA 3. *Let* $q > 0$. *The function* $\log g_q(|z|)$ *is subharmonic in* $z \in U$.

Proof. We use (2.8)-(2.10). We must show that $H(q) > 0$ for $q > 0$. We consider the function $h(q)$ for $q > 0$. We have

$$h(\infty) \equiv \lim_{q\to\infty} h(q) = -\log(1-s^2); \qquad s \in [0,1).$$

By a direct computation

$$h''(q) = -\frac{1}{q^3}(1-s^2)^{1/q}[\log(1-s^2)]^2 < 0.$$

Hence $h'(q)$ is decreasing for $q > 0$. But

$$h'(\infty) \equiv \lim_{q\to\infty} h'(q) = 0$$

and thus

$$h'(q) > h'(\infty) = 0.$$

Therefore, $h(q)$ is increasing and so $H(q)$ is decreasing for $q > 0$. Let

$$f(s) \equiv H(\infty) = 2s^2 \log\frac{1+s}{1-s} + \left[2s + (1-s^2)\log\frac{1+s}{1-s}\right]\log(1-s^2)$$

and thus

$$H(q) > H(\infty) = f(s), \qquad q > 0.$$

We now consider $f(s)$ on $[0,1)$. We have $f(0) = f'(0) = 0$ and by a direct computation

$$f''(s) = 2\log\left(\frac{1}{1-s^2}\right)\left[\frac{2s}{1-s^2} + \log\frac{1+s}{1-s}\right] + \frac{2s^2}{1-s^2}\log\frac{1+s}{1-s} + \frac{2}{1-s^2}\left[\log\frac{1+s}{1-s} - 2s\right].$$

This is nonnegative because

$$\log\frac{1+s}{1-s} - 2s = 2s\sum_{n=1}^{\infty}\frac{s^{2n}}{2n+1} \geq 0, \qquad s \in [0,1).$$

Therefore, $f'(s)$ is increasing in $s \in (0,1)$ and thus $f'(s) > f'(0) = 0$. Consequently, $f(s)$ is increasing in $s \in (0,1)$ and hence $f(s) > f(0) = 0$. This concludes the proof of the lemma.

PROPOSITION 2. *The function* $f_q(|z|)$ *is subharmonic in* $z \in U$ *for* $0 < q \leq 1$ *while* $g_q(|z|)$ *is subharmonic in* $z \in U$ *for all* $q > 0$.

Proof. This follows from Lemmas 2, 3 and (2.6).

The differential metric induced by ρ_q is given by

$$(2.11)\qquad ds_q(z) = \frac{q|z|}{f_q(|z|)} \cdot \frac{|dz|}{1-|z|^2}, \qquad z \in U,$$

which reduces to the Poincaré metric when $q = 1$. Of course, when $q \neq 1$ this is not a conformally invariant metric and therefore, its curvature is not a constant. Recall that for a metric $g(z, \bar{z})|dz|$ its Gaussian curvature is given by

$$\kappa(g) = -g^{-2}\Delta \log g, \qquad g = g(z, \bar{z}).$$

Using (2.7) we find that the curvature of the metric (2.11) is given by

$$\kappa_q(z) = -4\frac{f_q^{\,2}}{q^2|z|^2} + \frac{2}{q|z|^2}\frac{(1-|z|^2)^q}{f_q^{\,2}}(f_q^{\,2} - q|z|^2); \quad f_q = f_q(|z|), \ z \in U.$$

Therefore, as expected $\kappa_1(z) \equiv -4$ and also $\kappa_q(z) < 0$ whenever $q \geq 1$.

§3. *Metrics on manifolds*

Let M be a complex manifold in C^n and let $H(M:U)$ denote the family of holomorphic functions from M into the unit disk U. For a fixed $\zeta \in M$ we write $H_\zeta(M:U) = \{f \in H(M:U) : f(\zeta) = 0\}$. For $q > 0$ one introduces the "*q-Möbius pseudo-distance*"

$$(3.1)\qquad \delta_{M,q}(z,\zeta) = \sup\{\delta_q(f(z), f(\zeta)) : f \in H(M:U)\}$$

of two points $z, \zeta \in M$. A normal family argument shows that the supremum is indeed attained by a member of $H(M:U)$. This pseudo-distance satisfies all axioms of a distance except that $\delta_{M,q}(z,\zeta)$ can be zero even if $z \neq \zeta$. The triangle inequality axiom is a consequence of Theorem 7 which will be shown in §5. M is said to belong to class $\mathcal{B}$

if for any two distinct points $z, \zeta \in M$ there exists an $f \in H(M:U)$ with $f(z) \neq f(\zeta)$. Clearly, for a fixed $q > 0$, $\delta_{M,q}$ is a distance on M if and only if $M \in \mathcal{B}$.

The "*q-Carathéodory pseudo-distance*" is defined by

$$\rho_{M,q}(z,\zeta) = \sup\{\rho_q(f(z), f(\zeta)) : f \in H(M:\Delta)\}. \tag{3.2}$$

Again, a normal family argument shows that the supremum is attained. Therefore, using the monotonicity of $g(r)$ in (2.4), $0 \leq r < 1$, (2.2), (3.1) and (3.2) we obtain

$$\rho_{M,q}(z,\zeta) = \frac{1}{2}\log\frac{1+\delta_{M,q}(z,\zeta)}{1-\delta_{M,q}(z,\zeta)}; \quad z, \zeta \in M.$$

Also, by virtue of Theorem 1, $\rho_{M,q}$ satisfies the triangle inequality axiom if and only if $q \in (0,1]$ and therefore, for a fixed $q \in (0,1]$, $\rho_{M,q}$ is a distance on M if and only if $M \in \mathcal{B}$.

We write $\delta_M = \delta_{M,1}$ and $\rho_M = \rho_{M,1}$ and we call these pseudo-distances the Möbius and the Carathéodory pseudo-distances, respectively, of M.

Obviously, $\delta_{M,q}$ and $\rho_{M,q}$ are increasing functions of $q > 0$ and

$$\delta_{M,q} = \tanh \rho_{M,q}, \quad \operatorname{sech} \rho_{M,q} = \operatorname{sech}^q \rho_M; \quad 0 \leq \delta_{M,q} < 1. \tag{3.3}$$

LEMMA 4. *Let $q > 0$ and consider the functions f_q and g_q as in (2.3) and (2.4) respectively. Then*

$$\delta_{M,q}(z,\zeta) = \max\{f_q(|f(z)|) : f \in H_\zeta(M:U)\}$$

and

$$\rho_{M,q}(z,\zeta) = \max\{g_q(|f(z)|) : f \in H_\zeta(M:U)\}$$

for any $z, \zeta \in M$.

Proof. We, of course, only need to prove the first identity. Indeed,

$$\begin{aligned}\delta_{M,q}(z,\zeta) &\geq \max\{\delta_q(f(z), 0) : f \in H_\zeta(M:U)\} \\ &= \max\{f_q(|f(z)|) : f \in H_\zeta(M:U)\}.\end{aligned}$$

On the other hand

$$\delta_{M,q}(z,\zeta) = \max\left\{f_q(|g(z)|) : g(\eta) = \frac{f(\eta)-f(\zeta)}{1-\overline{f(\zeta)}f(\eta)}, \quad f \in H(M:U)\right\}$$

$$\leq \max\{f_q(|g(z)|) : g \in H_\zeta(M:U)\}$$

which concludes the proof.

If f is a C^1-function near the point $z = (z_1, \cdots, z_n) \in \mathbf{C}^n$, and $v = (v_1, \cdots, v_n) \in \mathbf{C}^n$ we write

$$\partial_v f(z) = \sum_{j=1}^{n} \frac{\partial f}{\partial z_j} v_j, \quad \bar{\partial}_v f(z) = \sum_{j=1}^{n} \frac{\partial f}{\partial \bar{z}_j} \bar{v}_j$$

and therefore

$$\Delta_v f(z) = 4\partial_v \bar{\partial}_v f(z)$$

represents the Hessian (or "Laplacian") of f at z along the direction v, whenever f is a C^2-function near z.

The "*Carathéodory-Reiffen metric*" is given by

$$C_M(\zeta:v) = \sup\{|\partial_v f(\zeta)| : f \in H(M:U)\},$$

where $\zeta \in M$ and $v \in \mathbf{C}^n - \{0\}$. Again, a normal family argument shows that the supremum is attained and evidently

$$(3.4) \qquad C_M(\zeta:v) = \max\{|\partial_v f(\zeta)| : f \in H_\zeta(M:U)\}.$$

One also shows that $C_M(\zeta:v)$ is a continuous and a locally Lipschitz function in (ζ, v) (see, for example, [10] for additional details). Let $\tilde{\rho}_M$ denote the integrated metric on M of C_M. Then (see, for example, [10])

$$(3.5) \qquad \rho_M(z,\zeta) \leq \tilde{\rho}_M(z,\zeta); \qquad z, \zeta \in M.$$

Clearly, $\tilde{\rho}_M$ is a pseudo-distance on M and when $M \in \mathcal{B}$ it is also a distance.

Obviously, in the case that M is the unit disk U we have

$$\delta_{\Delta,q} = \delta_q, \quad \rho_{\Delta,q} = \rho_q, \quad \tilde{\rho}_\Delta = \rho_1 = \rho, \quad C_\Delta = c\,.$$

This, with Proposition 1 and Lemma 1, leads to the following almost classical assertion:

THEOREM 2. *Let* $\phi: M \to M^*$ *be a holomorphic mapping of a complex manifold* M *of* $\mathbf{C}^n$ *into another complex manifold* M^* *of* $\mathbf{C}^m$. *Then, for* $q > 0$,

$$\delta_{M^*,q}(\phi(z), \phi(\zeta)) \le \delta_{M,q}(z,\zeta)\,,$$

$$\rho_{M^*,q}(\phi(z), \phi(\zeta)) \le \rho_{M,q}(z,\zeta)\,,$$

$$\tilde{\rho}_{M^*}(\phi(z), \phi(\zeta)) \le \tilde{\rho}_M(z,\zeta)$$

where, $z, \zeta \in M$. *Also,*

$$C_{M^*}(\phi(z) : \phi_*(v)) \le C_M(z:v)\,,$$

where $\phi_*(v) = (\partial_v \phi_1, \cdots, \partial_v \phi_m)$ *and* $\phi = (\phi_1, \cdots, \phi_m)$; $\phi_j = \phi_j(z_1, \cdots, z_n)$, $1 \le j \le m$.

An immediate consequence of this theorem is that $\delta_{M,q}$, $\rho_{M,q}$, $\tilde{\rho}_M$ and C_M are biholomorphically invariant. We should also note that the functions $\delta_{M,q}$, $\rho_{M,q}$ and $\tilde{\rho}_M$ are continuous on $M \times M$.

From Lemma 4 follows that there exists an $F \in H_\zeta(M:U)$, $F(z) \equiv F(z: \zeta, \eta)$, with $F(\zeta) = F(\zeta : \zeta, \eta) = 0$ such that

$$\delta_{M,q}(\eta,\zeta) = f_q(|F(\eta)|) \tag{3.6}$$

and

$$\rho_{M,q}(\eta,\zeta) = g_q(|F(\eta)|)\,. \tag{3.7}$$

Also, from (3.4) follows that there exists a function $G \in H_\zeta(M:U)$, $G(z) \equiv G(z: \zeta, v)$, with $G(\zeta) = G(\zeta : \zeta, v) = 0$ such that

$$C_M(\zeta : v) = |\partial_v G(\zeta)|\,. \tag{3.8}$$

§4. *The Hessian of metrics*

Let f be upper semi-continuous near $z \in C^n$ and $v \in C^n$. The generalized Hessian (or "Laplacian") of f at z along the direction v is defined by

$$\Delta_v f(z) = 4 \varliminf_{r \to 0} \frac{1}{r^2}\left(\frac{1}{2\pi}\int_0^{2\pi} f(z + re^{i\theta}v)\,d\theta - f(z)\right).$$

Note that, if f is a C^2-function near z, then $\Delta_v f(z)$ reduces to the usual Hessian $4\partial_v\bar{\partial}_v f(z)$. Clearly, if f assumes a local minimum at z then $\Delta_v f(z) \geq 0$ for each direction v. Also, f is said to be plurisubharmonic in a complex manifold M of C^n if f is upper semi-continuous in M and $\Delta_v f(z) \geq 0$ for each $z \in M$ and each direction v.

We now prove:

THEOREM 3. *Let* $0 < q \leq 1$ *and let* $\zeta \in M$ *be fixed. Then* $\log \delta_{M,q}(z,\zeta)$ *is a continuous plurisubharmonic function in* $z \in M$.

Proof. Let $\eta \in M$ and consider the function $F(z) = F(z: \eta, \zeta)$ as in (3.6). Define

$$g(z) = \frac{F(z) - F(\zeta)}{1 - \overline{F(\zeta)}F(z)}, \qquad z \in M.$$

Clearly, $g \in H_\zeta(M:U)$ and $g(\eta) = -F(\zeta)$. By Lemma 4,

$$\delta_{M,q}(z,\zeta) \geq f_q(|g(z)|)$$

for all $z \in M$ and by (3.6) we have equality at $z = \eta$. Near η, $f_q(|g(z)|)$ is positive and hence $\log \delta_{M,q}(z,\zeta)/f_q(|g(z)|)$ assumes a local minimum at $z = \eta$. Therefore,

$$\Delta_v \log \delta_{M,q}(\eta,\zeta) \geq \Delta_v \log f_q(|g(z)|)\big|_{z=\eta}$$

for each direction v. However,

$$\Delta_v \log f_q(|g(z)|) = |\partial_v g(z)|^2 \Delta \log f_q(t), \quad t = |g(z)|.$$

The theorem now follows from Lemma 2 and we also note that

$$\Delta_v \log f_q(|g(z)|)\big|_{z=\eta} = |\partial_v F(\eta)|^2 (1-|F(\zeta)|^2)^2 \Delta \log f_q(|F(\zeta)|) .$$

Similarly, using Lemma 3 and (3.7), we obtain the following theorem:

THEOREM 4. *Let* $q > 0$ *and let* $\zeta \in M$ *be fixed. Then* $\log \rho_{M,q}(z,\zeta)$ *is a continuous plurisubharmonic function in* $z \in M$.

The special case of $q = 1$ of this theorem was first proven by Vesentini [9] by using different methods. The paper of [9] also contains some applications of the special case of this theorem.

Let $v \in \mathbf{C}^n$ and consider the function $G(z) = G(z:\zeta, v)$ as in (3.8). For a direction u we define

$$d(z:u) = \frac{|\partial_u G(z)|}{1-|G(z)|^2} .$$

Therefore,

$$\Delta_w \log d(z:u) = 4[d(z:w)]^2$$

for each direction w and whenever $\partial_u G(z) \neq 0$. Especially,

$$\Delta_w \log d(\zeta:v) = 4[d(\zeta:w)]^2 . \tag{4.1}$$

Note that $d(\zeta:v) = C_M(\zeta:v)$.

The next theorem has also been proven in [3] (see also [10]).

THEOREM 5. *Let* $(\zeta, v) \in M \times \mathbf{C}^n$. *Then*

$$\Delta_w \log C_M(\zeta:v) \geq 4[d(\zeta:w)]^2$$

for each direction w *and thus* $\log C_M$ *is plurisubharmonic on* $M \times \mathbf{C}^n$.

Proof. Let

$$g(z) = \frac{G(z)-G(\eta)}{1-\overline{G(\eta)}\,G(z)}; \qquad z, \eta \in M .$$

Thus $g \in H_\eta(M:U)$ and therefore, $|\partial_u g(\eta)| \leq C_M(\eta:u)$ for each direction u. However, $|\partial_u g(\eta)| = d(\eta:u)$. Thus $d(\eta:u) \leq C_M(\eta:u)$ and we have equality at $(\eta, u) = (\zeta, v)$. Near (ζ, v), $d(\eta:u)$ is positive and hence $\log C_M(\eta:u)/d(\eta:u)$ assumes a local minimum at $(\eta, u) = (\zeta, v)$ and so

$$\Delta_w \log C_M(\zeta:v) \geq \Delta_w \log d(\zeta:v)$$

for each direction w. The theorem now follows from (4.1).

Let $v \in C^n$ and assume that $\mu(z:v)$ is a positive upper semi-continuous function near z. The "*curvature*" of $\mu(z:v)$ at z in the direction v is given by

$$\kappa(\mu: z, v) = -\frac{1}{[\mu(z:v)]^2} \Delta_v \log \mu(z:v) .$$

By (4.1) the curvature of $d(z:u)$ is -4 for every $z \in M$ and any direction u. We also have (see [3]):

THEOREM 6. *The curvature of* $C_M(\zeta:v)$ *is always* ≤ -4.

Proof. By Theorem 5, $\Delta_v \log C_M(\zeta:v) \geq 4[d(\zeta:v)]^2$ and, since $d(\zeta:v) = C_M(\zeta:v)$, the theorem follows.

It can be shown that when M is a ball or a polydisk in C^n then the curvature of $C_M(\zeta:v)$ is exactly -4 (see [3] for details).

§5. *Skwarczynski pseudo-distance*

Let $\mathcal{H}(M)$ be a Hilbert space of holomorphic functions or forms on a complex manifold M of C^n. We shall assume that point evaluations are bounded linear functionals on $\mathcal{H}(M)$. Therefore, $\mathcal{H}(M)$ possesses a reproducing kernel k_ζ, $\zeta \in M$, and convergence in the norm implies uniform convergence on compacta of M. It follows from the usual properties of reproducing kernels that for $\zeta \in M$

$$f(\zeta) = (f, k_\zeta)$$

for each $f \in \mathcal{H}(M)$. Here, (f, g) and $\|f\| = (f, f)^{1/2}$ denote the inner product and the norm, respectively of $f, g \in \mathcal{H}(M)$.

We also write $K(z, \bar{\zeta}) = k_\zeta(z)$, $z, \zeta \in M$, and note that $K(z, \bar{z})$ is a real analytic and nonnegative. We shall assume that $K(z, \bar{z}) > 0$ for every $z \in M$. This property is, of course, equivalent to the requirement that for any $z \in M$ there is an $f \in \mathcal{H}(M)$ with $f(z) \neq 0$. Therefore,

$$\|k_\zeta\| = \sqrt{K(\zeta, \overline{\zeta})} > 0$$

for every $\zeta \in M$.

Let $\zeta \in M$ and consider the circle

$$U_M(\zeta) = \{f \in \mathcal{H}(M) : f = e^{i\theta} \frac{k_\zeta}{\|k_\zeta\|}, \quad 0 \leq \theta \leq 2\pi\}.$$

The following pseudo-distance on M was first studied by Skwarczynski [8] when $K(z, \overline{\zeta})$ is the usual Bergman kernel of M. We set

$$\lambda_M(z, \zeta) = \frac{1}{\sqrt{2}} \operatorname{dist}(U_M(z), U_M(\zeta)); \qquad z, \zeta \in M.$$

This is clearly a pseudo-distance on M and we call it the "*Skwarczynski pseudo-distance*" of M. An alternative and useful expression for this pseudo-distance may be derived as follows:

$$\lambda_M(z, \zeta) = \frac{1}{\sqrt{2}} \min_{\phi,\theta} \left\| e^{i\phi} \frac{k_z}{\|k_z\|} - e^{i\theta} \frac{k_\zeta}{\|k_\zeta\|} \right\|$$

$$= \frac{1}{\sqrt{2}} \left\{ 2-2 \left| \left(\frac{k_z}{\|k_z\|}, \frac{k_\zeta}{\|k_\zeta\|} \right) \right| \right\}^{1/2}$$

and therefore,

$$(5.1) \qquad \lambda_M(z, \zeta) = \left[1 - \left(\frac{|K(z, \overline{\zeta})|^2}{K(z, \overline{z}) K(\zeta, \overline{\zeta})} \right)^{1/2} \right]^{1/2}; \qquad z, \zeta \in M.$$

We now prove:

PROPOSITION 3. *Suppose that* $1, f_1, \cdots, f_n \in \mathcal{H}(M)$ *where* $f_j(z) = z_j$, $1 \leq j \leq n$, *and* $z = (z_1, \cdots, z_n) \in M$. *Then* λ_M *is a distance on* M.

Proof. Assume that $\lambda_M(z, \zeta) = 0$ for $z, \zeta \in M$. Therefore, in view of (5.1), we have $|K(z, \overline{\zeta})|^2 = K(z, \overline{z}) K(\zeta, \overline{\zeta})$ which means that $k_\zeta = \alpha k_z$

for some $\alpha \in \mathbb{C}$. This shows that $f(\zeta) = \bar{\alpha} f(z)$ for every $f \in \mathcal{H}(M)$. Putting $f = 1$ and $f = f_j$, $j = 1, \cdots, n$, we obtain $\zeta = z$ which concludes the proof.

EXAMPLE. Let $q > 0$ and consider the space $\mathcal{H}_q(U)$ of holomorphic functions $f(z) = \sum_{n=0}^{\infty} a_n z^n$ in the unit disk U so that

$$\|f\|_q^2 \equiv \sum_{n=0}^{\infty} \frac{n!}{\Gamma(2q+n)} |a_n|^2 < \infty .$$

This is a Hilbert space with the inner product

$$(f, g)_q = \sum_{n=0}^{\infty} \frac{n!}{\Gamma(2q+n)} a_n \bar{b}_n ; \qquad f, g \in \mathcal{H}_q(U) ,$$

where

$$f(z) = \sum_{n=0}^{\infty} a_n z^n , \quad g(z) = \sum_{n=0}^{\infty} b_n z^n ; \qquad z \in U .$$

The space $\mathcal{H}_q(U)$ possesses the reproducing kernel

$$\text{(5.3)} \qquad k_\zeta(z) = K_q(z, \bar{\zeta}) = \Gamma(2q)(1 - z\bar{\zeta})^{-2q} ; \quad z, \zeta \in U .$$

Indeed, for any $f \in \mathcal{H}_q(U)$

$$f(\zeta) = \sum_{n=0}^{\infty} a_n \zeta^n = (f, k_\zeta) , \quad \zeta \in U ,$$

because

$$K_q(z, \bar{\zeta}) = \sum_{n=0}^{\infty} \frac{\Gamma(2q+n)}{n!} (z\bar{\zeta})^n ; \qquad z, \zeta \in U .$$

The Skwarczynski pseudo-distance in this case is

$$\lambda_{U,q}(z,\zeta) = \left[1 - \left(\frac{(1-|z|^2)(1-|\zeta|^2)}{|1-z\bar{\zeta}|^2}\right)^q\right]^{1/2}$$

or

$$\lambda_{U,q}(z,\zeta) = [1-(1-\delta^2)^q]^{1/2} ,$$

where $\delta = \delta(z,\zeta)$, $z,\zeta \in U$, is the previously defined Möbius distance on U. Therefore, using (2.1), we have that $\delta_q = \lambda_{U,q}$. This, with the aid of Proposition 3, proves the following result which was anticipated earlier.

THEOREM 7. *Let* $q > 0$. *Then the function* δ_q *defines a distance on* U.

Many other interesting examples of the pseudo-distance λ_M can be obtained by using different reproducing kernels on M. We shall however pursue the previous example a little further. Let D be a simply connected plane region such that $D \neq \mathbb{C}$ and let $K(z,\bar{\zeta})$ be its usual Bergman kernel function. Let $\phi: D \to U$ be a conformal mapping of D onto the unit disk U. One may define the space $\mathcal{H}_q(D)$, $q > 0$, by setting $\mathcal{H}_q(D) = T[\mathcal{H}_q(U)]$ where $Tf = f_o\phi(\phi')^q$, $f \in \mathcal{H}_q(U)$. Then $\mathcal{H}_q(D)$ is a linear space of holomorphic functions in D and T is a bijective linear transformation from $\mathcal{H}_q(U)$ onto $\mathcal{H}_q(D)$. The inverse of T is given by $T^{-1}g = g_o\psi(\psi')^q$, $g \in \mathcal{H}_q(D)$, where $\psi = \phi^{-1}$. We define

$$(f,g)_{D,q} = (T^{-1}f, T^{-1}g)_q ; \quad f,g \in \mathcal{H}_q(D) .$$

Then $\mathcal{H}_q(D)$ is a Hilbert space of holomorphic functions in D with the above defined inner product. Moreover, $\mathcal{H}_q(D)$ possesses a reproducing kernel

$$K_{D,q}(z,\bar{\zeta}) = K_q(\phi(z), \overline{\phi(\zeta)})\,\phi'(z)^q\,\overline{\phi'(\zeta)}^q ; \quad z,\zeta \in D .$$

Therefore,

$$K_{D,q}(z,\bar{\zeta}) = \Gamma(2q)\,\pi^q[K(z,\bar{\zeta})]^q ; \qquad z,\zeta \in D , \tag{5.4}$$

and thus

$$\lambda_{D,q}(z,\zeta) \equiv \delta_{D,q}(z,\zeta) = \left[1 - \left(\frac{K(z,\bar{\zeta})^2}{K(z,\bar{z})K(\zeta,\bar{\zeta})}\right)^{q/2}\right]^{1/2}$$

is a distance function on D for any $q > 0$.

For $q \geq 1/2$ the norm of $\mathcal{H}_q(D)$ can be realized in a form which appears frequently in function theory. In fact,

$$(5.5) \qquad \|f\|^2_{D,q} = \frac{\pi^{-q}}{\Gamma(2q-1)} \int_D |f(z)|^2 [K_D(z,z)]^{1-q} dm(z), \qquad q > 1/2$$

where m is the area Lebesgue measure and

$$(5.6) \qquad \|f\|^2_{D,1/2} = \frac{1}{2\pi} \int_{\partial D} |f(z)|^2 |dz| .$$

In the last integral f stands for the nontangential boundary values of the holomorphic function $f(z)$ in D.

§6. *The Bergman metric*

We consider the previously defined Hilbert space $\mathcal{H}(M)$ and we let $S(M) = \{f \in \mathcal{H}(M) : \|f\| \leq 1\}$. Let $\zeta \in M$ and define $S_\zeta(M) = \{f \in S(M) : f(\zeta) = 0\}$. For a direction v we define

$$(6.1) \qquad b_M(\zeta : v) = \sqrt{K(\zeta,\bar{\zeta})} \max\{|\partial_v f(\zeta)| : f \in S_\zeta(M)\} .$$

It follows, using standard Hilbert space arguments, that

$$(6.2) \qquad b^2_M(\zeta : v) = \frac{1}{4} \Delta_v \log K, \qquad K = K(\zeta,\bar{\zeta}) .$$

This is a Kahler metric which we label as the "*Bergman metric*." When $K(\zeta,\bar{\zeta})$ is the usual Bergman kernel $b^2_M(\zeta : v)$ is precisely the classical Bergman metric of M.

This metric is connected with the pseudo-distance λ_M. In fact, for a tangent vector $v = (dt_1, \cdots, dt_n) \in C^n$ we have

$$d^2 \lambda_M^2(t, \zeta)|_{t=\zeta} = b_M^2(\zeta : v) .$$

We also note that the first differential of λ_M^2 vanishes, i.e., $d\lambda_M^2(t, \zeta)|_{t=\zeta} = 0$, and that the critical points of λ_M are related to the Bergman representative domains (see [8] for more details).

Let β_M denote the integrated metric on M of b_M. This is a pseudo-distance on M and we call it the *"Bergman pseudo-distance"* of M. Again, if $K(\zeta, \bar{\zeta})$ is the Bergman kernel then β_M is the usual Bergman pseudo-distance of M.

It should be emphasized here that, in general, λ_M, β_M and b_M are not necessarily biholomorphically invariant. We, however, make the following observation: Let $\phi : M^* \to M$ be a biholomorphic mapping of M^* onto M with a non-vanishing Jacobian $J_\phi = \partial w/\partial z$, $w = \phi(z)$. Assume that there is an $\alpha \in R$ such that $[J_\phi(z)]^\alpha$ is holomorphic in M^*. We set $\mathcal{H}(M^*) = T[\mathcal{H}(M)]$ where $Tf = f_o \phi [J_\phi]^\alpha$, $f \in \mathcal{H}(M)$. Then $\mathcal{H}(M^*)$ is a linear space of holomorphic functions or forms in M^* and T is a bijective linear transformation from $\mathcal{H}(M)$ onto $\mathcal{H}(M^*)$. By insisting that T is an isometry, $\mathcal{H}(M^*)$ becomes a Hilbert space with the reproducing kernel

(6.3) $$K_{M^*}(z, \bar{\zeta}) = K_M(\phi(z), \overline{\phi(\zeta)}) [J_\phi(z)]^\alpha \overline{[J_\phi(\zeta)]}^\alpha .$$

Under these circumstances we also have

(6.4) $$\lambda_{M^*}(z, \zeta) = \lambda_M(\phi(z), \phi(\zeta)), \quad \beta_{M^*}(z, \zeta) = \beta_M(\phi(z), \phi(\zeta))$$

and

(6.5) $$b_{M^*}(z : v) = b_M(\phi(z) : \phi_*(v)) .$$

We now prove:

THEOREM 8. *Assume that* $1 \in \mathcal{H}(M)$ *with* $\|1\| \leq 1$. *Then* $\log b_M$ *is plurisubharmonic on* $M \times C^n$.

Proof. According to (6.1) there is an $h \in \mathcal{H}(M)$, $h(z) = h(z : \zeta, v)$, with $h(\zeta) = 0$, $\|h\| = 1$ and

$$[K(\zeta, \zeta)]^{-1} b_M^2(\zeta : v) = |\partial_v h(\zeta)|^2 .$$

Let $\eta \in M$ and let

$$g(\eta) = \|h - h(\eta)\|^2 = 1 + \alpha |h(\eta)|^2 - \beta h(\eta) - \overline{\beta}\,\overline{h(\eta)}$$

where

$$\alpha = \|1\|^2, \qquad \beta = (1, h) .$$

Define

$$f(z) = \frac{1}{\sqrt{g(\eta)}} [h(z) - h(\eta)], \qquad z \in M .$$

Then $f(\eta) = 0$, $\|f\| = 1$ and so $f \in S_\eta(M)$. Consequently, using (6.1), for any direction u we have

$$[K(\eta, \eta)]^{-1} b_M^2(\eta : u) \geq |\partial_u f(\eta)|^2 = \frac{1}{g(\eta)} |\partial_u h(\eta)|^2$$

where equality occurs when $(\eta, u) = (\zeta, v)$. Therefore, for any direction w,

$$\Delta_w \log b_M^2(\zeta : v) \geq \Delta_w \log K(\zeta, \overline{\zeta}) + \Delta_w \log |\partial_u h(\eta)|^2 \big|_{(\eta, u) = (\zeta, v)}$$
$$- \Delta_w \log g(\eta) \big|_{\eta = \zeta} .$$

Thus

$$\Delta_w \log b_M(\zeta : v) \geq 2[b_M^2(\zeta : w) - \overline{\partial}_w \partial_w \log g(\eta)\big|_{\eta = \zeta}] .$$

However,

$$\overline{\partial}_w \partial_w \log g(\eta)\big|_{\eta = \zeta} = |\partial_w h(\zeta)|^2 [\alpha - |\beta|^2]$$

and since

$$b_M^2(\zeta : w) \geq K(\zeta, \overline{\zeta}) |\partial_w h(\zeta)|^2$$

we obtain

$$\Delta_w \log b_M(\zeta ; v) \geq 2 |\partial_w h(\zeta)|^2 \{K(\zeta, \overline{\zeta}) - [\alpha - |\beta|^2]\} .$$

Let $r = 1-(1,h)h = 1-\beta h$. Then $r \in \mathcal{H}(M)$, $r(\zeta) = 1$ and

$$\|r\|^2 = \|1\|^2 - |(1,h)|^2 = \alpha - |\beta|^2 > 0 .$$

Therefore, using the reproducing property of $k_\zeta(z) = K(z,\bar{\zeta})$,

$$1 = |r(\zeta)|^2 = |(r, k_\zeta)|^2 \leq \|r\|^2 K(\zeta,\bar{\zeta})$$

and so

$$K(\zeta,\zeta) - [\alpha - |\beta|^2] = K(\zeta,\zeta) - \|r\|^2 \geq \|r\|^{-2} - \|r\|^2 = \|r\|^{-2}(1-\|r\|^4) .$$

Since $\|1\| \leq 1$ it follows that $\|r\| \leq 1$ and the theorem is proved.

This leads to the following result:

COROLLARY 1. *Let the assumptions of formula (6.3), with $\alpha \neq 0$, prevail and assume that $1 \in \mathcal{H}(M)$. Then $\log b_M$ is plurisubharmonic on $M \times C^n$. Especially, if M is a bounded domain in C^n and b_M is the classical Bergman metric then $\log b_M$ is plurisubharmonic on $M \times C^n$.*

Proof. In view of Theorem 8 we may assume that $\|1\|_M = \beta > 1$. Let $\gamma = \beta^{-1/n\alpha}$ and consider the biholomorphic mapping $\phi: M \to M^*$ given by $w = \phi(z) = \gamma z$. Then $Tf = f_o\phi[J_\phi]^\alpha = \beta^{-1} f_o\phi$, $f \in \mathcal{H}(M^*)$, defines an isometry from $\mathcal{H}(M^*)$ onto $\mathcal{H}(M)$ and therefore

$$\|1\|_{M^*} = \beta^{-1}\|1\|_M = 1 .$$

Consequently, $\log b_{M^*}$ is plurisubharmonic on $M^* \times C^n$. Let $v \in C^n$. By (6.5) we have $b_M(z{:}v) = \gamma b_{M^*}(w{:}v)$, $w = \phi(z) = \gamma z$, and thus

$$\log b_M(z{:}v) = \log b_{M^*}(w{:}v) + \log \gamma .$$

Let $Z = (z{:}v) = (Z_1, \cdots, Z_{2n})$ and $W = (w{:}v) = (W_1, \cdots, W_{2n})$. Here,

$$W_j = \gamma_j Z_j, \quad j = 1, \cdots, 2n; \quad (\gamma_1 = \cdots = \gamma_n = \gamma,\ \gamma_{n+1} = \cdots = \gamma_{2n} = 1) .$$

We write $g_M(Z) = \log b_M(z{:}v)$ and $g_{M^*}(W) = \log b_M(w{:}v)$. Let $u = (u_1, \cdots, u_{2n}) \in C^{2n}$. Then

$$\Delta_u g_M(Z) = \Delta_{u_\gamma} g_{M^*}(W)$$

where $u_\gamma = (\gamma u_1, \cdots, \gamma u_n, u_{n+1}, \cdots, u_{2n}) \in C^{2n}$. Since $\log b_{M^*}$ is plurisubharmonic on $M^* \times C^n$ we have $\Delta_{u_\gamma} g_{M^*}(W) \geq 0$ and thus $\Delta_u g_M(Z) \geq 0$. This shows that $\log b_M$ is plurisubharmonic on $M \times C^n$. Finally, in the case that $\mathcal{H}(M)$ is the usual Bergman space, formula (6.3) holds with $\alpha = 1$ and, of course, $1 \in \mathcal{H}(M)$ when M is a bounded domain in C^n. This concludes the proof.

§7. *On a theorem of Pick*

In this section we shall make an additional assumption on the Hilbert space $\mathcal{H}(M)$. Specifically, we assume that $\mathcal{H}(M)$ is the space of *all* holomorphic functions or forms f in M normed by

$$\|f\|_\mu^2 = \int_{M_o} |f(z)|^2 \, d\mu(z) < \infty .$$

Here μ is a positive measure acting on M_o, where M_o is either M or any part of the boundary ∂M which determines the holomorphic functions in M as, for example, the Šilov boundary of M. In the case that M_o is not M, f in the last integral stands for the nontangential boundary values of the holomorphic function $f(z)$, $z \in M$. In this way we may regard $\mathcal{H}(M) \equiv H_2(M:\mu)$ as a closed subspace of $L_2(M:\mu)$ in a natural manner. Examples of such spaces are the Bergman spaces and the Hardy-Szegö spaces to name only a few (see [4] for more details).

Let $f \in H(M:U)$ and consider the Hermitian kernel

$$B(z,\bar{\zeta}) = K(z,\bar{\zeta})[1 - f(z)\overline{f(\zeta)}]; \qquad z, \zeta \in M .$$

Obviously, for any $\zeta \in M$, $B(\cdot,\zeta)$ is in $H_2(M:\mu)$ and, in fact,

$$\|B(\cdot,\bar{\zeta})\|^2 \leq 4K(\zeta,\bar{\zeta}) .$$

We also write

$$R(z,\zeta) = K(\zeta,\bar{z})[f(z)-f(\zeta)]$$

and

$$S(z,\zeta) = K(\zeta,z)[1-|f(z)|^2]^{\frac{1}{2}} .$$

For any $\zeta \in M$ these functions belong to $L_2(M:\mu)$ and we have

$$\|R(\cdot,\zeta)\|^2 \leq 4K(\zeta,\bar{\zeta}) ,$$

and

$$\|S(\cdot,\zeta)\|^2 \leq 2K(\zeta,\bar{\zeta}) .$$

LEMMA 5. *For* $z,\zeta \in M$ *we have*

$$B(z,\bar{\zeta}) = (R(\cdot,z),R(\cdot,\zeta)) + (S(\cdot,z),S(\cdot,\zeta)) .$$

Proof. By the reproducing property

$$\begin{aligned} B(z,\bar{\zeta}) &= (B(\cdot,\bar{\zeta}),k_z) \\ &= (k_\zeta[1-\overline{f(\zeta)}f],k_z) . \end{aligned}$$

But

$$1-f(t)\overline{f(\zeta)} = 1-|f(t)|^2+f(t)[\overline{f(t)}-\overline{f(\zeta)}]$$

$$f(t)[\overline{f(t)}-\overline{f(\zeta)}] = [f(t)-f(z)][\overline{f(t)}-\overline{f(\zeta)}]+f(z)[\overline{f(t)}-\overline{f(\zeta)}] .$$

However,

$$\begin{aligned} &\overline{(k_\zeta f(z)[\bar{f}-\overline{f(\zeta)}],k_z)} \\ &= \overline{f(z)}(k_z[f-f(\zeta)],k_\zeta) \\ &= \overline{f(z)}\{k_z(\zeta)f(\zeta)-f(\zeta)k_z(\zeta)\} = 0 . \end{aligned}$$

Therefore,

$$\begin{aligned} B(z,\bar{\zeta}) &= (k_\zeta[f-f(z)][\bar{f}-\overline{f(\zeta)}],k_z)+(k_\zeta[1-|f|^2],k_z) \\ &= (\bar{k}_z[f-f(z)],\bar{k}_\zeta[f-f(\zeta)])+(\bar{k}_z[1-|f|^2]^{\frac{1}{2}},\bar{k}_\zeta[1-|f|^2]^{\frac{1}{2}}) \end{aligned}$$

and the assertion follows.

The following theorem is a generalization of a theorem of Pick (see, for example, Ahlfors [1, p. 3-4]) on the Szegö kernel of the unit disk $K_{1/2}(z,\bar{\zeta}) = (1-z\bar{\zeta})^{-1}$ (see (5.3) and (5.6)). The proof of our theorem is simpler than the classical one and can be further generalized in several directions (see [4, 5] for more details).

THEOREM 9. *Let* $f \in H(M{:}U)$. *Then the kernel* $B(z,\bar{\zeta})$ *is positive definite on* M. *That is, given any finite number of points* $z_1, \cdots, z_N \in M$ *and corresponding numbers* $\alpha_1, \cdots, \alpha_N \in \mathbb{C}$ *we have*

$$\sum_{m,k=1}^{N} B(z_m, \bar{z}_k)\alpha_m\bar{\alpha}_k \geq 0\,.$$

Proof. We use Lemma 5. Then

$$\sum_{m,k=1}^{N} B(z_m, \bar{z}_k)\alpha_m\bar{\alpha}_k$$

$$= \sum_{m,k=1}^{N} \alpha_m\bar{\alpha}_k\{(R(\cdot, z_m), R(\cdot, z_k)) + (S(\cdot, z_m), S(\cdot, z_k))\}$$

$$= \left\| \sum_{m=1}^{N} \alpha_m R(\cdot, z_m) \right\|^2 + \left\| \sum_{m=1}^{N} \alpha_m S(\cdot, z_m) \right\|^2 \geq 0$$

and the theorem follows.

COROLLARY 2. *Let* $f \in H(M{:}U)$. *Then*

$$\det\,[B(z_m, \bar{z}_k)]_{m,k=1}^{N} \geq 0$$

for any $z_1, \cdots, z_N \in M$.

Many interesting distortion theorems arise from this general result by putting various reproducing kernels. For example, since the space

$\mathcal{H}_q(D)$, $q \geq 1/2$, is of the form $H_2(D{:}\mu)$, as (5.5) and (5.6) show, we have, in view of (5.4)

$$\det[K_D{}^q(z_m, \bar{z}_k)(1-f(z_m)\overline{f(z_k)})]_{m,k=1}^N \geq 0, \qquad q \geq 1/2,$$

for any $f \in H(D{:}U)$ and any $z_1, \cdots, z_N \in D$. Here, $K_D(z, \bar{\zeta})$ is the Bergman kernel of the simply connected plane region D. Similar result holds for homogeneous domains in $\mathbf{C}^n$. Especially,

$$\det[(1-z_m\bar{z}_k)^{-2q}(1-f(z_m)\overline{f(z_k)})]_{m,k=1}^N \geq 0, \qquad q \geq 1/2$$

for any $f \in H(U{:}U)$ and any $z_1, \cdots, z_N \in U$. The case $q = 1/2$ is the classical theorem of Pick.

The following corollary may be viewed as a direct generalization of the classical Schwarz-Pick lemma as stated in Proposition 1.

COROLLARY 3. *Let* $f \in H(M{:}U)$. *Then*

$$\frac{|K(z,\bar{\zeta})|^2}{K(z,\bar{z})K(\zeta,\bar{\zeta})} \leq \frac{(1-|f(z)|^2)(1-|f(\zeta)|^2)}{|1-f(z)\overline{f(\zeta)}|^2}$$

for any $z, \zeta \in M$.

Proof. This follows from Corollary 2 by taking $N = 2$ and $(z, \zeta) = (z_1, z_2)$.

The next corollary establishes a relationship between the Skwarczynski and the Carathéodory pseudo-distances.

COROLLARY 4. *Let* δ_M, ρ_M *and* λ_M *be the Möbius, Carathéodory and Skwarczynski, respectively, pseudo-distances of* M. *Then*

$$\lambda_M \geq [1-(1-\delta_M^2)^{1/2}]^{1/2} = [1-\operatorname{sech}\rho_M]^{1/2}$$

and therefore $\delta_M \leq \sqrt{2}\,\lambda_M$.

Proof. This follows from Corollary 3, (3.1), (3.3) and (5.1).

The following corollary has been also proved in [2] (see also [6]).

COROLLARY 5. *Let* $(\zeta, v) \in M \times C^n$. *Then*

$$C_M(\zeta:v) \le b_M(\zeta:v)$$

and therefore $\rho_M \le \tilde{\rho}_M \le \beta_M$.

Proof. Let $f \in H_\zeta(M:U)$ and consider

$$g \equiv g(z,\bar{z}) = \log(1-|f(z)|^2) - \log \frac{|K(z,\bar{\zeta})|^2}{K(z,\bar{z})K(\zeta,\bar{\zeta})}, \quad z \in M .$$

By Corollary 3, $g \ge 0$ and g assumes a local minimum at $z = \zeta$. Thus $\Delta_v g \ge 0$ at $z = \zeta$. However,

$$\Delta_v g = -4(1-|f(z)|^2)^{-2}|\partial_v f(z)|^2 + 4\partial_v \bar{\partial}_v \log K(z,\bar{z})$$

and so

$$\partial_v \bar{\partial}_v \log K(\zeta,\bar{\zeta}) \ge |\partial_v f(\zeta)|^2 .$$

The corollary now follows by appealing to (3.4), (3.5), (6.2) and the fact that β_M is the integrated metric of b_M.

When γ_M is a distance function on M, M becomes a metric space (M, γ_M). This metric space is said to be complete (see also [7, p. 53]) if for each point $\zeta \in M$ and each $r > 0$, the closed ball $\{z \in M : \gamma_M(z,\zeta) \le r\}$ is a compact subset of M. If (M, γ_M) is complete in this sense then it is complete in the usual sense. The converse is not true in general, but it is true if γ_M is induced from a Riemannian metric.

We already know that, for $q > 0$, $\delta_{M,q}$ is a distance if and only if $M \in \mathfrak{B}$ and the same is true for $\rho_{M,q}$ provided $1 \ge q > 0$. This, together with the previous corollaries, leads to the following result:

THEOREM 10. *Let* $q > 0$. *Then*

$$\delta_M \le \rho_M \le \tilde{\rho}_M \le \beta_M, \quad \delta_{M,q} \le \rho_{M,q}, \quad \delta_M \le \sqrt{2}\,\lambda_M . \tag{7.1}$$

Moreover, if $M \in \mathcal{B}$ *then*:

(i) *All these pseudo-distances, where for* $\rho_{M,q}$ *we take* $q \in (0,1]$, *are in fact distances on* M.

(ii) *If for some* $q_o > 0$, (M, δ_{M,q_o}) *is complete then* $(M, \delta_{M,q})$ *and* $(M, \rho_{M,q})$ *are complete for every* $q > 0$ *and every* $q \in (0,1]$, *respectively. Also, if* (M, ρ_{M,q_o}) *is complete for some* $q_o \in (0,1]$ *then* $(M, \delta_{M,q})$ *and* $(M, \rho_{M,q})$ *are complete for every* $q > 0$ *and every* $q \in (0,1]$, *respectively.*

(iii) *If* (M, ρ_M) *is complete then* (M, λ_M) *is complete.*

(iv) (M, ρ_M) *complete* $\Rightarrow (M, \tilde{\rho}_M)$ *complete* $\Rightarrow (M, \beta_M)$ *complete.*

Proof. The inequalities in (7.1) follow from Corollaries 4, 5 and

$$\rho_{M,q} = \frac{1}{2} \log \frac{1+\delta_{M,q}}{1-\delta_{M,q}} = \sum_{n=0}^{\infty} \frac{\delta_{M,q}^{2n+1}}{2n+1} \geq \delta_{M,q} .$$

Item (i) follows from (7.1) and the fact that $\delta_{M,q}$ is a distance when $M \in \mathcal{B}$. A use of (3.3) shows that, for q_o, $q > 0$,

$$\delta_{M,q} = f_{q/q_o}(\delta_{M,q_o}), \qquad \rho_{M,q} = g \circ f_{q/q_o}(\delta_{M,q_o})$$

and

$$\delta_{M,q} = f_{q/q_o} \circ g^{-1}(\rho_{M,q_o}), \qquad \rho_{M,q} = g \circ f_{q/q_o} \circ g^{-1}(\rho_{M,q_o}),$$

where f_q and g are the increasing functions defined in (2.3) and (2.4). This proves item (ii). Items (iii) and (iv) follow from (7.1).

DEPARTMENT OF MATHEMATICS
UNIVERSITY OF PITTSBURGH
PITTSBURGH, PENNSYLVANIA 15260

BIBLIOGRAPHY

[1] Ahlfors, L. V., *Conformal Invariants: Topics in Geometric Function Theory*, McGraw-Hill, New York, 1973.

[2] Burbea, J., The Carathéodory metric and its majorant metrics, Can. J. Math. 29 (1977), 771-780.

[3] ________, On the Hessian of the Carathéodory metric, Rocky Mount. J. Math. 8 (1978), 555-559.

[4] ________, A generalization of Pick's theorem and its applications to intrinsic metrics, Ann. Polon. Math. 39 (1979), to appear.

[5] ________, Pick's theorem with operator-valued holomorphic functions, Kōdai J. Math., to appear.

[6] Hahn, K. T., On completeness of the Bergman metric and its subordinate metrics, Proc. Nat. Acad. Sci. U.S.A. 73 (1976), 4294.

[7] Kobayashi, S., *Hyperbolic Manifolds and Holomorphic Mappings*, Marcel Dekker, New York, 1970.

[8] Skwarczynski, M., The invariant distance in the theory of pseudo-conformal transformations and the Lu Qi Keng conjecture, Proc. Amer. Math. Soc. 22 (1969), 305-310.

[9] Vesentini, E., Variations on a theme of Carathéodory, Ann. Scuola Norm. Sup. Pisa (4) 6 (1979), 39-68.

[10] ________, Invariant distances and invariant differential metrics in locally convex spaces, to appear.

NECESSARY CONDITIONS FOR SUBELLIPTICITY AND HYPOELLIPTICITY FOR THE $\bar{\partial}$-NEUMANN PROBLEM ON PSEUDOCONVEX DOMAINS

David Catlin

In a recent paper [8], J. J. Kohn studied the question of when subelliptic estimates hold for the $\bar{\partial}$-Neumann problem for (p,q)-forms on a pseudoconvex domain. For a given boundary point z_0 he defined an increasing sequence of ideals $I_k^q(z_0)$, $k = 1, 2, \cdots$, over the ring $\mathcal{E}_{z_0}$ of germs of C^∞-functions at z_0. The ideals are associated with the order of contact of q-dimensional complex-analytic subvarieties with the boundary of the domain at the point z_0. He was able to show that if for some integer k, $I_k^q(z_0) = \mathcal{E}_{z_0}$, then a subelliptic estimate holds near z_0. By using the geometric description of real-analytic boundaries of Diederich and Fornaess [2], he obtained the following theorem. (In all of this we shall assume the reader is familiar with the basic terminology and set-up of the $\bar{\partial}$-Neumann problem as contained in [6] or in the introduction of [8].)

THEOREM. *Let z_0 be a boundary point of a pseudoconvex domain Ω in $\mathbb{C}^n$ with smooth boundary. Let U be a neighborhood of z_0 and assume that the boundary $b\Omega$ is real-analytic in U. Suppose that there is no germ of a q-dimensional subvariety V_q with $z_0 \in V_q$ and $V_q \subset b\Omega$. Then there exist a positive constant ε and a neighborhood U_0 of z_0 such that a subelliptic estimate of order ε holds for (p,q)-forms supported in $U_0 \cap \bar{\Omega}$, i.e.,*

Recent Developments in Several Complex Variables
0-691-08285-5/81 000093-08 $00.50/0 (cloth)
0-691-08281-2/81/000093-08 $00.50/0 (paperback)
For copying information, see copyright page.

$$|||u|||_\varepsilon^2 \le C(\|\bar\partial u\|^2 + \|\vartheta u\|^2), \quad u \in \mathcal{D}^{p,q}(U_0 \cap \bar\Omega).$$

For $q = n-1$, precise results were obtained for domains with smooth boundary. These results show that the order of the best possible subelliptic estimate near a given boundary point is inversely related to the order of contact with the boundary of (n–1)-dimensional varieties containing that point.

In this paper we shall give necessary conditions for subelliptic estimates for (p,q)-forms that are similar to but not as precise as the sufficient conditions known for $q = n-1$. The question of necessary conditions for hypoellipticity of the $\bar\partial$-Neumann problem is also considered. We now state precisely the results

THEOREM 1. *Suppose that* Ω *is a smoothly bounded pseudoconvex domain in* $\mathbb{C}^n$. *Let* U_0 *be a neighborhood of a boundary point* z_0 *and suppose that a subelliptic estimate of order* ε *holds for the* $\bar\partial$*-Neumann problem for* (p,q)*-forms supported in* U_0, *i.e.,*

$$|||u|||_\varepsilon^2 \le C(\|\bar\partial u\|^2 + \|\vartheta u\|^2), \quad u \in \mathcal{D}^{p,q}(U_0).$$

If V *is any* q*-dimensional variety containing* z_0, *then there is a sequence of points* $z_1, z_2, \cdots, \in V$ *such that* $z_k \to z_0$ *and*

$$r(z_k) \ge |z_k - z_0|^{(n+2)/\varepsilon}. \tag{1}$$

Here $r(z)$ *denotes the boundary-defining function of* $b\Omega$.

When $q = 1$ and $n = 2$, Greiner [7] has shown that the exponent $\frac{n+2}{\varepsilon}$ in (1) can be strengthened to $\frac{1}{\varepsilon}$. His method extends to $q = n-1$. Thus it is clear that our results are not the best possible. Necessary conditions along these lines for Lipschitz and L^p norms have also been given by Krantz [9]. In the case of arbitrary q, when the variety V is a manifold near z_0, Egorov has announced in [5] that (1) holds with the exponent of $\frac{1}{\varepsilon}$. The question of when a singular variety can be tangent to the boundary to a given order has been studied by D'Angelo in [3].

The method used to prove Theorem 1 is a quantitative version of the argument used to study the question of hypoellipticity of the $\bar{\partial}$-Neumann problem.

THEOREM 2. *Let* z_0 *be a boundary point of a smoothly bounded pseudo-convex domain* Ω *in* $\mathbb{C}^n$ *and suppose there is a* q*-dimensional complex manifold* V *passing through* z_0 *with* $V \subset b\Omega$. *Then there is a* $\bar{\partial}$*-closed* (p,q)*-form* $\alpha \in L^2_{(p,q)}(\Omega)$ *such that if* u *is any* $(p,q-1)$*-form with* $\bar{\partial}u = \alpha$, *then* sing supp u *strictly contains* sing supp α *(as distributions on* $\bar{\Omega}$*).*

REMARK. The author has recently learned that a similar proof of this theorem has been discovered by Diederich and Pflug [4].

Proof of Theorem 2. It suffices to prove the theorem for $p = 0$ since if I and J are multi-indices of length p and q, respectively, then

$$\bar{\partial}(f(z)\,dz_I \wedge d\bar{z}_J) = (-1)^p\, dz_I \wedge \bar{\partial}(f(z)\,d\bar{z}_J)\,.$$

By hypothesis we may choose coordinates on $\mathbb{C}^n$ and a neighborhood U around the point z_0 satisfying:

(i) $U = U' \times U''$, where

$$U' = \{z' \in \mathbb{C}^q;\ |z_1|^2 + \cdots + |z_q|^2 < 16\} \text{ and}$$

$$U'' = \{z'' \in \mathbb{C}^{n-q};\ |z_{q+1}|^2 + \cdots + |z_n|^2 < 4\}\,,$$

(ii) the point z_0 is given by the origin,

(iii) writing $z_n = x_n + iy_n$, we have $\frac{\partial r}{\partial x_n} > 0$ on $b\Omega \cap U$.

(iv) there is a mapping $h: U' \to U''$, $h = (h_{q+1}, \cdots, h_n)$ such that the h_j are holomorphic and satisfy $|h| < 1$, and $V \cap U = \{(z', h(z'));\ z' \in U'\}$.

Let $\zeta_k = (0, 0, \cdots, 0, -\frac{1}{k})$. Then by a theorem of Pflug [9], there is a holomorphic function $f \in L^2(\Omega)$ such that $\overline{\lim_{k\to\infty}}\, |f(\zeta_k)| = +\infty$. Let $\chi_1 : U' \to \mathbb{R}$ be defined by

$$\chi_1(z') = \phi(|z_1|^2 + \cdots + |z_q|^2)\,,$$

where ϕ is a smooth nonnegative function such that $\phi(t) = 0$ for $t \geq 2$ and $\phi(t) = 1$ for $t \leq 1$. Set

$$\alpha(z) = \chi_1(z')f(z)\,d\bar{z}_1 \cdots d\bar{z}_q \, .$$

Clearly $\bar{\partial}\alpha = 0$ and $\alpha \in L^2_{(0,q)}(\Omega)$. We shall show by way of contradiction that if u is any $(0, q-1)$-form with $\bar{\partial}u = \alpha$, then sing supp u strictly contains sing supp α. Thus assume that

$$\text{sing supp } \alpha = \text{sing supp } u \, .$$

Let $g_k : U' \to \Omega$ be defined by

$$g_k(z') = g_k(z_1, \cdots, z_q) = \left(z_1, \cdots, z_q, h_{q+1}(z'), \cdots, h_{n-1}(z'), h(z') - \frac{1}{k}\right).$$

Then for all k, $g_k(U') \subset U' \times U''$. Hence we may take the pullbacks of u and α through g_k. They satisfy $\bar{\partial}(g_k^* u) = g_k^*(\alpha)$.

We put the standard hermitian metric $\sum_{i=1}^{q} dz_i \otimes d\bar{z}_i$ on U'. This gives rise to an inner product on forms, which we denote by $< \, , \, >$. For $z' \in U'$, let $\chi_2(z')$ be defined by $\chi_2(z') = \phi\left(\frac{|z'|^2}{4}\right)$, and let w be the $(0,q)$-form on U' defined by

$$w = \chi_2 d\bar{z}_1 \wedge \cdots \wedge d\bar{z}_q \, .$$

Note that α is identically 0 on the set $\{z \in \bar{\Omega}; |z_1|^2 + \cdots + |z_q|^2 > 2\}$. Since sing supp u = sing supp α, it follows that u is smooth on this set. Hence there is a constant $M > 0$ such that $|g_k^* u| \leq M$ on the set $\{z' \in C^q; q \geq |z_1|^2 + \cdots + |z_q|^2 \geq 3\}$.

Integration by parts gives

$$\left| \int_{U'} < \bar{\partial}(g_k^* u), w > dV \right| = \left| \int_{U'} < g_k^* u, \vartheta w > dV \right| ,$$

Since ϑw is supported in the region where $|g_k^* u| \leq M$, one obtains the estimate

$$\left| \int_{U'} < \bar{\partial}(g_k^* u), w > dV \right| \leq \left| \int_{U'} dV \right| \cdot M \cdot \sup_{z' \in U'} |\vartheta w| \leq C, \tag{2}$$

where C is independent of k. On the other hand, since $\bar{\partial}(g_k^* u) = g_k^*(\alpha) = \chi_1(z') f(g_k(z')) d\bar{z}_1 \wedge \cdots \wedge d\bar{z}_q$, we have

$$\left| \int_{U'} < \bar{\partial}(g_k^* u), w > dV \right| = \left| 2^q \int_{U'} f(g_k(z')) \chi_1(z') dV \right| .$$

The factor 2^q in the right-hand side appears because $< d\bar{z}_i, d\bar{z}_j > = 2\delta_{ij}$. Note that $f(g_k(z'))$ is a holomorphic function of z' and therefore satisfies the mean-value equality:

$$\int_{|z'|=r} f(g_k(z')) dS = f(g_k(0)) \int_{|z'|=r} dS .$$

Integrating with respect to r gives

$$\left| 2^q \int_{U'} f(g_k(z')) \chi_1(z') dV \right| = \left| 2^q f(g_k(0)) \int_0^{\sqrt{2}} r^{2q-1} \phi(r^2) dr \int_{|z'|=1} dS \right|$$

$$= C |f(g_k(0))| = C |f(\zeta_k)| .$$

But since $\overline{\lim_{k \to \infty}} |f(\zeta_k)| = +\infty$, we have a contradiction of (2). This completes the proof.

If the $\bar{\partial}$-Neumann problem is hypoelliptic, then it is always possible to solve $\bar{\partial} u = \alpha$, with sing supp u = sing supp α. Thus Theorem 2 implies the following corollary.

COROLLARY. *Suppose that* Ω *is a smoothly bounded pseudoconvex domain in* $\mathbb{C}^n$ *and that there exists a q-dimensional complex manifold* V *with* $V \cap b\Omega$. *Then the* $\bar{\partial}$*-Neumann problem for* (p,q)*-forms on* $\bar{\Omega}$ *is not hypoelliptic.*

When $q = 1$ there are necessary conditions that are stronger than those of Theorem 2. These conditions are stated in terms of certain holomorphic convexity properties at the boundary. To state the results the following notation is needed.

$$A(\Omega) = \{u \in C^\infty(\Omega);\ u \text{ is holomorphic in } \Omega\}.$$

$$A^\infty(\Omega) = A(\Omega) \cap C^\infty(\bar{\Omega}).$$

For K a compact subset of $\bar{\Omega}$, we define

$$\hat{K} = \{z \in \bar{\Omega};\ |f(z)| \le \sup_K |f| \text{ for all } f \in A^\infty(\Omega)\}.$$

Using this terminology we have

THEOREM 3. *Let* U *be an open subset of* $\bar{\Omega}$ *in the relative topology of* $\bar{\Omega}$. *Suppose that for every* $\bar{\partial}$*-closed* (p,1)*-form* α *with coefficients in* $L^2(\Omega)$, *there is a* (p,0)*-form with*

$$\text{sing supp } u \cap U = \text{sing supp } \alpha \cap U.$$

Then for every compact subset K *of* $\bar{\Omega}$,

$$\hat{K} \cap b\Omega \cap U = K \cap b\Omega \cap U. \tag{3}$$

It is clear that if the above condition (3) is satisfied, then there can be no 1-dimensional complex manifold contained in $b\Omega \cap U$. For if there were such a manifold V containing the boundary point z_0, then one could choose holomorphic coordinates $z_1, \cdots, z_n$ in a neighborhood U_0 around z_0, such that z_0 is given by the origin and the manifold is given by $\{z \in U_0;\ z_1 = \cdots = z_{n-1} = 0\}$. Setting

$$K = \{z \in U_0;\ z_1 = \cdots = z_{n-1} = 0, |z_n| = \delta\},$$

then for sufficiently small positive δ, $K \subset b\Omega \cap U$. By applying the Maximum Modulus Theorem to a function $f \in A^{\infty}(\Omega)$ restricted to the manifold V, one concludes that $z_0 \in \hat{K}$. But this contradicts (3).

Finally we state two theorems concerned with the question of when (3) holds. It turns out that if the dimension of Ω is two, then (3) is equivalent to the hypothesis in Theorem 2 that there is no 1-dimensional complex manifold contained in $b\Omega$. If the dimension of Ω is greater than two, then this is no longer the case.

THEOREM 4. *Suppose Ω has dimension 2, and that in a given open subset U of $\bar{\Omega}$, there is no 1-dimensional complex manifold V contained in $b\Omega \cap U$. Then for all compact subsets K of $\bar{\Omega}$*

$$\hat{K} \cap b\Omega \cap U = K \cap b\Omega \cap U .$$

THEOREM 5. *There exists a smoothly bounded pseudoconvex domain $\Omega \subset \mathbb{C}^3$, such that*

(i) *$b\Omega$ contains no one-dimensional complex manifold,*

(ii) *there is a compact subset K of $b\Omega$ such that $\hat{K} \cap b\Omega$ strictly contains K.*

The proofs of both Theorem 4 and Theorem 5 are contained in [1]. The example in Theorem 5 is based on a well-known example of Stolzenberg [11] of a compact set K in $\mathbb{C}^2$ such that the polynomial hull $\hat{K}$ contains no one-dimensional complex manifold, and yet $\hat{K}$ strictly contains K.

REFERENCES

[1] Catlin, D., Boundary behavior of holomorphic functions on weakly pseudoconvex domains. Thesis, Princeton Univ. 1978.

[2] Diederich, K. and Fornaess, J. E., Pseudoconvex domains with real-analytic boundary. Ann. of Math., 107 (1978), 371-384.

[3] D'Angelo, J., Finite type conditions for real hypersurfaces. J. Differential Geometry, to appear.

[4] Diederich, K. and Pflug, P., Necessary conditions for hypoellipticity of the $\bar{\partial}$-problem. To appear in these proceedings.

[5] Egorov, Yu. V., Subellipticity of the $\bar{\partial}$-Neumann problem. Dokl. Akad. Nauk. SSSR, 235, No. 5 (1977), 1009-1012.

[6] Folland, G. B. and Kohn, J. J., The Neumann problem for the Cauchy-Riemann complex. Ann. of Math. Studies, No. 75, P. U. Press, 1972.

[7] Greiner, P. C., On subelliptic estimates of the $\bar{\partial}$-Neumann problem in $\mathbb{C}^2$. J. Differential Geometry, 9 (1974), 239-250.

[8] Kohn, J. J., Subellipticity of the $\bar{\partial}$-Neumann problem on pseudoconvex domains: sufficient conditions. Acta Math., 142 (1979), 79-122.

[9] Krantz, S. G., Characterization of various domains of holomorphy via $\bar{\partial}$-estimates and applications to a problem of Kohn. Illinois Journal of Math., 91 (1979), 267-285.

[10] Pflug, P., Quadratintegrable holomorphe Funktionen und die Serre-Vermutung. Math. Ann. 216 (1975), 285-288.

[11] Stolzenberg, G., A hull with no analytic structure, J. Math. Mech. 12 (1963), 103-111.

DAS FORMALE PRINZIP FÜR KOMPAKTE KOMPLEXE UNTERMANNIGFALTIGKEITEN MIT 1-POSITIVEM NORMALENBÜNDEL

Michael Commichau und Hans Grauert

Reinhold Remmert gewidmet

Einleitung

Sei X eine n-dimensionale zusammenhängende komplexe Mannigfaltigkeit und A eine d-codimensionale zusammenhängende komplexe Untermannigfaltigkeit von X, $\mathfrak{m}$ die Idealgarbe von A. Der komplexe Unterraum

$$A_{(\nu)} := (A, \mathcal{H}_\nu)$$

von X mit der Strukturgarbe $\mathcal{H}_\nu := \mathcal{O}_X/\mathfrak{m}^{\nu+1}|A$ heisst die ν-te infinitesimale Umgebung von A. Natürlich ist $A_{(0)} = A$ die komplexe Untermannigfaltigkeit selbst.

Wit setzen $\mathcal{H}_\infty = \lim_{\infty \leftarrow \nu} \mathcal{H}_\nu$ und nennen den formalen komplexen Raum $A_{(\infty)} := (A, \mathcal{H}_\infty)$ die formale Umgebung von A in X. Die Frage, zu der diese Arbeit einen Beitrag liefert, heisst: Wann bestimmt die Struktur von $A_{(\infty)}$ die Struktur des Umgebungskeims von A in X?

Bekanntlich ist dies nicht immer der Fall. V. I. Arnol'd [A] konnte 1976 zeigen, dass es im Fall $n=2$ komplexe Tori A mit topologisch trivialem Normalbündel N gibt, bei denen der Umgebungskeim von A nicht isomorph zum Umgebungskeim der Null-Schnittfläche $\mathfrak{O} \subset N$ ist, obgleich $A_{(\infty)} \cong \mathfrak{O}_{(\infty)}$ gilt. Andererseits ist die Antwort auf unsere

Recent Developments in Several Complex Variables
0-691-08285-5/81/000101-26 $01.30/0 (cloth)
0-691-08281-2/81/000101-26 $01.30/0 (paperback)

Frage positiv – wir sagen: es gilt das formale Prinzip –, wenn das Normalenbundel $N = N(A)$ schwach negativ ist ([GM] oder [HR]).

Für den Fall positiven Normalenbündels wurde die Frage von Nirenberg und Spencer [NS] gestellt. Für den Fall $\dim A \geq 3$ und "genügend positiven" Normalenbündels bewies Griffiths [G], dass das formale Prinzip gilt, dann folgten Verschärfungen von Hartshorne [Ha] und Gieseker [Gi]. Im Fall $n \geq 2$ und einer nur schwachen Positivitätseigenschaft des Normalenbündels gab Hirschowitz [Hi] eine bejahende Antwort.

In der vorliegenden Arbeit wird die Gültigkeit des formalen Prinzips nun auch für den Fall nachgewiesen, dass das Normalenbündel $N = N(A)$ 1-positiv ist. Das ist sicher jedenfalls dann der Fall, wenn N eine Hermitesche Form trägt, bei der die Krümmungsform in jedem Punkt von A wenigstens einen positiven Eigenwert hat.

§1. *Die Aufbereitung komplexer Untermannigfaltigkeiten*

1. Sei X eine n-dimensionale zusammenhängende komplexe Mannigfaltigkeit und $A \subset X$ eine kompakte d-codimensionale zusammenhängende komplexe Untermannigfaltigkeit. Unter einem ausgezeichneten Koordinatensystem auf X verstehen wir eine biholomorphe Abbildung

$$F : U \xrightarrow{\sim} G$$

zwischen einer offenen Menge $U \subset\subset X$ und einem Gebiet $G \subset \mathbb{C}^n = \{\mathfrak{z} = (z_1, \cdots, z_n)\}$, so dass gilt:

(i) $U \cap A = F^{-1}(G \cap \{z_1 = \cdots = z_d = 0\})$

(ii) Es gibt eine Zahl $r > 0$ und ein Gebiet $\underline{G} \subset \mathbb{C}^{n-d}$, s.d.
$G = H_r \times \underline{G}$
mit $H_r := \{\mathfrak{z}' = (z_1, \cdots, z_d) \in \mathbb{C}^d : |z_\nu| < r \text{ für } 1 \leq \nu \leq d\}$.

Natürlich gibt es um jeden Punkt $x \in A$ ein ausgezeichnetes Koordinatensystem.

Wir wählen nun eine Überdeckung $\mathfrak{W} = \{W_\iota, \iota = 1, \cdots, \iota_*\}$ von A mit ausgezeichneten Koordinatensystemen sowie offene Teilmengen $D_\iota \subset\subset W_\iota$ derart, dass die D_ι mit den Koordinaten von W_ι ebenfalls noch eine

Überdeckung von A mit ausgezeichneten Koordinatensystemen bilden. Diese Wahlen zu A und X seien ein für allemal fest getroffen.

Es gibt nun beliebig feine offene Überdeckungen $\mathfrak{U} = \{U_i, i = 1, \cdots, i_*\}$ von A mit ausgezeichneten Koordinatensystemen, so dass dabei noch Folgendes gilt:

(i) Zu jedem i gibt es ein ι mit $U_i \subset\subset D_\iota$. Ein solches ι sei ausgewählt und mit $\iota(i)$ bezeichnet.

(ii) Das Koordinatensystem auf U_i ist die Einschränkung desjenigen auf $D_{\iota(i)}$.

(iii) Wann immer $U_{ij} := U_i \cap U_j \neq \emptyset$ ist, folgt $U_i \cup U_j \subset\subset W_{\iota(i)}$. In U_{ij} werden wir die durch $F_{\iota(i)}$ gegebenen Koordinaten benutzen.

$\mathfrak{W}(\mathfrak{U}) := \{W_{\iota(i)}, i = 1, \cdots, i_*\}$ ist dann eine Überdeckung von A mit ausgezeichneten Koordinatensystemen. Manche W_ι kommen in $\mathfrak{W}(\mathfrak{U})$ eventuell mehrfach, andere möglicherweise gar nicht vor. Für $W_{\iota(i)}$ schreiben wir fortan einfach W_i.

2. Es sei $\underline{W}_i := W_i \cap A$, $\underline{U}_i := U_i \cap A$, $\underline{U}_{ij} := U_{ij} \cap A$ $\underline{\mathfrak{U}} := \{\underline{U}_i, i = 1, \cdots, i_*\}$ ist eine Überdeckung von A mit Koordinatensystemen. Ist $\mathfrak{z} = (z_1, \cdots, z_n) \in \mathbf{C}^n$, so sei $\mathfrak{z}' := (z_1, \cdots, z_d)$ und $\mathfrak{z}'' =: = (z_{d+1}, \cdots, z_n)$, also $\mathfrak{z} = (\mathfrak{z}', \mathfrak{z}'')$. Unter einer Produktmenge verstehen wir eine offene Menge $D = H_r \times \underline{D} \subset \mathbf{C}^n$, wo H_r ein d-dimensionaler Polyzylinder ist (s.o.) und $\underline{D} \subset \mathbf{C}^{n-d}$ ein Gebiet. Zu $\mathfrak{U}$ sei nun $\tilde{\mathfrak{U}} = \{\tilde{\mathfrak{U}}_i, i = 1, \cdots, i_*\}$ eine weitere offene Überdeckung von A mit ausgezeichneten Koordinatensystemen mit

(i) $\tilde{U}_i \subset\subset U_i$, die Koordinaten stimmen überein. Es seien Produktmengen D_i^* gewählt mit $\tilde{U}_i \subset\subset D_i^* \subset\subset U_i$.

(ii) Falls $U_{ij} \neq \emptyset$ ist, seien Produktmengen D_{ij}^* in Bezug auf die W_i-Koordinaten gegeben mit $\tilde{U}_{ij} \subset\subset D_{ij}^* \subset\subset U_{ij}$.

Da man die Überdeckung $\underline{\mathfrak{U}}$ von A schrumpfen lassen kann, ist es klar, dass es ein solches $\tilde{\mathfrak{U}}$ gibt.

3. Die oben gewählten ausgezeichneten Koordinaten in W_ι seien mit $F_\iota : W_\iota \xrightarrow{\sim} G_\iota \subset \mathbb{C}^n$ bezeichnet. Es wird auch die Bezeichnung $F_\iota(x) = : \mathfrak{z}_\iota(x)$ verwendet. In den Durchschnitten $W_{\iota\kappa}$ transformieren sich die Koordinaten wie folgt:

$$\mathfrak{z}_\iota(x) = F_\iota \circ F_\kappa^{-1}(\mathfrak{z}_\kappa(x)) = : \mathfrak{f}_{\iota\kappa}\ (\mathfrak{z}_\kappa(x)),\quad x \in W_{\iota\kappa}\,.$$

Dabei ist $\mathfrak{f}_{\iota\kappa} = (f_1^{(\iota,\kappa)}, \cdots, f_n^{(\iota,\kappa)})$ ein n-tupel von in $F_\kappa(W_{\iota\kappa})$ holomorphen Funktionen, $\mathfrak{f}_{\iota\kappa} : F_\kappa(W_{\iota\kappa}) \xrightarrow{\sim} F_\iota(W_{\iota\kappa})$ eine biholomorphe Abbildung. Da die Koordinaten ausgezeichnet sind, gilt

$$(f_1^{(\iota,\kappa)}, \cdots, f_d^{(\iota,\kappa)}) = 0 \quad \text{für} \quad \mathfrak{z}'_\kappa = 0\,.$$

In der Nähe von $F_\kappa(\underline{W}_{\iota\kappa})$ gilt daher die Potenzreihenentwicklung

$$f_\nu^{(\iota,\kappa)}(\mathfrak{z}_\kappa) = \sum_{|\underline{\mu}|=1}^{\infty} f_{\nu,\underline{\mu}}^{(\iota,\kappa)}(0, \mathfrak{z}''_\kappa) \cdot \frac{1}{\underline{\mu}!} \cdot (\mathfrak{z}'_\kappa)^{\underline{\mu}}$$

für $1 \le \nu \le d$; dabei ist $\underline{\mu} = (\mu_1, \cdots, \mu_d)$ ein d-tupel nicht negativer ganzer Zahlen, $|\underline{\mu}| : = \mu_1 + \cdots + \mu_d$, $\underline{\mu}! = \mu_1! \cdot \ldots \cdot \mu_d!$ und $(\mathfrak{z}')^{\underline{\mu}} = z_1^{\mu_1} \cdot \ldots \cdot z_d^{\mu_d}$.

Schreibt man für das d-tupel $\underline{\mu} = (0, \cdots, 0, 1, 0, \cdots, 0)$, in dem die 1 an der τ-ten Stelle steht, einfach τ, so sind die längs $\underline{W}_{\iota\kappa}$ definierten holomorphen Matrizen

$$((f_{\nu,\tau}^{(\iota,\kappa)}(0, \mathfrak{z}''_\kappa(x)))_{\tau=1,\cdots,d}^{\nu=1,\cdots,d}) \in GL(d, \mathbb{C})$$

gerade die Übergangsmatrizen des Normalenbündels $N = N(A)$ von A in X. Unter einer Aufbereitung der komplexen Untermannigfaltigkeit A von X verstehen wir das System

$$(W_i, U_i, \tilde{U}_i, D_i^*, D_{ij}^*, \mathfrak{f}_{ij})\,.$$

4. Sei Y eine weitere n-dimensionale zusammenhängende komplexe Mannigfaltigkeit und $B \subset Y$ eine d-codimensionale kompakte zusammenhängende komplexe Untermannigfaltigkeit.

Ist eine Zahl $s \in \mathbb{N}_0$ und eine Biholomorphie

$$E_{(s)}: A_{(s)} \xrightarrow{\sim} B_{(s)}$$

gegeben, so sind A und B isomorph, und Y kann als eine weitere Umgebung von A angesehen werden oder – genauer ausgedrückt – ergibt einen weiteren Umgebungskeim von A, der mit dem durch $A \subset X$ gegebenen vermöge $E_{(s)}$ bis zur Ordnung s einschliesslich isomorph ist. Man kann diese neue Umgebung von A durch Änderung des Zusammenheftens der Koordinatensysteme U_i realisieren. Dazu sei $\mathfrak{U}$ so fein gewählt, dass jede Menge $E_{(s)}(\bar{\underline{U}}_i)$ in einem lokalen Koordinatensystem von Y liegt. $E_{(s)}|(\bar{\underline{U}}_i)_{(s)}$ kann dann zu einer biholomorphen Abbildung $E^i_{(s)}: V \to Y$ fortgesetzt werden, wo $V = V(\bar{\underline{U}}_i)$ in X offen ist. Nur kann man i.a. nicht erwarten, dass sich die $E^i_{(s)}$ zu einer global definierten holomorphen Abbildung zusammenfügen.

Es sei jetzt U eine zusammenhängende hinreichend kleine offene Umgebung von A in X mit

(i) $U \subset\subset \cup \tilde{U}_i$

(ii) Falls $\tilde{U}_i = H_r \times \underline{\tilde{U}}_i$, so $\bar{U} \cap (\partial H_r \times \bar{\underline{\tilde{U}}}_i) = \emptyset$, d.h. U ragt nirgends bis zum Rand ∂H_r hinaus.

(iii) $E_{(s)}$ lässt sich zu einer biholomorphen Abbildung $E^i_{(s)}$ in eine Umgebung von $\overline{U_i \cap U}$ fortsetzen.

(iv) $(E^i_{(s)})^{-1} \circ E^j_{(s)}$ ist noch auf $U_{ij} \cap U$ definiert und bildet $U_{ij} \cap U$ nach W_i ab.

Es ist dann $(E^i_{(s)})^{-1} \circ E^j_{(s)}|A_{(s)} = \mathrm{id}$, und daher schreibt sich mit $E^{ij}_{(s)} := (E^i_{(s)})^{-1} \circ E^j_{(s)}$ in den Koordinaten von W_i:

$$F_i \circ E^{ij}_{(s)} \circ F_i^{-1}(\mathfrak{z}) = \mathfrak{z} + \mathfrak{g}^{ij}_{(s)}(\mathfrak{z})$$

für alle $\mathfrak{z} \in F_i(U_{ij} \cap U)$, wo $\mathfrak{g}^{ij}_{(s)}$ von der Ordnung (s+1) auf $F_i(A)$ verschwindet.

Verheften wir nun die offenen Mengen

$$[(E^{ij}_{(s)})^{-1}(U_i \cap U)] \cap (U_j \cap U) \quad \text{und}$$

$$(U_i \cap U) \cap E^{ij}_{(s)}(U_j \cap U)$$

vermöge $E^{ij}_{(s)}$ miteinander, so erhalten wir eine komplexe Mannigfaltigkeit $\hat{X} \supset A_{(s)}$, die durch $E^* = \{E^i_{(s)}\}$ biholomorph auf eine Umgebung von B in Y abgebildet wird: $A_{(s)}$ liegt in $\hat{X}$ genauso wie B in Y.

Falls nun $E_{(s)}$ zu einer Umgebungsäquivalenz fortgesetzt werden kann, d.h. dass es eine biholomorphe Abbildung $E : U \xrightarrow{\sim} V = V(B) \subset Y$ gibt mit $E|A_{(s)} = E_{(s)}$, so wählen wir eine Umgebung $\tilde{U} = \tilde{U}(A) \subset\subset U$ so klein, dass stets

$$E(\tilde{U}_j \cap \tilde{U}) \subset E^j_{(s)}(U_j \cap U)$$

$$E(\tilde{U}_{ij} \cap \tilde{U}) \subset E^j_{(s)}(U_{ij} \cap U)$$

gilt. Wir setzen dann $\tilde{E}^i_{(s)} := (E^i_{(s)})^{-1} \circ E|\tilde{U}_i \cap \tilde{U}$ und erhalten

$$(*)\begin{cases} \tilde{E}^i_{(s)} = E^{ij}_{(s)} \circ \tilde{E}^j_{(s)} \quad \text{auf } \tilde{U}_{ij} \cap \tilde{U} \\ \tilde{E}^i_{(s)} \,|\, A_{(s)} = \mathrm{id} \\ \tilde{E}^i_{(s)}(\tilde{U}_i \cap \tilde{U}) \subset U_i \cap U \\ \tilde{E}^i_{(s)}(\tilde{U}_{ij} \cap \tilde{U}) \subset U_{ij} \cap U\,. \end{cases}$$

Sind umgekehrt injektive holomorphe Abbildungen $\tilde{E}^i_{(s)} : \tilde{U}_i \cap \tilde{U} \to U_i$ mit den genannten Eigenschaften gegeben, so setzen sich die Abbildungen $E^i_{(s)} \circ \tilde{E}^i_{(s)} : \tilde{U}_i \cap \tilde{U} \to Y$ zu einer biholomorphen Abbildung $E : \tilde{U}(A) \xrightarrow{\sim} \tilde{V}(B)$ mit $E|A_{(s)} = E_{(s)}$ zusammen.

Die Keime von Umgebungsäquivalenzen $(X, A) \xrightarrow{\sim} (Y, B)$, die Fortsetzungen von $E_{(s)}$ sind, entsprechen also umkehrbar eindeutig den Systemen $\{\tilde{E}^i_{(s)}\}$ mit den Bedingungen $(*)$.

Analog zeigt man:

Ist $\hat{E} : A_{(\infty)} \xrightarrow{\sim} B_{(\infty)}$ ein Isomorphismus, so erhält man durch

$$\hat{E}^i_{(s)} := (E^i_{(s)})^{-1} \circ \hat{E} : \underline{U}_{i(\infty)} \xrightarrow{\sim} \underline{U}_{i(\infty)}\,.$$

Isomorphismen mit $\hat{E}^i_{(s)} = E^{ij}_{(s)} \circ \hat{E}^j_{(s)}$ in $\underline{U}_{ij(\infty)}$.

In dem W_i-Koordinaten hat man die Darstellung

$$\hat{F}_i \circ \hat{E}^i_{(s)} \circ \hat{F}_i^{-1}(\mathfrak{z}) = \mathfrak{z} + \mathfrak{g}^i_{(s)}(\mathfrak{z}),$$

wo $\mathfrak{g}^i_{(s)} = (g^i_{(s),1}, \cdots, g^i_{(s),n})$ auf $F_i(A)$ von der Ordnung $(s+1)$ verschwindet. Man hat die formale Potenzreihenentwicklung

$$g^i_{(s),\nu}(\mathfrak{z}) = \sum_{|\underline{\mu}|=s+1}^{\infty} g^i_{(s),\nu,\underline{\mu}}(\mathfrak{z}'') \cdot \frac{1}{\underline{\mu}!} \cdot (\mathfrak{z}')^{\underline{\mu}}$$

für $1 \leq \nu \leq n$. Die $g^i_{(s),\nu,\underline{\mu}}$ sind dabei holomorphe Funktionen auf $F_i(\underline{U}_i)$, i.a. aber nicht mehr partielle Ableitungen der Funktionen $g^i_{(s),\nu}$, da die Potenzreihe nicht zu konvergieren braucht.

Kann man zeigen, dass die Potenzreihen entlang der $\overline{\underline{U}}_i$ konvergieren, so kann man $\hat{E}$ zu einer Umgebungsäquivalenz fortsetzen.

§2. *1-positive Vektorbündel*

1. Sei M eine n-dimensionale komplexe Mannigfaltigkeit und $V \subset\subset M$ ein Teilgebiet mit glattem C^2-Rand. Ist $x_0 \in \partial V$, so gibt es also ein holomorphes Koordinatensystem $U = U(x_0)$ mit Koordinaten $\mathfrak{z} = (z_1, \cdots, z_n)$ und eine C^2-Funktion $\phi: U \to \mathbf{R}$ mit $d\phi(x) \neq 0$ für alle $x \in U$ und

$$U \cap V = \{x \in U : \phi(x) < 0\}.$$

In $x \in U$ sei $T_x = \left\{\xi = \sum_{\nu=1}^{n} \left(a_\nu \frac{\partial}{\partial z_\nu}\Big|_x + \bar{a}_\nu \frac{\partial}{\partial \bar{z}_\nu}\Big|_x\right);\ a_\nu \in \mathbf{C}\right\}$ der reelle Tangentialraum, der vermöge $c \cdot \xi = \Sigma\left(ca_\nu \frac{\partial}{\partial z_\nu} + \overline{ca}_\nu \frac{\partial}{\partial \bar{z}_\nu}\right)$ ein n-dimensionaler komplexer Vektorraum ist, und $T_{\phi,x}$ der (n−1)-dimensionale komplexe Untervektorraum $\{\xi \in T_x : \partial\phi(\xi) = 0\}$. Die Leviform

$$L(\phi) = \sum_{\nu,\mu=1}^{n} \phi_{z_\nu \bar{z}_\mu}(x)\, dz_\nu\, d\bar{z}_\mu$$

ist eine Hermitesche Form auf T_x, ihre Beschränkung eine Hermitesche Form auf $T_{\phi,x}$.

V heisst 1-konkav, wenn es zu jedem Punkt $x_0 \in \partial V$ ein (U, ϕ) gibt, so dass $L(\phi)$ auf einem 1-dimensionalen komplexen Untervektorraum von T_{ϕ,x_0} negativ definit ist. $L(\phi)$ und $T_{\phi,x}$ sind koordinaten-invariant festgelegt, also ist die 1-Konkavität unabhängig von den Koordinaten definiert. Sie ist auch unabhängig von der Wahl der Randfunktion ϕ, denn es gilt bekanntlich der folgende

SATZ 1. *V sei 1-konkav,* $x_0 \in \partial V$, $U = U(x_0)$ *eine Umgebung,* $\psi : U \to \mathbb{R}$ *eine* C^2*-Funktion mit* $d\psi(x) \neq 0\ \forall x \in U$ *und* $V \cap U = \{x \in U : \psi(x) < 0\}$. *Dann ist* $L(\psi)$ *in jedem* $x \in U \cap \partial V$ *auf-einem 1-dimensionalen* $\mathbb{C}$*-Untervektorraum von* $T_{\psi,x}$ *negativ definit.*

Zum Beweis vgl. [GR], chap. IX, A.

2. Sei nun A eine kompakte $m = (n-d)$-dimensionale komplexe Mannigfaltigkeit und N ein holomorphes Vektorbündel vom Rang d auf A.

DEFINITION 1. N heisst 1-positiv, wenn es eine 1-konkave Umgebung $V = V(\mathfrak{O}) \subset\subset N$ der Nullschnittfläche $\mathfrak{O} \subset N$ mit glattem C^2-Rand ∂V gibt, so dass für jedes $x \in A$ gilt:

(i) In der Faser N_x ist $V \cap N_x$ sternförmig bzgl. $O_x \in N_x$.

(ii) Jeder Strahl in N_x, der von $O_x \in N_x$ ausgeht, schneidet ∂V transversal.

HILFSSATZ 1. *Sei A eine m-dimensionale kompakte komplexe Mannigfaltigkeit, N ein Geradenbündel auf A.* $V = V(\mathfrak{O})$ *sei eine 1-konkave Umgebung des Nullschnitts* $\mathfrak{O}$ *in N mit Eigenschaften wie in Definition 1.* $U \subset A$ *sei eine offene Teilmenge mit holomorphen Koordinaten* $\mathfrak{z}'' = (z_2, \cdots, z_n)$, $n = m+1$. *Weiter sei ein Vektorbündel-Isomorphismus* $N|U \xrightarrow{\sim} \mathbb{C} \times U$ *gegeben und damit eine lineare holomorphe Koordinate* z_1 *auf den Fasern von* $N|U$. *Es sei noch eine offene Menge* $\tilde{U} \subset\subset U$ *gegeben.*

Dann gibt es Zahlen $\delta > 0$, $q > 0$ *mit* $\delta < \operatorname{dist}(\tilde{U}, \partial U)$ *und zu jedem Punkt* $\mathfrak{z}_0 \in \partial V$ *mit* $\mathfrak{z}_0'' \in \tilde{U}$ *einen Vektor* $\underline{\xi} = (\xi_2, \cdots, \xi_n) \in \mathbf{C}^m$, *der die euklidische Länge* 1 *hat, und eine holomorphe Funktion*

$$f(t) = z_1^o + t\xi_1 + t^2 a, \qquad a \in \mathbf{C},$$

so dass für

$$\mathfrak{z}''(t) := (z_2^o + t\xi_2, \cdots, z_n^o + t\xi_n) \text{ und}$$

$$t \in \mathbf{C}, \quad |t| \le \delta,$$

gilt:

$$(f(t), \mathfrak{z}''(t)) \in (1 - q \cdot |t|^2) \cdot \overline{V}.$$

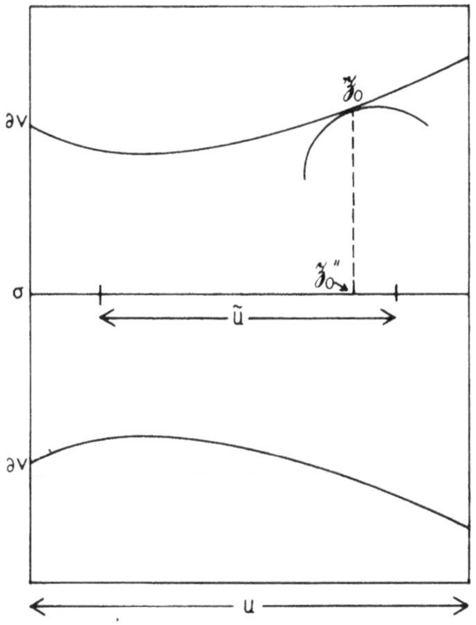

BEMERKUNGEN: 1) Die durch $(f(t), \mathfrak{z}''(t))$ gegebene komplexe Kurve ist also eine holomorphe Stützkurve an ∂V durch $\mathfrak{z}_0$ innerhalb von V. Dabei ist von grosser Bedeutung, dass das durch δ und q gegebene Ausmass, wie weit die Kurve an ihrem Rand $|t| = \delta$ ins Innere von V hereinragt, unabhängig von der Wahl des Randpunktes $\mathfrak{z}_0$ ist.

2) Es hat keinen Sinn, in Hilfssatz 1 mehr als eine lokale Situation zu betrachten, da später (Satz 3) mit einer kompakten Untermannigfaltigkeit A in X gearbeitet werden muss, für die es eine holomorphe Faserung von X über A nicht unbedingt zu geben braucht.

Beweis von Hilfssatz 1. Für die z_1-Koordinate verwenden wir Polarkoordinaten: $z_1 = r e^{i\vartheta}$.

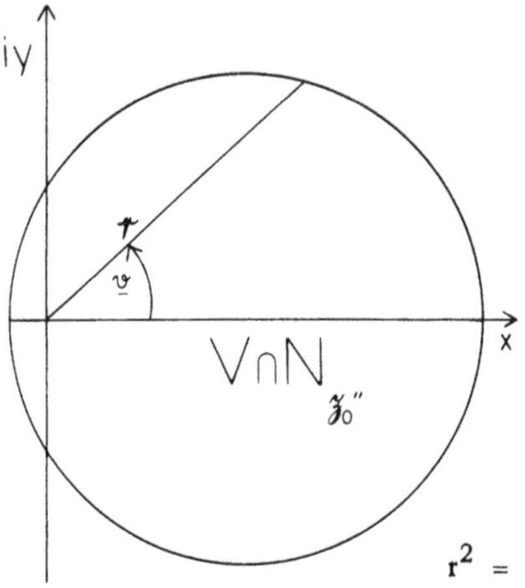

In jeder Faser $N_{\mathfrak{z}_o''}$ des Geradenbündels ist $V \cap N_{\mathfrak{z}_o''}$ gemäss Definition 1 ein C^2-glatt-berandetes bezüglich des Nullpunktes sternförmiges Gebiet in $\mathbb{C}$, dessen Rand von den vom Nullpunkt ausgehenden Strahlen transversal geschnitten wird. Über U lässt sich ∂V durch

$$r^2 = \psi(\vartheta, \mathfrak{z}'')$$

geben, wo ψ eine C^2-Funktion ist, die in ϑ periodisch von der Periode 2π ist. Setzt man

$$\alpha(\mathfrak{z}) := \psi\left(\frac{\log z_1 - \overline{\log z_1}}{2i}, \mathfrak{z}''\right) - z_1\overline{z_1},$$

was wegen der Periodizität von ψ eine wohldefinierte C^2-Funktion auf $N|U \setminus$ Nullschnitt ist, so wird ∂V über U durch $\alpha = 0$ gegeben. Dabei ist $V \setminus \mathfrak{O} = \{\alpha > 0\}$ und

$$\partial\alpha = \left[\frac{-i}{2z_1} \cdot \psi_\vartheta\left(\frac{\log z_1 - \overline{\log z_1}}{2i}, \mathfrak{z}''\right) - \overline{z_1}\right] dz_1 + \sum_{\nu=2}^{n} \psi_{z_\nu}(\cdots)\, dz_\nu .$$

Man errechnet leicht, dass $\alpha_{z_1} \neq 0$ in Punkten von ∂V ist. Wenn daher $(\xi_1, \underline{\xi}) \in T_{\alpha,\mathfrak{z}}$ ist (wobei $\mathfrak{z} \in \partial V$), also

$$0 = \xi_1 \alpha_{z_1}(\mathfrak{z}) + \sum_{\nu=2}^{n} \xi_\nu \alpha_{z_\nu}(\mathfrak{z}),$$

folgt $(\xi_1, \underline{\xi}) = 0$ oder $\underline{\xi} \neq 0$.

Entwicklung von α um $\mathfrak{z}_o$ ergibt (wobei $\mathfrak{z}_o \in \partial V$):

$$\alpha(\mathfrak{z}) = 2\,\mathrm{Re}\left[\sum_{\nu=1}^{n} a_\nu(z_\nu - z_\nu^{o}) + \sum_{\nu,\mu=1}^{n} a_{\nu\mu}(z_\nu - z_\nu^{o})(z_\mu - z_\mu^{o})\right.$$

$$\left. + \sum_{\nu,\mu=1}^{n} b_{\nu\mu}(z_\nu - z_\nu^{o})(\overline{z}_\mu - \overline{z_\mu^{o}})\right] + \beta(\mathfrak{z}),$$

wo $a_\nu = \alpha_{z_\nu}(\mathfrak{z}_o)$, $a_{\nu\mu} = \frac{1}{2}\alpha_{z_\nu z_\mu}(\mathfrak{z}_o)$, $b_{\nu\mu} = \alpha_{z_\nu \overline{z}_\mu}(\mathfrak{z}_o)$ als Ableitungen von $\alpha \in C^2$ stetig von $\mathfrak{z}_o$ abhängen und

$$\lim_{\mathfrak{z}\to\mathfrak{z}_o} \frac{\beta(\mathfrak{z})}{\|\mathfrak{z}-\mathfrak{z}_o\|^2} = 0$$

ist.

V ist 1 konkav, mit Satz 1 folgt daher: Zu jedem $\mathfrak{z} \in \partial V$ mit $\mathfrak{z}'' \in \tilde{U}$ gibt es $(\xi_1, \underline{\xi}) \in T_{\alpha,\mathfrak{z}}$ mit $\|\underline{\xi}\| = 1$ und

$$\sum_{\nu,\mu=1}^{n} b_{\nu\mu}\xi_\nu\overline{\xi}_\mu > K. \tag{1}$$

K kann dabei wegen $\tilde{U} \subset\subset U$ unabhängig von $\mathfrak{z}$ gewählt werden.

Zum Beweis von Hilfssatz 1 sind $f(t)$, $\mathfrak{z}''(t)$, q, δ so zu bestimmen, dass gilt

$$(f(t), \mathfrak{z}''(t)) \in (1-q|t|^2)\overline{V}, \quad \text{d.h.}$$

$$|f(t)|^2 \le (1-q|t|^2)\psi(\operatorname{arc} f(t), \mathfrak{z}''(t)) \quad \text{für } |t| \le \delta$$

oder

$$q|t|^2\psi(\operatorname{arc} f(t), \mathfrak{z}''(t)) \le \alpha(f(t), \mathfrak{z}''(t)).$$

Das ist wegen der lokalen Beschränktheit von ψ sicher dann möglich, wenn $K_2 > 0$, $\delta > 0$ existieren mit

$$\alpha(f(t), \mathfrak{z}''(t)) \ge K_2|t|^2 \quad \text{für } |t| \le \delta$$

bei geeigneter Wahl von f und $\mathfrak{z}''(t)$, weil auch $a = a(\mathfrak{z}_o)$ beschränkt gewählt werden kann. Geht man mit dem Ansatz

$$f(t) = z_1^o + \xi_1 t + t^2 a$$

$$\mathfrak{z}''(t) = \mathfrak{z}_o'' + t(\xi_2, \cdots, \xi_n)$$

in die Formel für die Entwicklung von α um $\mathfrak{z}_o$ —dabei sei $(\xi_1, \underline{\xi}) = (\xi_1, \underline{\xi})$ $(\mathfrak{z}_o)$ so bestimmt, dass (1) gilt in $\mathfrak{z}_o$ —, so führt dies zu

$$\alpha(f(t), \mathfrak{z}''(t)) = 2|t|^2 \sum_{\nu,\mu=1}^{n} b_{\nu\mu} \xi_\nu \overline{\xi_\mu} +$$

$$+ 2 \operatorname{Re}\left[t^2 \left(a_1 a + \sum_{\nu,\mu=1}^{n} a_{\nu\mu} \xi_\nu \xi_\mu\right)\right] + o(|t|^2) .$$

Da $a_1 = \alpha_{z_1}(\mathfrak{z}_o) \neq 0$ ist, können wir $a = -\frac{1}{a_1} \cdot \sum_{\nu,\mu=1}^{n} a_{\nu\mu} \xi_\nu \xi_\mu$ setzen und erhalten damit $[\quad] = 0$, also

$$\alpha(f(t), \mathfrak{z}''(t)) \geq K_2 |t|^2 \quad \text{für } t \in \mathbf{C}, |t| \leq \delta ,$$

wenn $\delta > 0$ klein genug ist.

Es muss noch gezeigt werden, dass K_2 und δ unabhängig von $\mathfrak{z}_o \in \partial V$ mit $\mathfrak{z}_o'' \in \tilde{U}$ gefunden werden können: Es ist ja $\|\underline{\xi}\| = 1$, $\sum_{\nu=1}^{n} a_\nu \xi_\nu = 0$, $a_\nu = \alpha_{z_\nu}$, a_1 stetig und ohne Nullstellen, ξ_1 ist also gleichgradig (für alle $\mathfrak{z}_o$ durch eine einheitliche Konstante) beschränkt. Gleiches ergibt sich für $a(\mathfrak{z}_o)$ aus der obigen Definition. Die gleichgradige Beschränktheit des Restglieds $\beta(\mathfrak{z})$ und damit der in $o(|t|^2)$ zusammengefassten Terme höherer Ordnung ergibt sich aus dem folgenden Hilfssatz 2. Hilfssatz 1 ist damit bewiesen.

HILFSSATZ 2. *Sei* $U \subset \mathbf{R}^n$ *offen,* $f: U \to \mathbf{R}$ *eine* C^k*-Funktion,* $\tilde{U} \subset\subset U$. *Dann gibt es eine Konstante* $K > 0$, *eine stetige monoton wachsende Funktion* $H: \mathbf{R}_o^+ \to \mathbf{R}_o^+$ *mit* $H(0) = 0$, *ausserdem zu jedem Multiindex* $\underline{\lambda} = (\lambda_1, \cdots, \lambda_n)$ *mit* $|\underline{\lambda}| = \lambda_1 + \cdots + \lambda_n = k$ *eine auf* $\tilde{U} \times \tilde{U}$ *definierte Funktion* $R_{\underline{\lambda}}(\mathfrak{p}, \mathfrak{q})$ *mit*

$$|R_{\underline{\lambda}}(\mathfrak{p},\mathfrak{q})| \leq H(\|\mathfrak{p}-\mathfrak{q}\|),$$

so dass gilt

$$f(\mathfrak{p}) = \sum_{|\underline{\lambda}|=0}^{k} \frac{f_{,\underline{\lambda}}(\mathfrak{p}_0)}{\underline{\lambda}!}(\mathfrak{p}-\mathfrak{p}_0)^{\underline{\lambda}} + \sum_{|\underline{\lambda}|=k} R_{\underline{\lambda}}(\mathfrak{p},\mathfrak{p}_0)(\mathfrak{p}-\mathfrak{p}_0)^{\underline{\lambda}}.$$

Der Beweis ergibt sich aus [GF].

BEMERKUNG: 1) Die Konstante a in Hilfssatz 1 ergibt sich als holomorphe Funktion in

$$\left(\xi_1, \cdots, \xi_n, \frac{a_{\nu\mu}}{a_1}\right) \in C^{n+\frac{n(n+1)}{2}}$$

2) In der Situation des Hilfssatzes 1 sind die $N_x \cap V$ für alle $x \in \bar{\tilde{U}}$ sternförmig bezüglich des Nullpunktes $0_x \in N_x$. Sie kömnen also biholomorph auf den Einheitskreis $E = \{|z| < 1\} \subset C$ so abgebildet werden, dass der Nullpunkt in den Nullpunkt übergeht. Die Menge $N_x \cap (1-q\delta^2)\bar{V}$ geht dabei stets auf eine kompakte Teilmenge von E. Sei nun $e > 0$ die kleinste Zahl, so dass alle diese Mengen noch in $\{|z| \leq e\}$ liegen. Da $N_x \cap V \xrightarrow{\sim} E$ stetig von x abhängt (Konstruktion mittels Dirichletschem Prinzip, sie Courant [C], chap. I,7), folgt $e < 1$.

SATZ 2. *Sei A eine m-dimensionale kompakte komplexe Mannigfaltigkeit, N ein 1-positves, L ein beliebiges holomorphes Vektorbündel auf A. Dann gibt es eine Zahl* $s \in N$, *so dass* $L \otimes (N^*)^p$, *wo* $(N^*)^p$ *die p-te symmetrische Tensorpotenz des Dualen* N^* *bezeichnet, für* $p \geq s$ *keine Schnittflächen ausser dem Nullschnitt mehr hat.*

Beweis. Wir nehmen zunächst an, dass N ein 1-positives Geradenbündel ist. Wir wählen eine endliche offene Überdeckung $\mathfrak{U} = \{U_i, i=1,\cdots,i_*\}$ von A mit Vektorbündelisomorphismen

$$N|U_i \xrightarrow[\phi_i]{\sim} C \times U_i, \qquad L|U_i \xrightarrow[\psi_i]{\sim} C^{\ell} \times U_i.$$

Vermöge ψ_i ist

$$L \otimes (N^*)^p|U_i = \underbrace{(N^*)^p \oplus \cdots \oplus (N^*)^p}_{\ell\text{-mal}} .$$

Eine Schnittfläche $h \in \Gamma(A, L \otimes (N^*)^p)$ wird über U_i durch ein ℓ-Tupel $\mathfrak{h}^{(i)} = (h_1^{(i)}, \cdots, h_\ell^{(i)})$ von Schnitten in $(N^*)^p$ gegeben. Über U_{ij} gibt es holomorphe nur von L abhängige Matrizen Q_{ij} mit $\mathfrak{h}^{(i)} = Q_{ij} \circ \mathfrak{h}^{(j)}$. Jedes $h_\lambda^{(i)}$ wiederum ist durch eine holomorphe Funktion: $N|U_i \to C$ gegeben, die faserweise in Bezug auf die durch ϕ_i eingeführte lineare Faserkoordinate $z_1 = z_1^{(i)}$ ein homogenes Polynom p-ten Grades ist, also kann man schreiben

$$h_\lambda^{(i)}(\mathfrak{z}'')(z_1) = : h_\lambda^{(i)}(z_1, \mathfrak{z}'') = a_\lambda^{(i)}(\mathfrak{z}'') \cdot z_1{}^p .$$

Wir lassen nun $\mathfrak{U}$ zu einer offenen Überdeckung $\tilde{\mathfrak{U}} = \{\tilde{U}_i, i = 1, \cdots, i_*\}$ mit $\tilde{U}_i \subset\subset U_i$ schrumpfen. Wir setzen $|\mathfrak{h}^{(i)}(\mathfrak{z})| : = \max |h_\nu^{(i)}(\mathfrak{z})|$. Es sei $V = V(\mathfrak{D}) \subset\subset N$ gemäss Definition 1 und $\mathfrak{z}_o \in \partial V$ mit $\mathfrak{z}_o'' \in \bar{\tilde{U}}_i$ so, dass $|\mathfrak{h}^{(i)}(\mathfrak{z}_o)|$ das Maximum der $|\mathfrak{h}^{(j)}(\mathfrak{z})|$ mit $\mathfrak{z} \in \partial V$, $\mathfrak{z}'' \in \bar{\tilde{U}}_j$ ist. Ausserdem wählen wir unabhängig von i und p die Zahlen $\delta > 0$, $q > 0$ gemäss Hilfssatz 1.

Die dort gefundene durch $\mathfrak{z}_o$ laufende Kurve sei durch $(f(t), \mathfrak{z}''(t))$, $|t| \leq \delta$, gegeben. $r(t) \geq 1$ sei so bestimmt, dass $(r(t) f(t), \mathfrak{z}''(t))$ in ∂V liegt. Sofern $\mathfrak{z}''(t) \in \tilde{U}_i$ ist, folgt $|\mathfrak{h}^{(i)}(r(t) f(t)), \mathfrak{z}''(t)| \leq |\mathfrak{h}^{(i)}(\mathfrak{z}_o)|$ nach Wahl von i und $\mathfrak{z}_o$. Ist jedoch $\mathfrak{z}''(t) \in \tilde{U}_j \setminus \tilde{U}_i$, so ergibt sich

$$\mathfrak{h}^{(i)}(\mathfrak{z}'') = Q_{ij}(\mathfrak{z}'') \circ \mathfrak{h}^j(\mathfrak{z}'')$$

$$\mathfrak{h}^{(i)}(\mathfrak{z}''), \mathfrak{h}^{(j)}(\mathfrak{z}'') \in ((N^*)^p \oplus \cdots \oplus (N^*)^p)_{\mathfrak{z}''}$$

$$\mathfrak{h}^{(i)}(\mathfrak{z}'')(r(t) f(t)) = Q_{ij}(\mathfrak{z}'') \circ \mathfrak{h}^{(j)}(\mathfrak{z}'')(r(t) f(t))$$

$$|\mathfrak{h}^{(i)}(r(t) f(t), \mathfrak{z}'')| \leq K |\mathfrak{h}^{(j)}(r(t) f(t), \mathfrak{z}'')| \leq K |\mathfrak{h}^{(i)}(\mathfrak{z}_o)|$$

mit einer nur von $\mathfrak{U}, \tilde{\mathfrak{U}}$ und den Q_{ij} abhängigen Konstanten $K \geq 1$. Es folgt für $|t| = \delta$

$$|\mathfrak{h}^{(i)}(f(t), \mathfrak{z}''(t))| \leq K\, e^p |\mathfrak{h}^{(i)}(\mathfrak{z}_0)|\ , \tag{2}$$

wo $e \in (0,1)$.

Es sei nun s so gross, dass $K\, e^p < 1$ ist für $p \geq s$. Die Ungleichung (2) widerspricht dem Maximumprinzip für die plurisubharmonische Funktion $\gamma(t) = |\mathfrak{h}^{(i)}(f(t), \mathfrak{z}''(t))|$ nur dann nicht, wenn $h = 0$ ist, was zu zeigen war.

Hat N höheren Rang als 1, so bläst man den Nullschnitt $\mathfrak{O} \cong A$ auf und erhält dadurch ein Geradenbündel $\tilde{N}$ über dem aufgeblasenen Nullschnitt $\tilde{A}$, dessen Fasern jeweils Geraden durch 0_x in einem N_x, $x \in A$, sind. Der Rand einer 1-konkaven Umgebung $V = V(\mathfrak{O}) \subset\subset N$ wird dabei nicht angetastet und bleibt daher 1-konkav. Man überzeugt sich leicht, dass auch die übrigen Bedingungen der Definition 1 erfüllt sind; $\tilde{N}$ ist daher ein 1-positives Geradenbündel über $\tilde{A}$.

Ist $L \to A$ ein beliebiges holomorphes Vektorbündel, so ist $L \times_A \tilde{A} \to \tilde{A}$ eins über $\tilde{A}$ (derselben Faserdimension wie L).

Ist nun $h \in \Gamma(A, L \otimes (N^*)^p)$, so schreibt sich $h(x)$ für $x \in A$ lokal als endliche Summe von Ausdrücken $h(x) = \sum_{i=1}^{r} a_i(x) \otimes p_i(x)$, wo $a_i(x) \in L_x$ und $p_i(x)$ homogene Polynome vom Grad p auf N_x sind und die $a_i(x)$ wie auch die $p_i(x)$ holomorph von x abhängen. Ist nun $y \in \tilde{A}$, etwa $y = g_x$ mit einer Geraden $g_x \subset N_x$, so definiert man durch

$$\tilde{h}(y) = \sum_{i=1}^{r} a_i(x) \otimes (p_i(x) | g_x)$$

einen holomorphen Schnitt

$$\tilde{h} \in \Gamma(\tilde{A}, (L \times_A \tilde{A}) \otimes (\tilde{N}^*)^p)\ .$$

Mit h ist auch $\tilde{h} \neq 0$. Dies ist aber für grosses p unmöglich, also ist die Behauptung von Satz 2 nunmehr vollständig bewiesen.

3. Sei nun wieder X eine n-dimensionale zusammenhängende komplexe Mannigfaltigkeit und $A \subset X$ eine 1-codimensionale kompakte zusammenhängende komplexe Untermannigfaltigkeit mit 1-positivem Normalenbündel N. Es gibt dann eine Umgebung $V = V(\mathfrak{D}) \subset\subset N$ gemäss Definition 1. Die Untermannigfaltigkeit A sei aufbereitet nach §1. In $N|\underline{W}_i$ wird dann ∂V gegeben durch eine C^2-Gleichung

$$|z_1^{(i)}| = \psi_i(\vartheta_i, \mathfrak{z}_i'') .$$

Wir setzen

$$\phi_i(\mathfrak{z}_i) := |z_1^{(i)}|^3 / \psi_i^3(\vartheta_i, \mathfrak{z}_i'') \geq 0 \text{ in } N|\underline{W}_i .$$

Die ϕ_i sind C^2-Funktionen, verschwinden genau auf dem Nullschnitt und haben den Wert 1 genau auf ∂V. Für $a \geq 0$ gilt:

$$\phi_i(a z_1, \mathfrak{z}'') = a^3 \phi_i(\mathfrak{z}) .$$

Die Flächen $\{\phi_i(\mathfrak{z}_i) = a^3\}$ sind gleich $a \circ \partial V = \{(az_1, \mathfrak{z}''), \mathfrak{z} \in \partial V\}$ und deswegen 1-konkave Ränder von $a \circ V$.

Wir wählen zu $\mathfrak{U}$ eine "Teilung der Eins." Das ist hier ein System $\{p_i, i = 1, \cdots, i_*\}$ von reellen C^2-Funktionen auf X mit:

a) $p_i(x) \geq 0 \quad \forall x \in X$

b) $\{p_i(x) \neq 0\} \subset\subset U_i$

c) $\sum_{i=1}^{i_*} p_i(x) \equiv 1$ in einer Umgebung von A.

Das Normalenbündel N wird nach §1 durch die Übergangsfunktionen $f_{1,1}^{(i,j)}$ gegeben (bezüglich der Überdeckung $\mathfrak{W}$ von A).

Um zwischen den Koordinaten von N und denen von X zu unterscheiden, schreiben wir fortan auf N statt $\mathfrak{z}_i = (z_1^{(i)}, \cdots, z_n^{(i)})$ einfach $\hat{\mathfrak{z}}_i = (\hat{z}_1^{(i)}, \cdots, \hat{z}_n^{(i)})$. Es ist also $\hat{\mathfrak{z}}_i'' = \mathfrak{z}_i''$ auf $A \cong \mathfrak{D}$. Über $\underline{W}_{ij}$ hat man die Transformationen $\hat{\mathfrak{z}}_i = \hat{\mathfrak{f}}_{ij}(\hat{\mathfrak{z}}_j)$ mit

$$\hat{z}_1^{(i)} = f_{1,1}^{(i,j)}(\hat{\mathfrak{z}}_j'') \cdot \hat{z}_1^{(j)} =: f_{ij}(\hat{\mathfrak{z}}_j'') \cdot \hat{z}_1^{(j)}$$

$$\hat{\mathfrak{z}}_i'' = \mathfrak{f}_{ij}''(0, \hat{\mathfrak{z}}_j'') =: \hat{\mathfrak{f}}_{ij}''(0, \hat{\mathfrak{z}}_j'') ,$$

wo $\mathfrak{f}_{ij}''$ (bzw. $\hat{\mathfrak{f}}_{ij}''$) die letzten (n–1) Komponenten von $\mathfrak{f}_{ij}$ (bzw. $\hat{\mathfrak{f}}_{ij}$) bezeichnet. Man rechnet leicht nach, dass sich die ϕ_i zu einer Funktion $\hat{\phi} : N \to R_0^+$ zusammensetzen. Wir transportieren die Funktionen ϕ_i nach X:

$$\tilde{\phi}_i(x) := \phi_i(\mathfrak{z}_i(x)), \; x \in U_i \subset X .$$

Auf X definieren wir in einer Umgebung $U(A)$

$$\phi(x) := \sum_{i=1}^{i_*} p_i(x) \tilde{\phi}_i(x) \geq 0$$

und erhalten damit eine C^2-Funktion, die auf A und – wenn U klein genug – nur auf A verschwindet.

Da die $\hat{\mathfrak{f}}_{ij}$ umkehrbar sind, kann man über einer Umgebung von $\underline{U}_{ij}$ in W_{ij} schreiben:

$$\mathfrak{f}_{ij} = \hat{\mathfrak{f}}_{ij}(\mathrm{id} + \mathfrak{f}^*_{ij}) ;$$

dabei verschwindet die erste Komponente von $\mathfrak{f}^*_{ij}$ von 2. Ordnung auf A und die übrigen von 1. Ordnung. Ist nun $U(A)$ klein, so gilt dort $\Sigma p_i \equiv 1$, und die Abbildung $\mathrm{id} + \mathfrak{f}^*_{ij}$ wirft $\overline{U_{ij}} \cap U$ in W_i. Für $x \in U_j \cap U$ gilt

$$\phi(x) = \sum_{i=1}^{i_*} p_i(x) \tilde{\phi}_i(x) = \sum_{i=1}^{i_*} p_i(x) \phi_i(\mathfrak{z}_i(x))$$

$$= \sum_{i=1}^{i_*} p_i(x) \cdot (\phi_i \circ \mathfrak{f}_{ij})(\mathfrak{z}_j(x))$$

$$= p_j(x) \phi_j(\mathfrak{z}_j(x)) + \sum_{\substack{i=1 \\ i \neq j}}^{i_*} p_i(x) \cdot (\phi_j \circ (\mathrm{id} + \mathfrak{f}^*_{ij}))(\mathfrak{z}_j(x)) ,$$

da $\phi_i \circ \hat{f}_{ij} = \phi_j$. Wegen $p_j(x) = 1 - \sum_{i \neq j} p_i(x)$ folgt $\phi(x) = \phi_j(\mathfrak{z}_j(x)) + \sum_{i \neq j} p_i(x)(\phi_j \circ (\mathrm{id} + f^*_{ij}) - \phi_j)(\mathfrak{z}_j(x))$. Für $a \geq 0$ und $\mathfrak{z} = (z_1, \cdots, z_n) \in \mathbf{C}^n$ sei wieder $a \circ \mathfrak{z} := (az_1, z_2, \cdots, z_n)$. Die Funktion $\frac{1}{a} \circ f^*_{ij}(a \circ \mathfrak{z}_j)$ von $\mathfrak{z}_j$ strebt einschliesslich ihrer Ableitungen für $a \to 0$ auf einer Umgebung von $\frac{1}{a} \circ (\overline{U_{ij}}) \cap U$ von mindestens 1. Ordnung lokal gleichmässig gegen 0. Es ist

$$\frac{1}{a^3} \phi(\mathfrak{z}_j^{-1}(a \circ \mathfrak{z}_j(x))) = \phi_j(\mathfrak{z}_j(x)) + \sum_{i \neq j} p_i(\mathfrak{z}_j^{-1}(a \circ \mathfrak{z}_j(x))$$

$$\cdot \frac{1}{a^3} \cdot (\phi_j \circ (\mathrm{id} + f^*_{ij}) - \phi_j)(a \circ \mathfrak{z}_j(x)).$$

Für $a \to 0$ konvergiert

$$p_i(\mathfrak{z}_j^{-1}(a \circ \mathfrak{z}_j(x))) \to p_i(\mathfrak{z}_j^{-1}(0, \mathfrak{z}_j''(x))).$$

Weiter gilt

$$\frac{1}{a^3} (\phi_j \circ (\mathrm{id} + f^*_{ij}) - \phi_j)(a \circ \mathfrak{z}_j(x)) = \phi_j(\mathfrak{z}_j(x) + \frac{1}{a} \circ f^*_{ij}(a \circ \mathfrak{z}_j(x))) - \phi_j(\mathfrak{z}_j(x)).$$

Daher konvergiert für $a \to 0$

$$\frac{1}{a^3} \phi(\mathfrak{z}_j^{-1}(a \circ \mathfrak{z}_j(x))) \to \phi_j(\mathfrak{z}_j(x)) \text{ auf } \left(\frac{1}{a} \circ U_j\right) \cap U.$$

Die in den letzten Zeilen erhaltenen Konvergenzen erstrecken sich jeweils auch noch auf die ersten und zweiten Ableitungen der angegebenen Funktionen nach den $z_\nu^{(j)}$, und alle Konvergenzen sind lokal gleichmässig.

Für die ϕ_j und $\{x \in U_j : \phi_j(\mathfrak{z}_j(x)) < \varepsilon\}$ gilt der Hilfssatz 1, wenn $\varepsilon > 0$ klein genug ist. Er gilt wegen der gleichmässigen Konvergenz auch für $\frac{1}{a^3} \cdot \phi(\mathfrak{z}_j^{-1}(a \circ \mathfrak{z}_j(x)))$, wenn $a > 0$ nur genügend klein gewählt ist, und damit auch für ϕ selbst. Wir haben also gezeigt:

SATZ 3. *Es sei* X *eine* n*-dimensionale zusammenhängende komplexe Mannigfaltigkeit,* $A \subset X$ *eine* 1*-codimensionale kompakte zusammen-*

hängende komplexe Untermannigfaltigkeit, deren Normalenbündel 1-positiv ist. Dann gibt es eine Umgebung $U = U(A) \subset \cup \tilde{U}_i$ *und eine* C^2*-Funktion* $\phi : U \to R_o^+$ *und ein* $\varepsilon_o > 0$, *so dass gilt:*

1) $\phi(x) > 0$ *für* $x \in U \setminus A$, $\phi(x) = 0$ *für* $x \in A$
2) $\{x \in U : \phi(x) \leq \varepsilon_o\} \subset\subset U$
3) *Ist* $x \in U_i$ *gegeben und* S *irgendein von* 0 *ausgehender Strahl in* $\{\mathfrak{z} \in C^n : \mathfrak{z}'' = \mathfrak{z}_i''(x)\}$, *so schneidet* $\mathfrak{z}_i^{-1}(S)$ *jede Menge* $\{y \in U : \phi(y) = \varepsilon\}$, *genau einmal, und zwar transversal, sofern* $0 < \varepsilon \leq \varepsilon_o$ *ist.*
4) *Es gibt feste Zahlen* $\delta > 0$, $q > 0$ *und zu jedem* $\varepsilon \in (0; \varepsilon_o]$ *und jedem* $x_o \in \tilde{U}_i \cap \partial V_\varepsilon$, *wo* $V_\varepsilon := \{x \in U : \phi(x) < \varepsilon\}$, *einen Vektor* $\underline{\xi} = (\xi_2, \cdots, \xi_n) \in C^{n-1}$ *mit* $\|\underline{\xi}\| = 1$ *und eine holomorphe Funktion*

$$f(t) = z_1^o + t\xi_1 + t^2 a, \quad wo \quad \mathfrak{z}_o := \mathfrak{z}_i(x_o),$$

so dass für $\mathfrak{z}''(t) = \mathfrak{z}_o'' + t\underline{\xi}$, $|t| \leq \delta$, *gilt:*

$$\mathfrak{z}''(t) \in \mathfrak{z}_i(\underline{U}_i), (f(t), \mathfrak{z}''(t)) \in (1 - q|t|^2) \circ \mathfrak{z}_i(\overline{V}_\varepsilon).$$

BEMERKUNG. Für $x \in \underline{U}_i$ bilden wir $[\varepsilon \leq \varepsilon_o]$

$$R(\varepsilon, x, i) := \sup \{|z_1^{(i)}(y)| : y \in U_i \cap \partial V_\varepsilon, \mathfrak{z}_i''(y) = \mathfrak{z}_i''(x)\}$$

$$r(\varepsilon, x, i) := \inf \{|z_1^{(i)}(y)| : y \in U_i \cap \partial V_\varepsilon, \mathfrak{z}_i''(y) = \mathfrak{z}_i''(x)\}$$

und

$$R(\varepsilon) := \sup_{x, i} R(\varepsilon, x, i), \quad r(\varepsilon) := \inf_{x, i} r(\varepsilon, x, i) \quad R_o := R(\varepsilon_o), r_o := r(\varepsilon_o).$$

Es gilt dann $0 < r(\varepsilon) \leq R(\varepsilon) < \infty$, und $\frac{R(\varepsilon)}{r(\varepsilon)}$ ist gleichgradig für alle $\varepsilon \in (0, \varepsilon_o]$ beschränkt.

§3. *Der Hauptsatz*

In diesem Paragraphen wird das Hauptresultat der Arbeit hergeleitet.

SATZ 4. *Seien* X *und* Y *n-dimensionale zusammenhängende komplexe Mannigfaltigkeiten;* $A \subset X$, $B \subset Y$ *seien d-codimensionale zusammen-*

hängende kompakte komplexe Untermannigfaltigkeiten. Das Normalenbündel N(A) *sei 1-positiv und* $\hat{E}: A_{(\infty)} \xrightarrow{\sim} B_{(\infty)}$ *ein Isomorphismus. Dann gibt es Umgebungen* U(A), V(B) *und eine biholomorphe Abbildung* $E: U \xrightarrow{\sim} V$ *mit* $E|A_{(\infty)} = \hat{E}$.

ZUSATZ. *Es gibt eine Zahl* $s \in \mathbf{N}$, *so dass* E *schon durch* $\hat{E}|A_{(s)}$ *eindeutig bestimmt ist.*

Beweis. Zunächst soll der Zusatz bewiesen werden. Es gebe ausser E noch $\tilde{E}$ mit $\tilde{E}|A_{(s)} = E|A_{(s)}$. Wir nehmen an, $\tilde{E}$ wäre in jeder Umgebung von A von E verschieden. Es gibt dann eine grösste Zahl $p \geq s$, so dass $E|A_{(p)} = \tilde{E}|A_{(p)}$ gilt. Für $F := \tilde{E}^{-1} \circ E$ gilt dann $F|A_{(p)} = \mathrm{id}$, aber $F|A_{(p+1)} \neq \mathrm{id}$. Es bezeichne Aut(p) die über A erklärte Garbe der Keime von lokalen Automorphismen von $\underline{V} \cap A_{(p+1)}$, welche auf $\underline{V} \cap A_{(p)}$ die Identität induzieren ($\underline{V} \subset A$ offen). Lokal kann man F zu einer holomorphen Abbildung $\hat{F}_x$ in eine Umgebung $V(x) \subset X$, $x \in A$, fortsetzen. Ist dann g eine holomorphe Funktion in V, die auf A überall von der Ordnung ≥ 1 verschwindet, so ist $g \circ \hat{F}_x = g + h$ in der Nähe von x, wobei h auf A von der Ordnung $\geq p+1$ verschwindet. Die Zuordnung $g \mapsto h$ liefert ein Element von

$$\mathrm{Hom}\,(\mathfrak{m}/\mathfrak{m}^2, \mathfrak{m}^{p+1}/\mathfrak{m}^{p+2}),$$

und dadurch erhält man auch einen Garbenepimorphismus

$$\mathrm{Aut}(p) \to \mathrm{Hom}\,(\mathfrak{m}/\mathfrak{m}^2, \mathfrak{m}^{p+1}/\mathfrak{m}^{p+2}).$$

Der Kern ist isomorph zu $\mathrm{Hom}\,(\Omega(A), \mathfrak{m}^{p+1}/\mathfrak{m}^{p+2})$, wobei $\Omega(A)$ die Garbe der Keime der Pfaffschen Formen auf A ist. Man hat also eine exakte Sequenz

$$0 \to \Theta \otimes (N^*)^{p+1} \to \mathrm{Aut}(p) \to N \otimes (N^*)^{p+1} \to 0,$$

wobei Θ die Garbe der Keime holomorpher Vektorfelder auf A bezeichnet [vgl. [GM], §4].

F liefert einen nicht trivialen Schnitt in Aut(p). Nach Satz 2 aus §2 existieren aber für grosses p in $\Theta \otimes (N^*)^{p+1}$ wie auch in $N \otimes (N^*)^{p+1}$

keine nicht trivialen Schnitte. Dasselbe gilt wegen der exakten Sequenz auch für $\mathrm{Aut}(p)$. Ist also s gross genug, so folgt aus $E|A_{(s)} = \tilde{E}|A_{(s)}$ schon $E = \tilde{E}$ (als Abbildungen von Umgebungskeimen). Der Zusatz ist bewiesen.

Beweis von Satz 4. Es werde $s \in \mathbf{N}$ entsprechend dem Zusatz gewählt. $\hat{E}$ induziert einen Isomorphismus $E_{(s)} : A_{(s)} \xrightarrow{\sim} B_{(s)}$. Dazu seien die $E^i_{(s)}$, $E^{ij}_{(s)}$, $\hat{E}^i_{(s)}$ wie in §1 bestimmt, ebenso die Aufbereitung von $A \hookrightarrow X$. In den W_i-Koordinaten hatten wir die Darstellung

$$\hat{F}_i \circ \hat{E}^i_{(s)} \circ \hat{F}_i^{-1}(\mathfrak{z}) = \mathfrak{z} + \mathfrak{g}^i_{(s)}(\mathfrak{z})$$

gewonnen. Es braucht jetzt nur noch gezeigt zu werden, dass die formalen Potenzreihen $\mathfrak{g}^i_{(s)}$ entlang der $\underline{\bar{U}}_i$ konvergieren. Dazu wählen wir ein für allemal Mengen $\tilde{\tilde{U}}_i \subset\subset \tilde{U}_i$, die noch A überdecken und mit den Koordinaten von U_i ebenfalls ausgezeichnete Koordinatensysteme bilden. Alle in Zukunft betrachteten Umgebungen U von A seien so klein, dass

$$F_i(U \cap \tilde{\tilde{U}}_i) \cap (F_i(\underline{\tilde{\tilde{U}}}_i) \times \partial H_r) = \emptyset \text{ ist, wobei}$$

$$F_i(\tilde{\tilde{U}}_i) = F_i(\underline{\tilde{\tilde{U}}}_i) \times H_r .$$

Wir führen eine Norm ein. Es sei $E_r = \{|z| < r\} \subset \mathbf{C}$ und $f(z) = \sum_{\nu=0}^{\infty} a_\nu z^\nu$ eine auf E_r erklärte beschränkte Funktion. Für $z \in E_r$ wird definiert:

$$\|f, z\| := \sum_{\nu=0}^{\infty} |a_\nu|\,|z|^\nu < \infty .$$

Es gilt: $|f(z)| \le \|f, z\|$ für alle $z \in E_r$; ist $z = |z| e^{i\phi}$, so ist $\|f, z\|$ von ϕ unabhängig;

$$\|f, z\| \le K_\rho \sup\{|f(\tilde{z})|, \tilde{z} \in E_r\} \text{ für } |z| \le \rho ,$$

wobei $0 < \rho < r$ und K_ρ eine nur von $\frac{\rho}{r}$ abhängige Konstante ist. Ist

$\mathfrak{z} = (z_1, \cdots, z_n) = (z_1, \mathfrak{z}'')$ und $f = \sum_{\nu=0}^{\infty} a_\nu(\mathfrak{z}'') z_1^{\nu}$, so sei $\|f, \mathfrak{z}\| := \sum_{\nu=0}^{\infty} |a_\nu(\mathfrak{z}'')| \cdot |z_1|^{\nu}$. Es gilt eine Produktregel

$$\|f \cdot g, \mathfrak{z}\| \leq \|f, \mathfrak{z}\| \cdot \|g, \mathfrak{z}\| .$$

Ist $\mathfrak{f} = (f_1, \cdots, f_n)$ ein n-tupel holomorpher Funktionen, so sei

$$\|\mathfrak{f}, \mathfrak{z}\| := \max_{\nu=1}^{n} \|f_\nu, \mathfrak{z}\| .$$

Es muss noch erwähnt werden, dass für $\| \quad \|$ ein Schwarzsches Lemma gilt.

Nach §1 waren die $\mathfrak{g}^{ij}_{(s)}(\mathfrak{z})$ in $F_i(U_{ij} \cap U)$ holomorphe Funktionen. Ist $U = U(A)$ klein genug und $\mathfrak{h}(\mathfrak{z})$ ein n-tupel holomorpher Funktionen, so gilt für

$$\mathfrak{z}, (\mathfrak{z} + \mathfrak{h}(\mathfrak{z})) \in F_i(D^*_{ij} \cap U)$$

die Darstellung

$$\mathfrak{g}^{ij}_{(s)}(\mathfrak{z} + \mathfrak{h}(\mathfrak{z})) = \mathfrak{g}^{ij}_{(s)}(\mathfrak{z}) + \vartheta^{ij}_{(s)}(\mathfrak{z}, \mathfrak{h}(\mathfrak{z})) ,$$

sofern $\|\mathfrak{h}, \mathfrak{z}\| \leq |z_1|$ ist. Da $\mathfrak{g}^{ij}_{(s)}$ auf $\{z_1 = 0\}$ von mindestens (s+1)-ter Ordnung verschwindet, hat man auf $F_i(D^*_{ij} \cap U)$ eine Abschätzung

$$\|\mathfrak{g}^{ij}_{(s)}, \mathfrak{z}\| \leq K_s \cdot |z_1|^{s+1}$$

mit einer von s abhängigen Konstanten K_s. Ebenso erhält man nach Schrumpfung von U –und die sei o.E. schon vorher bei U durchgeführt–:

$$\|\vartheta(\mathfrak{z}, \mathfrak{h}(\mathfrak{z})), \mathfrak{z}\| \leq \|\mathfrak{h}, \mathfrak{z}\| \cdot |z_1|^{s} \cdot K_s$$

für $\mathfrak{z}, (\mathfrak{z} + \mathfrak{h}(\mathfrak{z})) \in F_i(D^*_{ij} \cap U)$, $\|\mathfrak{h}, \mathfrak{z}\| \leq |z_1|$. Dabei wird ϑ abkürzend für $\vartheta^{ij}_{(s)}$ verwendet.

Im Folgenden sei $f^{(p)}(\mathfrak{z}) := \sum_{\nu=0}^{p} a_\nu(\mathfrak{z}'') z_1^{\nu}$ der Abschnitt bis zur Ordnung p einer Potenzreihe $f(\mathfrak{z}) = \sum_{\nu=0}^{\infty} a_\nu(\mathfrak{z}'') z_1^{\nu}$. Rechnet man die Gleichung

$$\hat{E}^{i}_{(s)} = E^{ij}_{(s)} \circ \hat{E}^{j}_{(s)} \quad \text{in} \quad \underline{U}_{ij(\infty)}$$

ins W_i-Koordinatensystem um, so erhält man mod $(z_1)^{p+1}$, $p \geq s$, die Kongruenz

$$\mathfrak{z} + (\mathfrak{g}^{i}_{(s)})^{(p)}(\mathfrak{z}) \equiv \mathfrak{z} + (\mathfrak{h}^{j}_{i})^{(p)}(\mathfrak{z}) + (\mathfrak{g}^{ij}_{(s)})^{(p)}(\mathfrak{z}) + \vartheta(\mathfrak{z}, (\mathfrak{h}^{j}_{i})^{(p)}(\mathfrak{z}))$$

in $F_i(\bar{\tilde{U}}_i \cap \bar{\bar{\tilde{U}}}_j \cap \tilde{U})$. Dabei ist

$$\mathfrak{z} + \mathfrak{h}^{j}_{i}(\mathfrak{z}) := \mathfrak{f}_{ij} \circ (\mathrm{id} + \mathfrak{g}^{j}_{(s)}) \circ \mathfrak{f}_{ji}(\mathfrak{z}) .$$

Es sei $\tilde{U}$ so klein, dass

$$\|(\mathfrak{g}^{j}_{(s)})^{(p)}, \mathfrak{z}\| \leq |z_1| \quad \text{in} \quad F_j(\bar{\tilde{U}}_j \cap \tilde{U}) \quad \text{und}$$

$$\|(\mathfrak{h}^{j}_{i})^{(p)}, \mathfrak{z}\| \leq |z_1| \quad \text{in} \quad F_i(\bar{\tilde{U}}_i \cap \bar{\bar{\tilde{U}}}_j \cap \tilde{U})$$

ist. In der obigen Kongruenz enthält höchstens der Term ϑ Glieder von höherer als p-ter Ordnung. Daher folgt

$$\|(\mathfrak{g}^{i}_{(s)})^{(p)}, \mathfrak{z}\| \leq \|(\mathfrak{h}^{j}_{i})^{(p)}, \mathfrak{z}\| + K_s \cdot |z_1|^{s+1} + K_s \cdot |z_1|^{s} \cdot \|(\mathfrak{h}^{j}_{i})^{(p)}, \mathfrak{z}\|$$

für $\mathfrak{z} \in F_i(\bar{\tilde{U}}_i \cap \bar{\bar{\tilde{U}}}_j \cap \tilde{U})$.

Es seien nun $\phi : U \to R^{+}_{0}$, q, δ gemäss Satz 3 in §2 gewählt. Als $\tilde{U} = \tilde{U}(A) \subset\subset U$ nehmen wir immer ein

$$V_\varepsilon = V_\varepsilon(A) = \{x \in U : \phi(x) < \varepsilon\}$$

mit $\varepsilon \in (0, \varepsilon_0]$. Es soll nun $\|(\mathfrak{h}^{j}_{i})^{(p)}\|$ durch $\|(\mathfrak{g}^{j}_{(s)})^{(p)}\|$ abgeschätzt werden. Für die Koordinatenwechsel $\mathfrak{f}_{\iota\kappa}$ kann

$$|\mathfrak{f}_{\iota\kappa}(x) - \mathfrak{f}_{\iota\kappa}(y)| \leq K \cdot |x-y|$$

für $x, y \in F_\kappa(W_{\iota\kappa})$ mit einer Konstanten K angenommen werden. Beachtet man die gegenseitige Abschätzbarkeit der Normen $\| \quad \|$ und $| \quad |$ bei Schrumpfung in z_1-Richtung, so erhält man bei kleinem ε_0:

$$\|(\mathfrak{h}_i^j)^{(p)}, \mathfrak{z}\| \leq K' \cdot \sup\{\|(\mathfrak{g}^j_{(s)})^{(p)}, \tilde{\mathfrak{z}}\|, \tilde{\mathfrak{z}} \in F_j(\tilde{U}_j \cap \partial V_\varepsilon)\}$$

für $\mathfrak{z} \in F_i(U_i \cap \tilde{\tilde{U}}_j \cap \overline{V_{\varepsilon_1}})$ mit $\varepsilon_1 = \frac{1}{2}(\varepsilon + (1-q\delta)\varepsilon)$. (Man benutzt dabei, dass $\|\mathfrak{f}, \mathfrak{z}\|$ vom Winkel ϕ der komplexen Zahl $z_1 = |z_1|e^{i\phi}$ unabhängig ist. Beim Beweis der Abschätzung kann man daher die (ungefähr) ellipsenförmigen Fasern von V_ε jeweils durch die kleinsten diese Fasern umfassenden Kreisscheiben ersetzen. Schliesslich benutzt man noch, dass die Koordinatenwechsel $\mathfrak{f}_{ij}$ lokal durch C-lineare Abbildungen approximiert werden können, also kleine Kreisscheiben wieder ungefähr in kleine Kreisscheiben abbilden.)

Die Zahl K' ist eine neue Konstante ≥ 1, die von s, p, ε_0, ε unabhängig ist.

Zu jedem $p \geq s$ existieren beliebig kleine Zahlen $\varepsilon \in (0, \varepsilon_0]$, so dass

$$\|(\mathfrak{g}^j_{(s)})^{(p)}, \mathfrak{z}\| \leq \frac{1}{K'} \cdot |z_1| \quad \text{in} \quad F_j(\tilde{U}_j \cap V_\varepsilon) \quad \text{für alle } j$$

ist. Daraus ergibt sich

$$\|(\mathfrak{h}_i^j)^{(p)}, \mathfrak{z}\| \leq |z_1| \quad \text{in} \quad F_i(U_i \cap \tilde{\tilde{U}}_j \cap V_{\varepsilon_1}).$$

Wir wählen nun i, $\mathfrak{z}_0 \in F_i(\bar{\tilde{U}}_i \cap \partial V_\varepsilon)$ in Abhängigkeit von s, p, ε so, dass

$$\|(\mathfrak{g}^i_{(s)})^{(p)}, \mathfrak{z}_0\| \quad \text{maximal}$$

unter allen

$$\|(\mathfrak{g}^j_{(s)})^{(p)}, \mathfrak{z}\| \quad \text{mit} \quad \mathfrak{z} \in F_j(\bar{\tilde{U}}_j \cap \partial V_\varepsilon)$$

ist. Da $U_i \cap \overline{V_{\varepsilon_1}}$ von den Mengen $U_i \cap \tilde{\tilde{U}}_j \cap \overline{V_{\varepsilon_1}}$ überdeckt wird, gilt für alle $\mathfrak{z} \in U_i \cap V_{\varepsilon_1}$

$$\|(\mathfrak{g}^i_{(s)})^{(p)}, \mathfrak{z}\| \leq K' \cdot (1+K_s R_0^s) \cdot \|(\mathfrak{g}^i_{(s)})^{(p)}, \mathfrak{z}_0\| + K_s R_0^s R(\varepsilon).$$

Wir legen nun die Kurve $\mathfrak{z}(t) = (f(t), \mathfrak{z}''(t))$, $|t| \leq \delta$, durch $F_i(\mathfrak{z}_0)$. Es gilt $F_i^{-1}(\mathfrak{z}(t)) \in U_i$ für $|t| \leq \delta$. Für $|t| = \delta$ erhalten wir

$$\|(g^i_{(s)})^{(p)}, \mathfrak{z}(t)\| \leq e^{s+1} \cdot \sup\{\|(g^i_{(s)})^{(p)}, \tilde{\mathfrak{z}}\|, \tilde{\mathfrak{z}}'' = \mathfrak{z}''(t), \tilde{\mathfrak{z}} \in F_i(\partial V_{\varepsilon_1})\}.$$

Dabei ist $e \in (0,1)$ ein von $p, s, \varepsilon, \varepsilon_o$ unabhängige Konstante. Es gilt also für $|t| = \delta$ die Abschätzung

$$\|(g^i_{(s)})^{(p)}, \mathfrak{z}(t)\| \leq e^{s+1}[K'(1+K_s R_o^s) \cdot \|(g^i_{(s)})^{(p)}, \mathfrak{z}_o\| + K_s R_o^s R(\varepsilon)].$$

Für die Funktion $t \mapsto \|(g^i_{(s)})^{(p)}, \mathfrak{z}(t)\|$, $|t| \leq \delta$, welche als endliche Summe von Beträgen holomorpher Funktionen plurisubharmonisch ist, gilt ein Maximumprinzip. Es ist $\mathfrak{z}_o = \mathfrak{z}(0)$ und daher

$$\|(g^i_{(s)})^{(p)}, \mathfrak{z}_o\| \leq \sup_{|t|=\delta} \|(g^i_{(s)})^{(p)}, \mathfrak{z}(t)\|.$$

Mithin folgt

$$\|(g^i_{(s)})^{(p)}, \mathfrak{z}_o\| \leq \frac{e^{s+1} K_s R_o^s R(\varepsilon)}{1 - e^{s+1} K'(1+K_s R_o^s)}.$$

Die Zahl e ist von ε, ε_o, s, p unabhängig, K' ebenfalls. Es sei nun also $s \in \mathbf{N}$ so gewählt, dass $e^{s+1}K' < \frac{1}{2}$ ist, dann $\varepsilon_o > 0$ so klein, dass $e^{s+1}K'(1+K_s R_o^s) < \frac{1}{2}$ und der ganze Bruch $\leq \frac{1}{2K'} \cdot r(\varepsilon)$ ist. Es folgt

$$\|(g^i_{(s)})^{(p)}, \mathfrak{z}_o\| \leq \frac{1}{2K'} \cdot r(\varepsilon) \text{ und damit}$$

$$\|(g^i_{(s)})^{(p)}, \mathfrak{z}\| \leq \frac{1}{2K'} \cdot |z_1| \text{ in } F_j(\tilde{U}_j \cap V_\varepsilon) \text{ für alle } j.$$

Es sei $\varepsilon_p \in (0, \varepsilon_o]$ nun maximal mit

$$\|(g^j_{(s)})^{(p)}, \mathfrak{z}\| \leq \frac{1}{K'} \cdot |z_1| \text{ in } F_j(\tilde{U}_j \cap V_{\varepsilon_p}) \text{ für alle } j.$$

Die eben erhaltene Ungleichung zeigt, dass $\varepsilon_p = \varepsilon_o$ sein muss für alle p. Daher gilt

$$\|(g^j_{(s)})^{(p)}, \mathfrak{z}\| \leq \frac{1}{2K'} \cdot |z_1| \leq \frac{1}{2K'} \cdot R_o$$

für alle p und $\mathfrak{z} \in F_j(\tilde{U}_j \cap V_{\varepsilon_0})$. Die formalen Potenzreihen $\mathfrak{g}^j_{(s)}$ konvergieren also in $F_j(\tilde{U}_j \cap V_{\varepsilon_0})$. Das war zu zeigen.

Der Beweis des Satzes für $d = \operatorname{codim}(A \hookrightarrow X) > 1$ folgt wieder durch Zurückführung auf den Fall $d = 1$ durch Aufblasen von A.

LITERATUR

[A] Arnol'd, V. I.: Bifurcations of invariant manifolds of differential equations and normal forms in neighborhoods of elliptic curves. Funct. Anal. and appl. 10 (1976), 249-259.

[C] Courant, R.: Dirichlet's principle, conformal mapping, and minimal surfaces. 3. Auflage. Interscience Publishers, New York 1967.

[G] Griffiths, Ph. A.: The Extension Problem in Complex Analysis II; Embeddings with positive normal bundle. Amer. J. Math. 88 (1966), 366-446.

[GF] Grauert, H. und Fischer, W.: Differential- und Integralrechnung II, 3. Auflage. Springer Heidelberg 1978.

[Gi] Gieseker, D.: On two theorems of Griffiths about embeddings with ample normal bundle. Amer. J. Math. 99 (1977), 1137-1150.

[GM] Grauert, H.: Über Modifikationen und exzeptionelle analytische Mengen. Math. Ann. 146 (1962), 331-368.

[GR] Gunning, R. und Rossi, H.: Analytic functions of several complex variables. Prentice-Hall, Englewood Cliffs, N. J., 1965.

[Ha] Hartshorne, R.: Cohomological dimension of algebraic varieties. Ann. Math. 88 (1968), 403-450.

[Hi] Hirschowitz, A.: On the convergence of formal equivalence between embeddings. To appear Ann. of Math.

[HR] Hironaka, H. und Rossi, H.: On the Equivalence of Inbeddings of Exceptional Complex Spaces. Math. Ann. 156 (1964), 313-333.

[NS] Nirenberg, L. und Spencer, D. C.: On rigidity of holomorphic imbeddings. In: Contributions to Function Theory. Tata Institute, Bombay 1960, 133-137.

PERTURBATIONS OF ANALYTIC VARIETIES

John P. D'Angelo

Introduction

Let D be an open domain in $\mathbb{C}^n$ with smooth boundary M. A basic problem in complex analysis is relating the function theory in D to geometric properties of M. In turn, we can describe many of the useful local geometric properties of M by considering the orders of contact of complex analytic varieties in the ambient space with M. One such property is the notion that the Levi form of M have $n-q$ eigenvalues of one sign at a point p. In reference [1], the author introduces condition F_q which generalizes this idea to degenerate situations. See Definition 1 below. Stated imprecisely, the problem is to determine when a Taylor polynomial of a defining function for M prevents any q-dimensional complex analytic variety from having high order of contact with M. The methods of reference [1] reduce this to algebraic questions about families of ideals of holomorphic polynomials. This leads to the following question about dimensions of analytic varieties. Let f be a p-tuple of holomorphic functions, and write V(f) for the variety defined by f. We assume that 0 lies in V(f). Given an integer s, we would like to know when f is s-stable. That is, when is

$$\text{dimension } V(f+u) \leq \text{dimension } V(f)$$

for every p-tuple u of holomorphic functions which all vanish to order s

Recent Developments in Several Complex Variables
0-691-08285-5/81/000127-06 $00.50/0 (cloth)
0-691-08281-2/81/000127-06 $00.50/0 (paperback)
For copying information, see copyright page.

at 0. For condition F_q we need to relate s-stability for a p-tuple of polynomials to their maximum degree. This question in general is quite difficult. The purpose of this paper is to give the basic definitions, and then to consider an example in detail. Namely we let $f = (z^a - w^b, z^c w^d)$, and determine the necessary and sufficient condition on the integers so that f is s-stable for $s = \max(a, b, c+d)$. This condition is several complicated inequalities on the integers and the greatest common divisor of a and b.

Section I

Let $\mathcal{O}$ denote the local ring of germs of analytic functions at 0 in $\mathbf{C}^n$ for some fixed n. Later we will take $n = 2$. Let $\mathfrak{m} \subset \mathcal{O}$ be the maximal ideal. We write $\mathfrak{m}^k$ for its k-th power, so that $\mathfrak{m}^k$ consists of those elements of $\mathcal{O}$ which vanish to order at least k. We write $\mathcal{O}^p$ for the module of p-tuples of elements of $\mathcal{O}$. If f is an element of $\mathcal{O}^p$, we write (f) for the ideal generated by f, and V(f) for the variety (germ) defined by f. Finally, if r is a smooth function defined near the origin in $\mathbf{C}^n$, we write $j_k r$ for its k-th order Taylor polynomial at 0. We also use this notation componentwise for elements of $\mathcal{O}^p$.

Recall that our motivation is the following definition:

1. CONDITION F_q. Let r be a smooth function defined near 0 in $\mathbf{C}^n$. We say $j_k r$ satisfies condition F_q if there is no q-dimensional complex analytic variety through 0 that lies in the zero set of r', where r' is any smooth function for which $j_k r = j_k r'$.

2. EXAMPLE. If the Levi form of r has n–q positive eigenvalues at 0, then condition F_q holds for $j_2 r$. (See reference [1].)

Notice that $j_k r$ is a real valued polynomial. The methods of reference [1] show how to reduce question F_q to questions about familes of ideals of holomorphic polynomials. In particular, one needs to know if the dimension of an analytic variety can increase if the defining functions are changed at high orders. This motivates the following definition:

3. DEFINITION. Let $f \in \mathcal{O}^p$. We say f is k-stable if, for every $g \in \mathcal{O}^p$ with $j_k g = 0$, we have $\dim V(f+g) \leq \dim V(f)$.

4. PROPOSITION. *Let* $f \in \mathcal{O}^p$. *Then* f *is* k*-stable for some integer* k.

This proposition is proved in reference [1], but we are mainly interested in better information on k. Namely, suppose the components of f are polynomials, and μ is the maximum degree of the components. When is f μ-stable? If it is not, there is some $g \in \mathcal{O}^p$ with $j_\mu g = 0$, and $\dim V(f+g) > \dim V(f)$. We call g a perturbation of f.

To illustrate the complications involved in this definition, we consider $f = (z^a - w^b, z^c w^d)$. Here a, b, c, d are non-negative integers, and $a, b, c+d$ are all positive. We write (a, b) for the greatest common divisor of a and b. We may also assume that $a \leq b$.

5. THEOREM. *Let* $f = (z^a - w^b, z^c w^d)$. *Let* $\mu = \max(b, c+d)$. *Then* f *is* μ*-stable if and only if one of the following three conditions fails to hold. (Without loss of generality we assume that* $a \leq b$.)

1. $a < b$
2. $a/(a, b) \leq c$
3. *Let* M *be the largest integer with* $c \geq Ma/(a, b)$. *If* $b > c+d$, *then* $b < c+d+M(b-a)/(a, b)$. *If* $b \leq c+d$, *there is no third condition.*

Proof. Certainly $V(f) = \{0\}$. We will determine when there are elements g_i in $\mathcal{O}$ with $j_\mu g_i = 0$, and $V(f+g) \neq \{0\}$.

Notice that V contains a non-trivial holomorphic curve if and only if $\dim V(f) \geq 1$. We think of such a curve as the image of a map $\phi: C \to C^2$. Let $\phi^* f = f(\phi(t))$ denote the pullback map. We may suppose that $\phi(t) = (ut^m, vt^n)$ where u, v are units or are identically zero. We divide the proof into four cases.

Case 1. $a = b$. Then f is μ-stable.

Suppose that $\phi^*(f+g) = 0$. Then $0 = \phi^*(z^a - w^b + g_1) = u^a t^{ma} - v^a t^{na} + \phi^* g_1$. Since $j_b g_1 = 0$, we must have either $ma = na = 0$ or $u = v = 0$. In the former case, $m = n$, and $0 = \phi^*(z^c w^d + g_2) = u^c v^d t^{m(c+d)} + \phi^*(g_2)$.

Since $j_{c+d}g_2 = 0$, we have $uv = 0$. It is simple to check that if only one of u and v is 0, then $\phi^*(f+g) \neq 0$. Therefore $u = v = 0$ in either case, and therefore $V(f+g) = \{0\}$.

Case 2. $a < b$, and $c < a/(a,b)$. Then f is μ-stable.

Suppose again that $\phi^*(f+g) = 0$. We want to show that $\phi = 0$. We have $0 = \phi^*(z^a - w^b + g_1) = t^{ma}u^a - t^{nb}v^b + \phi^*(g_1)$. Assuming that $\phi = 0$, we claim that this implies that $ma = nb$. Suppose first that $ma < nb$. We will get a contradiction by showing $j_{ma}(\phi^* g_1) = 0$. Let $z^k w^s$ be a term in the Taylor expansion of g_1. We have $k + s > b$. Also $\phi^*(z^k w^s) = t^{mk+ns}u'$ for some unit u'. We are assuming that $n/m > a/b$. Therefore

$$k + sn/m > k + sa/b .$$

The right side is smallest when $k = 0$ and $s = b+1$, since $a < b$. Thus

$$k + sn/m > (b+1)a/b = a + a/b > a .$$

This gives $mk + ns > ma$. Hence $ma < nb$ is impossible. Now suppose that $ma > nb$. Similarly,

$$s + km/n > s + kb/a .$$

Again the right side is smallest when $k = 0$ and $s = b+1$. Therefore

$$s + km/n > b+1 > b .$$

Finally $ns + mk > nb$. Hence $ma > nb$ is also impossible. Now we suppose $ma = nb$, and use the second equation

$$0 = t^{mc+nd}u^c v^d + \phi^*(g_2) .$$

Again consider $z^k w^s$ where $k+s > c+d$. As before, if $\phi \neq 0$, we have

$$mc + nd = mk + ns .$$

Solving for s gives

$$** \qquad s = (c-k)b/a + d .$$

Suppose that $k-c \geq 0$. Then ** gives

$$s-d = (c-k)b/a > c-k ,$$

contradicting $c+d > k+s$. Therefore $k-c < 0$. By hypothesis then $k < c \leq a/(a,b)$. Since ** holds we must have $(c-k)b/a$ is an integer. This implies that $a/(a,b)$ divides $c-k$. Since $c-k < a/(a,b)$ we get $k = c$, again a contradiction. The only way out of all these contradictions is that $\phi = 0$. Thus $V(f+g) = \{0\}$, and f is μ-stable.

Case 3. Suppose $a < b$, and that $a/(a,b) \leq c$. Then if $b \leq c+d$, f is perturbable.

Put $g_1 = 0$, and $g_2(z,w) = -z^{c-a/(a,b)}w^{d+b/(a,b)}$. Put $\phi(t) = (t^b, t^a)$. This gives $\phi^*(f+g) = 0$. We need only check that $j_\mu g = 0$. But $c-a/(a,b)+d+b/(a,b) = c+d+(b-a)/(a,b) > c+d \geq b$. Therefore g is a perturbation of f.

Case 4. Suppose $a < b$, and that $a/(a,b) \leq c$. Also say $b > c+d$. Let M be the largest integer for which $Ma/(a,b) \leq c$. Then we can perturb f if and only if $c+d+M(b-a)/(a,b) > b$.

Put $g_1 = 0$, and $g_2(z,w) = -z^{c-Ma/(a,b)}w^{d+Mb/(a,b)}$. Then, as in Case 3, g is a perturbation if $c+d+M(b-a)/(a,b) > b$.

Now suppose the inequality fails. We assume $\phi^*(f+g) = 0$. As in Case 2 we see that $ma = nb$, and that $g_2 \neq 0$. Again consider a term $z^k w^s$. ** still holds, and $k-c < 0$. Let $R = c-k$. Using ** we see that $R = Na/(a,b)$ for some integer N. Therefore

$$\begin{aligned} k+s &= c+d+R(b/a-1) \\ &= c+d+N(b-a)/(a,b) . \end{aligned}$$

Since $R \leq c$, we get $Na/(a,b) \leq c$, so that $N \leq M$. This gives

$$k+s = c+d+N(b-a)/(a,b) > b$$

as a necessary condition for a perturbation. This completes the proof of Case 4, and therefore completes the proof of the theorem.

6. REMARKS. Suppose $f \in \mathcal{O}^n$, where n is also the dimension. By the Nullstellensatz, $V(f) = \{0\}$ if and only if there is some k with $\mathfrak{m}^k \subset (f)$. It is easy to show that such an f is k-stable. Even in the examples of Theorem 5 it is possible for f to be s-stable for much smaller integers s. We can interpret Theorem 5 as giving necessary and sufficient conditions for the solvability of a partial differential congruence. By a theorem of Grothendieck (see Reference [2]) $V(f) = \{0\}$ if and only if $\det(df) \neq 0$ mod (f). Therefore f is k-stable if and only if there is no solution to the congruence

$$\det(d(f+g)) \equiv 0 \bmod (f+g) \quad \text{and} \quad j_k g = 0 .$$

7. *A simple application to real hypersurfaces*

For simplicity we consider an especially simple case; however, the general case exhibits the same basic ideas. See reference [1]. Suppose $M = r^{-1}(0)$ where $j_{2k}r = 2\mathrm{Re}(z_n) + \Sigma|f_j(z_1, \cdots, z_{n-1})|^2$. Notice that $V(z_n, f_1, \cdots, f_p)$ lies in the zero set of $j_{2k}r$, but that no complex variety of any larger dimension does. However if f is not k-stable, we can find g with $j_k g = 0$, and $\dim V(f+g) > \dim V(f)$. Let $r' = 2\mathrm{Re}(z_n) + \Sigma|f_j|^2 - \Sigma|g_j|^2$. Then $j_{2k}r' = j_{2k}r$, but $r'^{-1}(0)$ contains a complex variety of strictly higher dimension. This is most easily seen by letting ϕ be a map satisfying $\phi^*(f+g, z_n) = 0$. Then $\phi^* r' = 0$ whether or not $\phi^* f = 0$.

DEPARTMENT OF MATH
UNIVERSITY OF ILLINOIS
URBANA, ILLINOIS 61801

REFERENCES

[1] D'Angelo, John, "Orders of Contact of Real and Complex Subvarieties," (to appear in Illinois J. Math).

[2] Eisenbud, D. and Levine, H., "An algebraic Formula for the Degree of a C^∞ Map Germ," Annals of Math., 106, 1977.

BIHOLOMORPHIC MAPPINGS BETWEEN TWO-DIMENSIONAL HARTOGS DOMAINS WITH REAL-ANALYTIC BOUNDARIES

Klas Diederich and John Erik Fornaess

(Dedicated to H. Grauert and R. Remmert)

§1. *Introduction*

Let $\Omega^1, \Omega^2 \subset\subset \mathbb{C}^n$ be two domains with, for instance, smooth boundaries and $\Phi: \Omega^1 \to \Omega^2$ a biholomorphic mapping. Does Φ have a continuous (homeomorphic) extension $\hat{\Phi}: \bar{\Omega}^1 \to \bar{\Omega}^2$? Which additional regularity properties does $\hat{\Phi}$ eventually have? It is well known that for $n = 1$ the answer is in the affirmative and $\hat{\Phi}$ is a diffeomorphism, which is even real analytic if $b\Omega^1$ and $b\Omega^2$ are real analytic. For $n > 1$ the answer to this question is, in general, not known. But in the last years many partial results have been proved, all pointing into the affirmative direction (see [1], [6], [7], [8], [9], [10], [11], [12], [13], [14], [15], [16], [17], [18], [20], [21], [22]). Counterexamples are not known up to now.

In almost all known results, however, global pseudoconvexity or even strict pseudoconvexity plays an essential role.

W. Kaup showed already in [10] that it is not necessary to assume pseudoconvexity for the existence of an extension $\hat{\Phi}$, which is even holomorphic in some neighborhood of $\bar{\Omega}^1$, if one knows already that the automorphism groups of Ω^j, $j = 1, 2$, contain large subgroups of maps with nice boundary behavior, more precisely, if Ω^j are Cartan domains o

Recent Developments in Several Complex Variables
0-691-08285-5/81/000133-18$00.90/0 (cloth)
0-691-08281-2/81/000133-18$00.90/0 (paperback)
For copying information, see copyright page.

certain Reinhardt domains. On the other hand, recently the authors showed in [7] that one has a continuous extension $\hat{\Phi}$ if $\Omega^j \subset\subset \mathbb{C}^2$, $j=1,2$, are domains with smooth real-analytic boundaries which are not necessarily pseudoconvex, but satisfy the following additional condition:

(*) The "border" between the pseudoconvex and the pseudoconcave part of the boundary $b\Omega^j$ is a totally real manifold (for a precise statement see Theorem 1 of [7]).

The main result of this paper also shows the existence of a continuous extension $\hat{\Phi}$ for certain non-pseudoconvex domains. It is

THEOREM 1. *Let* $\Omega^j \subset\subset \mathbb{C}^2$, $j = 1,2$, *be two Hartogs domains (not necessarily complete nor pseudoconvex) with smooth real-analytic boundaries* $b\Omega^j$. *Then every biholomorphic mapping* $\Phi: \Omega^1 \to \Omega^2$ *has a homeomorphic extension* $\Phi: \bar{\Omega}^1 \to \bar{\Omega}^2$.

We want to point out that this theorem is *not* a simple consequence of H. Cartan's classification of Hartogs domains [2], [3]. In our proof we will make essential use of the results of [6] and [7] on the boundary behavior of the Kobayashi metric at pseudoconvex, real-analytic boundaries resp. on the "size" of the envelope of holomorphy of bounded domains with smooth real-analytic boundaries. The proof will be given by distinguishing between the following two cases:

I) The mapping Φ does not map all fiber components of the Hartogs domain Ω^1 into fiber components of Ω^2.

II) The mapping Φ preserves the fiber components. (According to H. Cartan this happens for most Hartogs domains.)

We will show that in the first case the whole "border" between the pseudoconvex and the pseudoconcave part of $b\Omega^j$ lies inside the envelope of holomorphy $\tilde{\Omega}^j$ of Ω^j, $j = 1,2$. But since Φ is known to extend to a biholomorphic mapping $\tilde{\Phi}: \tilde{\Omega}^1 \to \tilde{\Omega}^2$, this means that the only part where one still has to find the wanted extension $\hat{\Phi}: \bar{\Omega}^1 \to \bar{\Omega}^2$ of Φ is a compact subset of the pseudoconvex part of $b\Omega^1$. In this situation the methods of [6] give the desired result.

In the case II) we construct at first suitable "base" Riemann surfaces $\mathcal{R}^j$ together with holomorphic projections $\Pi^j : \Omega^j \to \mathcal{R}^j$, $j = 1, 2$, such that Ω^j is a Hartogs domain over $\mathcal{R}^j$, and such that there is a conformal "base" mapping ϕ making the following diagram commutative:

$$\begin{array}{ccc} \Omega^1 & \xrightarrow{\ \Phi\ } & \Omega^2 \\ \Big\downarrow \Pi^1 & & \Big\downarrow \Pi^2 \\ \mathcal{R}^1 & \xrightarrow{\ \phi\ } & \mathcal{R}^2 \end{array}$$

(see §4). The map ϕ extends to a Hölder continuous map $\phi : \bar{\mathcal{R}}^1 \to \bar{\mathcal{R}}^2$ (§5).

The main difficulty of the proof will then consist in finding the continuous extension $\hat{\Phi}$ (resp. $\hat{\Phi}^{-1}$) over the points of $b\mathcal{R}^1$ (resp. $b\mathcal{R}^2$) forming the "edge" of Ω^1 (resp. Ω^2). We will do this by giving at first an estimate for the boundary behavior of the Kobayashi metric of Ω^j at the points of the edge (see §6) and then essentially apply Henkin's method from [9] (§7).

Part I: Mappings not preserving all fiber components

§2. *The structure of the envelopes of holomorphy*

1. Let $\Omega^j \subset\subset \mathbf{C}^2 = \{(z_1^j, z_2^j)\}$ be Hartogs domains with smooth real-analytic boundaries, $j = 1, 2$. We may assume that the coordinate axis $\{z_1^j = 0\}$ is the symmetry plane of Ω^j. We denote by π^j the projection of $\mathbf{C}^2 = \{(z_1^j, z_2^j)\}$ on the z_2^j-axis and call the domain $B^j := \pi^j(\Omega^j) \subset\subset \mathbf{C}$ the (unreduced) basis of Ω^j. Notice that since we do not suppose Ω^j to be complete the fibers $(\pi^j)^{-1}(z_2^j) \cap \Omega^j$ for $z_2^j \in B^j$ consist of a finite number of pairwise disjoint annuli and possibly a disc all centered at 0.

Let $\Phi : \Omega^1 \to \Omega^2$ be a biholomorphic mapping. In this part I we always make the following additional assumption:

(A) There is a connected component D of a fiber $(\pi^1)^{-1}(z_2^1) \cap \Omega^1$, $z_2^1 \in B^1$, such that $\Phi(D)$ is not contained in a fiber $(\pi^2)^{-1}(z_2^2) \cap \Omega^2$, $z_2^2 \in B^2$, of Ω^2.

2. In this §2 we will show that (A) forces the envelope of holomorphy $\tilde{\Omega}^j$ of Ω^j to contain a rather large part of $b\Omega^j$. In order to be precise we define at first

a) $\hat{M}^j := \{p \in b\Omega^j :$ the Levi form of $b\Omega^j$ vanishes identically at $p\}$.

Notice that $\hat{M}^j$ is a closed real-analytic set in $b\Omega^j$ of dimension $\leqq 2$.

b) The set $M_+^j \subset b\Omega^j$ resp. $M_-^j \subset b\Omega^j$ is the interior (in $b\Omega^j$) of the set of pseudoconvex resp. pseudoconcave points of $b\Omega^j$. The rest

$$M^j := b\Omega^j \setminus (M_+^j \cup M_-^j)$$

is called the border (between the pseudoconvex region M_+^j and the pseudoconcave region M_-^j).

The border M^j is a closed subset of $\hat{M}^j$, which, in general, is not equal to $\hat{M}^j$, as can be seen from simple examples.

3. The main result of this section is

PROPOSITION. *Under the assumption* (A) *one has* $M^j \subset \tilde{\Omega}^j$ *for* $j = 1, 2$.

4. For the proof of Proposition 2.3 we need the following

PROPOSITION. *Let* $\Omega \subset\subset \mathbf{C}^2$ *be a Hartogs domain with smooth real-analytic boundary. Suppose that for some point* $p \in M$, *the border on* $b\Omega$, *there is an open neighborhood* U *such that* $U \cap M \subset U$ *is a closed real-analytic manifold. Then there is a non-empty relatively open subset* $M_0 \subset U \cap M$ *which is contained in the envelope of holomorphy* $\tilde{\Omega}$ *of* Ω.

REMARK. In general, p is not contained in M_0.

Proof. One has $\dim_{\mathbb{R}} M \cap U = 2$ since otherwise $M \cap U$ would not locally separate $b\Omega$. Furthermore, outside a lower dimensional real-analytic subset, $M \cap U$ has to be totally real since it and therefore also $b\Omega$ otherwise would contain an open piece of a 1-dimensional complex-analytic manifold. This would be a contradiction to Theorem 4 of [5]. Now Proposition 2.4 is an immediate consequence of Theorem 4 of [7]. (This theorem was formulated as a global statement. But a glance at its proof shows at once that it can be localized.)

5. The main step in the proof of Proposition 2.3 consists in showing the following

LEMMA. *Let Ω^j, $j = 1,2$, and Φ be as in the theorem and let Φ satisfy (A). Assume that for $j = 1$ or $j = 2$ and for a certain point $p \in M^j$ there is an open neighborhood U such that $M^j \cap U \subset U$ is a connected closed real-analytic manifold. Then there is a closed real-analytic connected manifold $D^j \subset b\Omega^j$ together with a neighborhood V^j of D^j such that $D^j = M^j \cap V^j$ and $M^j \cap U \subset D^j \subset \tilde{\Omega}^j$.*

Proof. 1) We may assume that $j = 1$. According to Proposition 2.4 and since Φ and therefore also Φ^{-1} do not preserve all fiber components of Ω^1 resp. Ω^2 (assumption (A)) one can choose a point $p^1 = (z^1_{10}, z^1_{20}) \in U \cap M^1$ together with a neighborhood $U^1 \subset U$ of p^1 satisfying the following properties:

a) $U^1 \cap M^1 \subset U^1$ is a connected closed real-analytic, totally real manifold.
b) $U^1 \cap M^1 \subset \tilde{\Omega}^1$.
c) If we define $p^2 := \Phi^{-1}(p^1), U^2 := \Phi^{-1}(U^1)$, then p^2 and U^2 have properties analogous to a), b).
d) With the notation $p^2 = (z^2_{10}, z^2_{20})$ and

$$\Phi^{-1}(z^2_1, z^2_2) = (z^1_1(z^2_1, z^2_2), z^1_2(z^2_1, z^2_2))$$

one has $z_{10}^2 \neq 0$ and

$$\frac{\partial}{\partial\theta} z_2^1(e^{i\theta} z_{10}^2, z_{20}^2) \neq 0 \text{ at } \theta = 0 .$$

We denote

$$\eta^2(\theta) := (e^{i\theta} z_{10}^2, z_{20}^2) .$$

Notice that because of the rotational invariance of Ω^2 the border M^2 has the same properties as stated in a) at all points of the circle η^2 and that $\eta^2 \subset \tilde{\Omega}^2$. We, therefore, can define

$$\eta^1(\theta) := \Phi^{-1}(\eta^2(\theta)) \subset \tilde{\Omega}^1$$

and M^1 also looks like in a) at all points of $\eta^1(\theta)$. We want to show that the compact, rotationally invariant set

$$D^1 := \{(e^{i\tau} z_1^1, z_2^2) : \tau \in \mathbf{R},\ (z_1^1, z_2^2) \in \eta^1(\theta)\}$$

satisfies the claim of our lemma. For this it is obviously enough to show that there is an open neighborhood V^1 of D^1 such that $D^1 \subset V^1$ is a closed real-analytic manifold with $D^1 = M^1 \cap V^1$.

2. Since M^1 is rotationally invariant in the z_1^1-direction and a totally real, real-analytic manifold near all points $q^1 = (z_1^1, z_2^1) \in \eta^1$, there is an open neighborhood $W(q^1)$ of q^1 such that

$$\pi^1(W(q^1) \cap M^1) = \gamma^1 \subset \mathbf{C}$$

is an open piece of a real-analytic 1-dimensional submanifold if $z_1^1 \neq 0$. But the same is also true if $z_1^1(q^1) = 0$ as follows again easily from total reality of M^1 near q^1.

3. Next, we want to show that, in fact, $z_1^1 \neq 0$ at each point of η^1. Namely, at the points $(0, z_2^1) \in b\Omega^1$ the real tangent plane $Tb\Omega^1$ is of the form $\mathbf{C} \times \mathbf{R}$, where $\mathbf{C}$ represents the z_1^1-direction and $\mathbf{R}$ is a real

direction in the z_2^1-plane. If $q = (0, z_2^1) \in \eta^1$ then M^1 would be a totally real manifold near q and $TM^1 \subset C \times R$. Therefore, TM^1 would have to contain a real direction in the z_1^1-plane. But, because of the rotational invariance of M^1 in the z_1^1-plane, this would imply that $\{z_2^1 = 0\} \subset TM^1$, a contradiction to M^1 being totally real.

4. We define the projections

$$\hat{\pi}: C^2 = \{(z_1^1, z_2^1)\} \to R \times C$$

and

$$\tilde{\pi}: R \times C \to C$$

by $\hat{\pi}(z_1^1, z_2^1) := (|z_1^1|, z_2^1)$ and by $\tilde{\pi}(r, z_2^1) := z_2^1$ and put

$$\hat{\eta} := \hat{\pi}(\eta^1) .$$

Notice that $\hat{\pi}^{-1}(\hat{\eta}) = D^1$ and that $\hat{\eta}(\theta)$ is a piecewise smooth real-analytic curve in R^3. Because of the rotational invariance of Ω^1 in the z_1^1-direction it is an easy consequence of 2) that for each point $\hat{p} \in \text{trace } \hat{\eta}$ there is a 1-dimensional real-analytic closed submanifold γ^1 in an open neighborhood $W(p)$ of $\tilde{\pi}(\hat{p})$ such that for each $\theta_o \in R$ with $\hat{\eta}(\theta_o) = \hat{p}$ one has

$$\hat{\eta}(\theta) \in \gamma^1 \text{ for all } \theta, \ |\theta - \theta_o| < \varepsilon$$

for some small $\varepsilon > 0$. As an easy consequence of this fact one obtains by reparametrizing $\hat{\eta}$ the following statement:

The trace of $\hat{\eta}$ is homeomorphic either to a closed interval $I \subset R$ or to the unit circle $b\Delta$.

5. We now want to exclude the possibility of trace $\hat{\eta}$ being homeomorphic to I. In that case, trace $\hat{\eta}$ and therefore also trace $\pi^1(\eta^1)$ would be contractible. Hence, one has for the winding number of $\pi^1(\eta^1)$

$$I(w_o; \pi^1(\eta^1)) = 0 \text{ for all } w_o \notin \pi^1(\eta^1) .$$

On the other hand, the function

$$f(z_1^2) := z_2^1(z_1^2, z_{20}^2)$$

is holomorphic and non-constant in a neighborhood of the disc $\overline{\Delta^2} := \{|z_1^2| \leqq |z_{10}^2|\}$. Choose $w_o \in f(\Delta^2) \setminus f(b\Delta^2)$. We get

$$I(w_o; \pi^1(\eta^1)) = \frac{1}{2\pi i} \int_{\pi(\eta^1)} \frac{d\zeta}{\zeta - w_o}$$

$$= \frac{1}{2\pi i} \int_{b\Delta^2} \frac{f'(\zeta)}{f(\zeta) - w_o} \, d\zeta \neq 0 ,$$

a contradiction.

6. In 4. and 5. together we showed that trace $\hat{\eta}$ is homeomorphic to the unit circle. Consequently, if we choose for $\hat{p} \in \hat{\eta}$ the neighborhood $W(\hat{p})$ of $\tilde{\pi}(\hat{p})$ as in 4. small enough there is a closed 1-dimensional real-analytic submanifold $\gamma^1 \subset W(\hat{p})$ with

$$\gamma^1 \subset \tilde{\pi}(\hat{\eta}) .$$

Hence, trace $\hat{\eta} \subset \mathbf{R} \times \mathbf{C}$ is a compact, 1-dimensional real-analytic submanifold, thereby showing that also

$$D^1 = \{(re^{i\theta}, z_2^1) : (r, z_2^1) \in \hat{\eta}, \theta \in \mathbf{R}\}$$

is a compact real-analytic connected manifold. Furthermore, we have because of 2. for each $q^1 \in \eta^1$ an open neighborhood $W(q^1)$ of q^1 such that

$$M^1 \cap W(q^1) = \{p \in W(q^1) : \tilde{\pi}(p) \in \hat{\eta}\} = D^1 \cap W(q^1) .$$

This proves the lemma. □

7. *Proof of Proposition 2.3.* We may assume that $j = 1$. We call all 2-dimensional connected real-analytic submanifolds $D^1 \subset b\Omega^1$ as arising from Lemma 2.5 submanifolds of type I. Any two of them are obviously disjoint. Therefore, we can choose a stratification

$$S_1, S_2, \cdots, S_m$$

of M_o^1 by real-analytic connected closed submanifolds $S_k \subset b\Omega^1 \setminus \bigcup_{j=k+1}^{m} S_j$ such that all submanifolds of type I occur as some of the S_j's. We will show by induction:

(*) For all $k = 1, \cdots, m$ one has $S_k \cap M^1 = \phi$ or S_k is of type I and therefore $S_k \subset M^1$.

Proposition 2.3 is then of course an immediate consequence.

Let us now fix a k, $1 \leqq k \leqq m$, and suppose that (*) has been proved for all $S_1, \cdots, S_{k-1}$.

Let $p \in S_k$ be arbitrary. There exists an open neighborhood U of p such that

$$U \cap S_\ell = \emptyset \text{ for all } \ell > k \text{ and } U \cap S_k \text{ is connected.}$$

Furthermore, if for some $j < k$ one has $p \in \overline{S_j}$ then $S_j \cup M^1 = \emptyset$ and hence $S_j \subset M_+^1$ or $S_j \subset M_-^1$.

As a consequence, we have

$$M^1 \cap U \subset S_k \cap U .$$

If now $\dim_{\mathbf{R}} S_k = 1$, S_k does not separate $b\Omega^1$ locally showing that $p \notin M^1$. If, on the other hand, $\dim_{\mathbf{R}} S_k = 2$ and $p \in M^1$, we get because of $\dim_{\mathbf{R}} M_o^1 = 2$ eventually after shrinking U

$$M^1 \cap U = S_k \cap U .$$

According to Lemma 2.5 and the choice of the stratification S_k therefore must be of type I. □

§3. *End of proof*

1. We want to show that Theorem 1 holds under the additional assumption A. We know that $\Phi: \Omega^1 \to \Omega^2$ extends to a biholomorphic mapping $\Phi: \tilde{\Omega}^1 \to \tilde{\Omega}^2$. Furthermore, because of Proposition 2.3 we have

$$M^j \cup M^j_- = \overline{M^j_-} \subset \tilde{\Omega}^j, \quad j = 1, 2 .$$

Therefore, there are compact subsets $K^j \subset M^j_+$ such that Theorem 1 is proved if we can extend Φ continuously to $\Omega^1 \cup K^1$ and Φ^{-1} to $\Omega^2 \cup K^2$.

2. The authors proved in [4] and [6]

PROPOSITION. *Let* $\Omega \subset\subset \mathbf{C}^n$ *be a domain with real-analytic smooth boundary. Let* $M_+ \subset b\Omega$ *be the pseudoconvex region on* $b\Omega$. *Then each compact subset* $K \subset M_+$ *has an open neighborhood such that one has:*

a) *For certain constants* $\varepsilon = \varepsilon(K) > 0$, $C = C(K) > 0$ *the following lower estimate for the Kobayashi differential metric* F_Ω *of* Ω *holds:*

$$F_\Omega(p; X) \geqq C \frac{|x|}{d^\varepsilon_\Omega(p)}$$

for all $p \in W \cap \Omega$, $X \in \mathbf{C}^n$. *Here* d_Ω *denotes the Euclidean boundary distance of* p *in* Ω.

b) *For a certain constant* $\eta = \eta(K) > 0$ *and a suitable defining function* ρ *of* Ω *defined in a neighborhood of* $\overline{\Omega}$ *the function*

$$\sigma := -(-\rho)^\eta$$

is strictly plurisubharmonic on $\Omega \cap W$.

3. The Proposition 3.2 contains everything needed for Henkin's method to extend Φ (resp. Φ^{-1}) in a Hölder continuous way to $\Omega^1 \cup K^1$ (resp. $\Omega^2 \cup K^2$). For more details see for instance [7].

Part II: Fiber preserving maps

§4. *Construction of a base Riemann surface*

4.1. We assume in this part that the biholomorphic map $\Phi : \Omega^1 \to \Omega^2$ is fiber component preserving. If we write Φ in coordinates as $\Phi(z_1^1, z_2^1) = (z_1^2(z_1^1, z_2^1), z_2^2(z_1^1, z_2^1))$, then this means that $\partial z_2^2/\partial z_1^1 \equiv 0$.

For each fixed z_2^j, the set $\Omega^j(z_2^j) = \{z_1^j \in \mathbb{C}; (z_1^j, z_2^j) \in \Omega^j\}$ is a finite union of annuli and at most one disc, all centered at the origin. Therefore, Φ is either of the form $\Phi(z_1^1, z_2^1) = (z_1^1 g^1(z_2^1), z_2^2(z_2^1))$ or of the form $\Phi(z, w) = (f^1(z_2^1)/z_1^1, z_2^2(z_2^1))$ for (multiple-valued) holomorphic functions. The latter case is only possible if Ω^j, $j = 1, 2$ contain no discs in any fiber. In particular, this means that $\overline{\Omega}^j$ does not intersect the z_2^j-axis. Hence, composing Φ with the map $(z_1^2, z_2^2) \to (1/z_1^2, z_2^2)$, we have reduced this to the former case.

4.2. We can describe the inverse mapping Φ^{-1} by $\Phi^{-1}(z_1^2, z_2^2) = (z_1^1 g^2(z_2^2), z_2^1(z_2^2))$. Using the coordinate functions of Φ and Φ^{-1}, we define equivalence relations on Ω^j.

Let p_1^j and p_2^j be in Ω^j. Then we define $p_1^j \sim p_2^j$ if $\pi^j(p_1^j) = \pi^j(p_2^j)$ and the germs of g^j and $z_2^{3-j}(z_2^j)$ are the same at p_1^j and p_2^j.

We define $\mathcal{R}^j = \Omega^j/\sim$ and let $[p]$ be the equivalence class of $p \in \Omega^j$. Since $p_1^1 \sim p_2^1$ if and only if $\Phi(p_1^1) \sim \Phi(p_2^1)$, Φ induces a bijective map $\phi : \mathcal{R}^1 \to \mathcal{R}^2$.

4.3. The natural projections $\Pi^j : \Omega^j \to \mathcal{R}^j$ induce a topology on $\mathcal{R}^j$, namely the weakest topology for which Π^j becomes open (and hence continuous).

These topologies are Hausdorff, and the map $\phi : \mathcal{R}^1 \to \mathcal{R}^2$ becomes a homeomorphism.

4.4. The maps $\rho^j : \mathcal{R}^j \to \mathbf{C}$; $\rho^j([(z_1^j, z_2^j)]) = z_2^j$, $j = 1, 2$, are local homeomorphisms which induce on the topological spaces $\mathcal{R}^j$ the additional structure of Riemann surfaces. Then $\phi : \mathcal{R}^1 \to \mathcal{R}^2$ becomes a biholomorphism. We will call $\mathcal{R}^j$ the (reduced) base of Ω^j and call $\phi : \mathcal{R}^1 \to \mathcal{R}^2$ the base mapping.

§5. *Extension of the base mapping to the boundary*

5.1. We write $z_\ell^j = x_\ell^j + iy_\ell^j$, $j, \ell = 1, 2$. The set $U^j \subset \mathbf{R} \times \mathbf{C}$, $U^j = \{(x_1^j, z_2^j) \in \Omega\}$ has a smooth real-analytic boundary, given by $bU^j = \{r^j(x_1^j, z_2^j) = 0\}$ for some real-analytic defining function r^j of Ω^j.

Let $\Sigma^j \subset bU^j$ be the real-analytic subset of bU^j on which $\partial r^j / \partial x_1^j = 0$.

We denote by $\partial \mathcal{R}^j$ the ideal boundary of $\mathcal{R}^j$. The projection ρ^j extends to a continuous map $\rho^j : \mathcal{R}^j \cup \partial \mathcal{R}^j \to \mathbf{C}$. Since $b\Omega^j$ is smooth the projection map $\Pi^j : \Omega^j \to \mathcal{R}^j$ extends in a continuous way to $\Pi^j : \overline{\Omega}^j \to \mathcal{R}^j \cup \partial \mathcal{R}^j$ with finite fibers.

For every point $p \in \partial \mathcal{R}^j$, we define $S_p^j := (\Pi^j)^{-1}(p) \cap bU^j \cap \{x_1^j \geq 0\}$.

Fix a point $(x_{10}^j, z_{20}^j) \in S_p^j$. Then Π^j is a one-to-one holomorphic map on the germ at (x_{10}^j, z_{20}^j) of $\gamma^- := \{(x_{10}^j, z_2^j); r^j(x_{10}^j, z_2^j) < 0\}$. This germ has a smooth real-analytic boundary γ in the $\{x_1^j = x_{10}^j\}$-plane. We can identify, via Π^j, the sets γ and γ^- with their projections in $\mathcal{R}^j \cup \partial \mathcal{R}^j$.

The projection of Σ^j in the z_2^j-plane is an at most 1-dimensional real-analytic subset. Inside this set one can find two real-analytic open curves γ_1 and γ_2 lying exterior to γ^- in the obvious sense, going in opposite directions from $\rho^j(p)$, such that $\gamma_1 \cup \{\rho^j(p)\} \cup \gamma_2$ is a C^1-curve. But since ρ^j is schlicht on $\Pi^j(W \cap \overline{\Omega}^j)$ if W is a small neighborhood of (x_{10}^j, z_{20}^j), the C^1-curve $\gamma_1 \cup \{\rho^j(p)\} \cup \gamma_2$ can be lifted via ρ^j to a curve $\hat{\gamma}_1 \cup \{p\} \cup \hat{\gamma}_2$ bounding $\Pi^j(W \cap \overline{\Omega}^j)$.

5.2. We can now describe more precisely the boundary of $\mathcal{R}^j$. There could be a finite number of isolated points. At these points the Riemann surface extends across as a (possibly branched) Riemann surface over $\mathbb{C}$. The rest of the boundary consists of a finite collection of closed curves. Each of these consists of a finite number of points and open arcs which are lifted via ρ^j from open real-analytic arcs contained in the projection of Σ^j in $\mathbb{C}$. This induces the analogous structure on $\partial\mathcal{R}^j$. In particular, at each singular point of $\partial\mathcal{R}^j$, there are two arcs in $\partial\mathcal{R}^j$ coming together in a well-defined angle $\geqq \pi$ (possibly larger than 2π). Hence, one can realize $\mathcal{R}^j$ as a relatively compact open subset in some Riemann surface $\tilde{\mathcal{R}}^j$ in which $b\mathcal{R}^j$ consists of a finite number of isolated points and a finite number of C^1, piecewise real-analytic closed curves.

5.3. The biholomorphic map $\phi: \mathcal{R}^1 \to \mathcal{R}^2$ extends holomorphically across the isolated boundary points of $b\mathcal{R}^1$ and maps these in a bijective fashion to the isolated boundary points of $\mathcal{R}^2$.

The map $\phi: \mathcal{R}^1 \to \mathcal{R}^2$ extends to a Hölder homeomorphism [19], also called $\phi: \overline{\mathcal{R}}^1 \to \overline{\mathcal{R}}^2$. This implies also that $\phi: \mathcal{R}^1 \to \mathcal{R}^2$ extends to a homeomorphism $\phi: \mathcal{R}^1 \cup \partial\mathcal{R}^1 \to \mathcal{R}^2 \cup \partial\mathcal{R}^2$.

§6. *An estimate for the Kobayashi metric*

6.1. Since Ω^1 need not be pseudoconvex, we cannot hope that the Kobayashi metric will blow up everywhere at the boundary. However, we will make an estimate from below near $b\overline{\mathcal{R}}^1$. Since we will work in this section only on Ω^1 we will drop all superscripts 1.

Fix a $p \in b\mathcal{R}$ and a $(x_{10}, z_{20}) \in S_p$ with $x_{10} > 0$. Then, as in 5.1, the curve $\Gamma := \hat{\gamma}_1 \cup \{p\} \cup \hat{\gamma}_2 \subset \overline{\mathcal{R}}$ bounds that part of $\mathcal{R}$ coming from the germ of Ω at (x_{10}, z_{20}).

Let $c_\Omega(q; \xi)$ denote the infinitesimal Kobayashi metric of Ω at the point $q = (z_1, z_2) \in \Omega$ in the direction of the tangent-vector ξ to $\mathbb{C}^2$ at

(z_1, z_2). For (z_1, z_2) in B, with $(|z_1|, z_2)$ close to (x_{10}, z_{20}), let $d(\Pi(z_1, z_2), \Gamma)$ denote distance measured on $\mathcal{R}$ using the projection $\rho: \mathcal{R} \to \mathbf{C}$.

LEMMA 6.1. *There exists an* $\varepsilon > 0$ *and a* $C > 0$ *such that*

$$c_\Omega((z_1, z_2); \xi) \geqq C|\xi|/d(\Pi(z_1, z_2), \Gamma)^\varepsilon$$

for all $((z_1, z_2); \xi)$ *with* $(z_1, z_2) \in \Omega$ *and* $(|z_1|, z_2)$ *close enough to* (x_{10}, z_{20}).

The rest of §6 is devoted to the proof of Lemma 6.1. We need at first some preliminary estimates.

6.2. Let $n \geqq 1$ be any integer.

LEMMA 6.2. *Let* $W = \{(z_1, z_2) \in \mathbf{C}^2;\ \|(z_1, z_2)\| < 1$ *and* $(\operatorname{Re} z_1)^{2n} - \operatorname{Re} z_2 < 0\}$. *Then there exist constants* C, $\varepsilon > 0$ *such that*

$$c_W((z_1, z_2); \xi) \geqq C|\xi|/d^\varepsilon((z_1, z_2), bW)$$

for all $(z_1, z_2) \in W$ *and* $\xi \in \mathbf{C}^2$.

Since W is bounded, and is the intersection of two real-analytic pseudoconvex domains without positive-dimensional complex analytic varieties in their boundaries, this is an immediate consequence of Diederich-Fornaess [6].

6.3. LEMMA 6.3. *Let* $V = \{(z_1, z_2) \in \mathbf{C}^2;\ \|(z_1, z_2)\| < \frac{1}{4}$ *and* $(\operatorname{Re} z_1)^{2n} < x_2 + |y_2|\}$. *Then there exist* C, $\varepsilon > 0$ *such that*

$$c_V((z_1, z_2); \xi) \geqq C|\xi|/|z_2|^\varepsilon$$

on V.

Proof. Let $\psi: V \to \mathbf{C}^2$ be the 1-1, regular holomorphic map $\psi(z_1, z_2) = (z_1, \sqrt{z_2})$. Then direct computation shows that $\psi(V) \subset W$ where W is as in Lemma 6.2. Therefore the result follows from the estimate in Lemma 6.2.

6.4. We may assume, to simplify notation, that $z_{20} = 0$, that Γ is parallel to the y_2-axis at p and that the part of $\mathfrak{R}$ coming from the germ of U at (x_{10}, z_{20}) lies to the right of Γ. Let us also suppose that γ_1 goes in the positive y_2-direction from p and that γ_2 goes in the negative y_2-direction.

Let $\lambda_1, \cdots, \lambda_r$ be the germs of curves in Σ lying over γ_1 starting at (x_{10}, z_{20}). Also let $\mu_1, \cdots, \mu_s$ be the curves in Σ lying over γ_2 starting at (x_{10}, z_{20}).

To each point (z_1', z_2') with $(|z_1'|, z_2') \in \lambda_1 \cup \cdots \cup \lambda_r \cup \mu_1 \cup \cdots \cup \mu_s$ one can attach a copy of a domain V of the type of Lemma 6.3 via the mapping

$$\Phi_{z_1', z_2'}(z_1, z_2) := (z_1' \exp z_1, z_2' + z_2)$$

where the n in Lemma 6.3 is independent of (z_1', z_2'). This maps V to a domain $V_{(z_1', z_2')}$. If only n is large enough, $V_{(z_1', z_2')}$ will contain Ω near (z_1', z_2') decreasing at most polynomially with $|z_2'|$. Hence one obtains from Lemma 6.3 the following:

LEMMA 6.4. *There exist constants* $C, \varepsilon > 0$ *and an integer* $m > 1$ *such that*

$$C_\Omega((z_1, z_2); \xi) \geqq C|\xi|/d(\Pi(z_1, z_2), \Gamma)^\varepsilon$$

for all $((z_1, z_2); \xi)$ *with* $(z_1, z_2) \in \Omega$, $(|z_1|, z_2)$ *close enough to* $(x_{10}, 0)$ *and* $d(\Pi(z_1, z_2), \Gamma) \leqq |z_2|^m$.

Using again Lemma 6.3 at points (z_1, z_{20}) with $|z_1| = x_{10}$ one obtains that there exist constants $C, \varepsilon > 0$ such that $C_\Omega((z_1; z_2); \xi) \geqq C|\xi|/|z_2|^\varepsilon \geqq C|\xi|/d(\Pi(z_1, z_2), \Gamma)^{\varepsilon/m}$ when $(z_1, z_2) \in \Omega$, $(|z_1|, z_2)$ is close enough to $(x_{10}, 0)$ and $d(\Pi(z_1, z_2), \Gamma) > |z_2|^m$. Lemma 6.1 is an immediate consequence of this and Lemma 6.4.

§7. *End of proof*

7.1. We will show that $\Phi: \Omega^1 \to \Omega^2$ has a continuous extension to $\bar{\Omega}^1$. The same argument will then apply to Φ^{-1}.

Fix a $q_o^1 = (z_{10}^1, z_{20}^1) \in b\Omega^1$. We will show that the cluster set of q_o^1, $Cl(q_o^1)$, in $\overline{\Omega}^2$ is a singleton. If $\Pi(z_{10}^1, z_{20}^1) \in \mathcal{R}^1$, this is immediate because of the special shape of Φ (see 4.1), so we may assume $\Pi(z_{10}^1, z_{20}^1) =: p^1 \in b\mathcal{R}^1$. Then $(|z_{10}^1|, z_{20}^1) =: (x_{10}^1, z_{20}^1)$ is in $S_{p^1}^1$.

Since $\phi: \mathcal{R}^1 \to \mathcal{R}^2$ extends continuously to $\overline{\mathcal{R}}^1$, there exists a $p^2 \in b\mathcal{R}^2$ such that $Cl(q_o^1)$ is contained in $\{(z_1^2, z_2^2) \in \overline{\Omega}^2; (|z_1^2|, z_2^2) \in S_{p^2}^2\}$. Since $Cl(q_o^1)$ is connected, there exists $(x_{10}^2, z_{20}^2) \in S_{p^2}^2$ such that $Cl(q_o^1)$ is contained in $\{(z_1^2, z_{20}^2); |z_1^2| = x_{20}^2\}$.

If $x_{10}^2 = 0$, this is a singleton, so we will assume from now on that $x_{10}^2 > 0$. There are then two cases: $x_{10}^1 = 0$ and $x_{10}^1 > 0$.

7.2. Assume at first that $x_{10}^1 = 0$. Then there are fibers arbitrarily close to (z_{10}^1, z_{20}^1) which are discs. Since all fibers near (x_{10}^2, z_{20}^2) are annuli, this case cannot happen.

7.3. It remains to consider the case $x_{10}^1 > 0$. Let Γ_-^1 be the (germ of a) subset of $\mathcal{R}^1$ coming from (the germ of) Ω^1 at (z_{10}^1, z_{20}^1) with boundary Γ^1. Also let Γ^2 be the subset of $\mathcal{R}^2$ coming from Ω^2 at (x_{10}^2, z_{20}^2) with boundary Γ^2. Then $\phi: \overline{\mathcal{R}}^1 \to \overline{\mathcal{R}}^2$ restricts to a Hölder homeomorphism between $\Gamma^1 \cup \Gamma_-^1$ and $\Gamma^2 \cup \Gamma_-^2$. Hence there exists $\varepsilon' > 0$ such that $d(\phi(\tau), \Gamma^2) \leqq d(\tau, \Gamma^1)^{\varepsilon'}$ for all $\tau \in \Gamma_-^1$.

Let C, ε be as in Lemma 6.1 applied to Ω^2 near $(x_{10}^2, z_{20}^2) \in S_{p^2}^2$.

Let $\tau \in \Gamma_-^1$. Then there exists a $(z_1^1, z_2^1) \in \Omega^1$ near (z_{10}^1, z_{20}^1) such that $\Pi^1(z_1^1, z_2^1) = \tau$ and $d(\zeta, \Gamma^1)$ equals the distance between (z_1^1, z_2^1) and $b\Omega^1$. Then if $\xi = (0, 1)$, we have the following estimates for the Kobayashi metric of Ω^1:

$$\frac{1}{d(\tau,\Gamma^1)} \geqq C_{\Omega^1}((z_1^1, z_2^1); (0,1))$$

$$= C_{\Omega^2}(\Phi(z_1^1, z_2^1); (z_1^1(g^1)'(z_2^1)), (z_2^2)'(z_2^1))$$

$$\geqq C|z_1^1|\,|(g^1)'(z_2^1)|/d(\phi(\tau),\Gamma^2)^{\varepsilon}$$

$$\geqq C\cdot\frac{x_{10}^1}{2}\cdot|(g^1)'(z_2^1)|/d(\tau,\Gamma^1)^{\varepsilon\varepsilon'}.$$

From this it follows that $|(g^1)'(z_2^1)| \leqq (2/(C x_{10}^1))/d^{1-\varepsilon\varepsilon'}(\tau,\Gamma^1)$. Or since g^1 naturally is a function on $\mathfrak{R}^1$, we can write

$$|(g^1)'(\tau)| \leqq (2/(C x_{10}^1))/d^{1-\varepsilon\varepsilon'}(\tau,\Gamma^1) \text{ on } \Gamma_-^1.$$

Then it is seen that g^1 is Hölder $\varepsilon\varepsilon'$-continuous on Γ_-^1. Hence $Cl(q_o^1)$ can only contain one point on $\{(z_1^2, z_{20}^2); |z_{10}^2| = x_{10}^2\}$. This completes the proof that Φ extends continuously up to $b\Omega^2$.

KLAS DIEDERICH
GESAMTHOCHSCHULE WUPPERTAL
FACHBEREICH MATHEMATIK
GAUSSSTR. 20
D-5600 WUPPERTAL 1

JOHN ERIK FORNAESS
PRINCETON UNIVERSITY
DEPARTMENT OF MATHEMATICS
FINE HALL – BOX 37
PRINCETON, N. J. 08544
U.S.A.

REFERENCES

[1] Bedford, E., Fornaess, J.E.: Biholomorphic maps of weakly pseudoconvex domains. Duke Math. J. 45(1978), 711-719.

[2] Cartan, H.: Les fonctions de deux variables complexes.... J. Math. IX, *10*(1931),

[3] ———: Sur les transformations analytiques des domaines cerclés et semi-cerclés bornés. Math. Ann. *106*(1932),

[4] Diederich, K., Fornaess, J. E.: Pseudoconvex domains: Bounded strictly plurisubharmonic exhaustion functions. Inv. math. *39*(1977), 129-141.

[5] ———: Pseudoconvex domains with real-analytic boundary. Ann. of Math. *107*(1978), 371-384.

[6] ———: Proper holomorphic maps onto pseudoconvex domains with real-analytic boundary. Ann. of Math.

[7] ———: Biholomorphic mappings between certain real-analytic domains in $\mathbb{C}^2$.

[8] Fefferman, C.: The Bergman kernel and biholomorphic mappings of pseudoconvex domains. Inv. math. *26*(1974), 1-65.

[9] Henkin, G. M.: An analytic polyhedron is not holomorphically equivalent to a strictly pseudoconvex domain. Dokl. Akad Nauk SSSR *210*(1973), 1026-1029; Soviet Math. Dokl. *14*(1973), 858-862.

[10] Kaup, W.: Über das Randverhalten von holomorphen Automorphismen beschränkter Gebiete. Manuscripta math. *3*(1970), 257-270.

[11] Lewy, H.: On the boundary behavior of holomorphic mappings. Acad. Naz. dei Lincei *35*(1977).

[12] Naruki, I.: On extendability of isomorphisms of Cartan connections and biholomorphic mappings of bounded domains. Tôkoku Math. J. *28*(1976), 117-122.

[13] Nirenberg, L., Webster, S., Yang, P.: Local boundary regularity of holomorphic mappings. Preprint 1979.

[14] Pinčuk, S.: On proper holomorphic mappings of strictly pseudoconvex domains. Siberian Math. J. *15*(1975), 644-649.

[15] ________: On the analytic continuation of holomorphic mappings. Math. USSR Sbornik *27*(1975), 373-392; Math. Sbornik *98*(1975), 416-435.

[16] ________: Proper holomorphic mappings between strongly pseudoconvex domains (Russian). Dokl. Akad. Nauk SSSR *241*(1978), 416-435.

[17] Range, R. M.: The Carathéodory metric and holomorphic maps on a class of weakly pseudoconvex domains. Pac. J. Math.

[18] Vormoor, N.: Topologische Fortsetzung biholomorpher Funktionen auf den Rand bei beschränkten streng pseudokonvexen Gebieten im $\mathbb{C}^n$ mit C^∞-Rand. Math. Ann. *204*(1973), 239-261.

[19] Warschawski, S. E.: On differentiability at the boundary in conformal mapping. Proc. Amer. Math. Soc. *12*(1961), 614-620.

[20] Webster, S.: On the mapping problem for algebraic real hypersurfaces. Inv. math. *43*(1977), 53-68.

[21] ________: On the reflection principle in several complex variables. Proc. Amer. Math. Soc. *71*(1978), 26-28.

[22] ________: Biholomorphic mappings and the Bergman kernel off the diagonal. Inv. math.

NECESSARY CONDITIONS FOR HYPOELLIPTICITY OF THE $\bar{\partial}$-PROBLEM

Klas Diederich and Peter Pflug

J. J. Kohn proved in [1] the following theorem on hypoellipticity of the $\bar{\partial}$-problem:

THEOREM. *Let $\Omega \subset\subset \mathbf{C}^n$ be a smooth pseudoconvex domain. Suppose $\partial\Omega \cap V$ is real analytic for an open neighborhood V of a point $z \in \partial\Omega$ and $\partial\Omega \cap V$ does not contain any complex analytic variety of dimension q. Then for every $L^2_{p,q}$-form β on Ω with $\bar{\partial}\beta = 0$ there is an $L^2_{p,q-1}$-form α on Ω with*

1) $\bar{\partial}\alpha = \beta$ *and*

2) $W \cap \text{sing supp}\, \alpha \subset \text{sing supp}\, \beta$ *for some neighborhood W of z.*

REMARK. J. J. Kohn proves in fact, that the $\bar{\partial}$-Neumann-problem for (p,q)-forms is subelliptic under the conditions of the theorem which is even stronger than the stated hypoellipticity.

Kohn observed already in some special cases that the nonexistence of analytic varieties in the boundary is also necessary for hypoellipticity of $\bar{\partial}$. It was conjectured that this is true in general for pseudoconvex boundaries. In this note we shall show that the above assumption is always necessary for the hypoellipticity of the $\bar{\partial}$-problem; in detail we prove the following theorem applying an existence theorem of good L^2-holomorphic functions [3].

Recent Developments in Several Complex Variables
0-691-08285-5/81/000151-04 $00.50/0 (cloth)
0-691-08281-2/81/000151-04 $00.50/0 (paperback)

THEOREM.* *Let* Ω *be a* C^2*-smoothly bounded pseudoconvex domain in* C^n *and assume that every equation* $\bar{\partial}\alpha = \beta$ *with a* $\bar{\partial}$*-closed* $L^2_{p,q}$*-form* β *on* Ω $(1 \le q \le n-1)$ *can be solved by an* $L^2_{p,q-1}$*-form* α *with* sing supp $\alpha \subset$ sing supp β. *Then no complex analytic variety of dimension larger or equal to* q *can be contained in the boundary of* Ω.

Proof. The proof will be given by a contradiction argument. Therefore we assume that the boundary $\partial\Omega$ does contain a q-dimensional complex analytic variety M $(1 \le q \le n-1)$.

Without loss of generality we suppose that the coordinates are chosen such that one has:

1) $O \in M \subset \partial\Omega$ and O is a regular point of M,
2) the outer normal in O at $\partial\Omega$ points in the x_n-direction,
3) M is given as a holomorphic graph in $\tilde{V} \times V' \subset C^q \times C^{n-q}$:

$$M = \{(\tilde{z}, h_{q+1}(\tilde{z}), \cdots, h_n(\tilde{z})) : \tilde{z} = (z_1, \cdots, z_q) \in \tilde{V} = \tilde{V}(O) \subset C^q\},$$

4) $(z_1, \cdots, z_n) \xrightarrow{\phi} (z_1, \cdots, z_q, z_{q+1} - h_{q+1}(\tilde{z}), \cdots, z_n - h_n(\tilde{z}))$ is a biholomorphic map from $\tilde{V} \times V' = V$ onto $U = U(O) \subset C^n$.

Choose a C^∞-function χ on C^q with the following properties

1) $\chi \equiv 1$ for $|\tilde{z}| < \delta_1$,
2) $\chi \equiv 0$ for $|\tilde{z}| > \delta_2 > \delta_1$,

where the δ_j's are sufficiently small.

Using the boundary sequence $\phi^{-1}\left(0, \cdots, 0, -\frac{1}{\nu}\right)$ $(\nu >> 1)$ there exists an L^2-holomorphic function f on Ω which is unbounded on that sequence as was shown in [3]. Now we are able to define the following (q, q−1)-form α on $\Omega \setminus \{z : \tilde{z} = 0\}$

---

*The case $q = 1$ was also known before to D. Catlin with a slightly different argument. The general case has also been established by D. Catlin independently; it appears in these proceedings.

$$\alpha = \frac{(q-1)!}{(2\pi i)^q} \sum_{j=1}^{q} f(z)\chi(z_1,\cdots,z_q) \frac{\bar{z}_j}{\left(\sum_{v=1}^{q} z_\nu \bar{z}_\nu\right)^q} dz_1 \wedge d\bar{z}_1 \wedge \cdots \wedge dz_j \wedge \widehat{d\bar{z}_j} \wedge \cdots \wedge dz_q \wedge d\bar{z}_q .$$

We calculate

$$\bar{\partial}\alpha = \frac{(q-1)!}{(2\pi i)^q} \sum_{j=1}^{q} f(z) \frac{\partial \chi}{\partial \bar{z}_j}(z_1,\cdots,z_q) \frac{-\bar{z}_j}{\left(\sum_{\nu=1}^{q} z_\nu \bar{z}_\nu\right)^q} dz_1 \wedge d\bar{z}_1 \wedge \cdots \wedge dz_q \wedge d\bar{z}_q .$$

According to our choice of χ the form $\bar{\partial}\alpha$ is a $\bar{\partial}$-closed $L^2_{q,q}$-form on Ω which is smooth up to the boundary if $|\tilde{z}| < \delta_1$ or $|\tilde{z}| > \delta_2$. Hence by the assumed hypoellipticity of the $\bar{\partial}$-problem, there exists an $L^2_{q,q-1}$-form β on Ω with

1) $\bar{\partial}\beta = \alpha$

2) sing supp $\beta \subset$ sing supp α.

Using the biholomorphic map ϕ we can now study the $\bar{\partial}$-closed $(q,q-1)$-form $(\phi^{-1})^*(\alpha-\beta)$ on $\phi(V \cap \Omega)$:

$$(\phi^{-1})^*(\alpha-\beta) = \frac{(q-1)!}{(2\pi i)^q} \sum_{j=2}^{q} f \circ \phi^{-1}(w)\, \chi(w_1,\cdots,w_q) \frac{\bar{w}_j}{\left(\sum_{\nu=1}^{q} w_\nu \bar{w}_\nu\right)^q}$$

$$dw_1 \wedge d\bar{w}_1 \wedge \cdots \wedge dw_j \wedge \widehat{d\bar{w}_j} \wedge \cdots \wedge dw_q \wedge d\bar{w}_q - \beta^*$$

where the form β^* is uniformly bounded on $\phi(V \cap \Omega) \cap \{w : |\tilde{w}_j| < \delta_1$ or $|\tilde{w}| > \delta_2, |w| < \varepsilon\}$ for a certain $\varepsilon > 0$. Hence by Stokes theorem and Martinelli's integral formula we obtain the following equalities

$$\int_{|\tilde{w}|=2\delta_2} (\phi^{-1})^*(\alpha-\beta)\Big|_{\substack{w_{q+1}=0 \\ \vdots \\ w_{n-1}=0 \\ w_n=-\frac{1}{\nu}}} + \int_{|\tilde{w}|=(1/2)\delta_1} \beta^*\Big|_{\substack{w_{q+1}=0 \\ \vdots \\ w_{n-1}=0 \\ w_n=-\frac{1}{\nu}}} = f \circ \phi^{-1}\left(0,\cdots,0,-\frac{1}{\nu}\right) \qquad (\nu >> 1)$$

whose left sides remain bounded for $\nu \to \infty$ contradicting the choice of the function f which is unbounded along $\phi^{-1}(0, \cdots, 0, -1/\nu)$.

REMARK. As the result in [3] and the above proof show one can further weaken the smoothness assumption on $\partial\Omega$ in our theorem.

KLAS DIEDERICH AND PETER PFLUG
FACHBEREICH MATHEMATIK
GESAMTHOCHSCHULE WUPPERTAL
GAUSSSTR 20
D-5600 WUPPERTAL 1

REFERENCES

[1] Kohn, J. J.: Subellipticity of the $\bar{\partial}$-Neumann problem on pseudoconvex domains: Sufficient conditions; Acta math. 142 (1979), 79-121.

[2] Diederich, K. and Fornaess, J. E.: Pseudoconvex domains with real-analytic boundary; Ann. of Math. 107 (1978), 371-384.

[3] Pflug, P.: Quadratintegrable holomorphe Funktionen und die Serre-Vermutung; Math. Ann. 216 (1975), 285-288.

THE EDGE-OF-THE-WEDGE THEOREM FOR PARTIAL DIFFERENTIAL EQUATIONS

Leon Ehrenpreis

1. *Extension problems and methods of solution*

The simplest form of the edge-of-the-wedge theorem goes as follows: We consider complex variables $z_j = \xi_j + i\eta_j$, $j = 1, 2, \cdots, n$. Let Γ^+ be a proper convex cone in the imaginary (that is, η) space and set $\Gamma^- = -\Gamma^+$. Let $f^{\pm}$ be functions which are analytic in interior $R + i\Gamma^{\pm}$. Suppose $f^{\pm}$ have continuous boundary values on R which are equal. Then there is an entire function g whose restrictions to $R + i\Gamma^{\pm}$ are $f^{\pm}$.

The above form of the edge-of-the-wedge theorem can be sharpened in several ways:

(a) We can consider two different proper cones Γ_1, Γ_2 where Γ_2 is not necessarily equal to $-\Gamma_1$. We begin with f_j holomorphic on $R + i\Gamma_j$ with boundary values equal on R. The conclusion is that there exists g which is holomorphic on the convex hull of $(R+i\Gamma_1) \cup (R+i\Gamma_2)$ with $g = f_j$ on $R + \Gamma_j$.

(b) We do not need to require that f_j have continuous boundary values on R. The existence of boundary values in a much weaker sense suffices. Results of this nature are discussed in [3].

(c) (Martineau [8]) Instead of considering two functions f_j and two cones Γ_j we could consider several f_j and several Γ_j. We assume

Recent Developments in Several Complex Variables
0-691-08285-5/81/000155-15$00.75/0 (cloth)
0-691-08281-2/81/000155-15$00.75/0 (paperback)
For copying information, see copyright page.

that the sum of the values of f_j on R vanishes. The conclusion is that each f_j is of the form

(1) $$f_j = \sum_{k \neq j} g_{jk}$$

where g_{jk} is holomorphic in the tube domain over the convex hull of $\Gamma_j \cup \Gamma_k$ and $g_{ij} = -g_{ji}$.

We shall not consider Martineau's extension of the edge-of-the-wedge, though it appears that our methods can be used for such results.

The edge-of-the-wedge theorem is an example of the general *extension problem* for holomorphic functions: Let us be given a region Ω (that is, an open set plus part of its boundary) and a function f holomorphic on interior Ω and continuous on all of Ω. Find regions $\tilde{\Omega} \supset \Omega$ so that f has an extension to a function g which is holomorphic on interior $\tilde{\Omega}$, g continuous on $\tilde{\Omega}$.

In this form we can ask the question for the case where "holomorphic" is replaced by "solution of a system of differential equations"

(2) $$\vec{D}f = 0$$

that is

(3) $$D_1 f = D_2 f = \cdots = D_m f = 0 .$$

As we shall soon see, the hypothesis of continuity on Ω or on $\tilde{\Omega}$ will have to be modified in order to obtain meaningful results.

We shall consider regions $\Omega \subset R^n$. We use variables $x = (x_1, \cdots, x_n)$ in R^n with dual variables $\hat{x}$.

The first general results on this extension problem are due to Bochner [1]; these results are completed in [2], p. 326, Theorem 11.5. We studied the case $\Omega = \Omega_1 - \Omega_2$ where Ω_1 is an open bounded domain, and Ω_2 is a compact subset and the D_j have constant coefficients. Under suitable assumptions on Ω_1 (which are always satisfied if Ω_1 is convex) we can choose $\tilde{\Omega} = \Omega_1$ if the system $D_j f = 0$ is truly overdetermined, that is, the codimension of the complex algebraic variety $V : \hat{D}_j(\hat{x}) = 0$ is ≥ 2. Here $\hat{D}_j$ is the Fourier transform of D_j so $\hat{D}_j$ is a polynomial.

For the Cauchy-Riemann system in more than one complex variable this result is due to Hartogs, so we refer to our result as the general Hartogs extension theorem.

Some extensions to operators with analytic coefficients were obtained by Kawai [7].

There are two types of proof of the general Hartogs extension theorem and, as well as I can determine, these two types of proof are the basis for all proofs of extension theorems (including the edge-of-the-wedge theorem).

A. Extend f in some manner to some function $\tilde{f}$ on all of $\tilde{\Omega}$. Of course, $\tilde{f}$ will not necessarily satisfy $\vec{D}\tilde{f} = 0$. We compute $\vec{D}\tilde{f} = \vec{h}$. Thus $\vec{h}$ represents the "error" we made in choosing the extension $\tilde{f}$. We now correct $\tilde{f}$ by solving an inhomogeneous problem involving $\vec{h}$.

B. We pick a point $x^0 \in \tilde{\Omega} - \Omega$. We then find a formula of the form

$$g(x^0) = S_{x^0} \cdot g \tag{4}$$

which holds for all solutions on $\tilde{\Omega}$ of $\vec{D}g = 0$. Here S_{x^0} is a distribution of compact support on Ω. (4) is proven by finding suitable fundamental solutions of $\vec{D}$. Using (4) and the geometry of $\tilde{\Omega}$, Ω we prove that if $\vec{D}f = 0$ and g is defined by

$$g(x^0) = S_{x^0} \cdot f \tag{5}$$

for all $x^0 \in \tilde{\Omega}$ then $\vec{D}g = 0$.

In the present work we shall introduce a new method of extension:

C. *Parametrization Method.* We seek to parametrize solution of $\vec{D}f = 0$ by the values of f and some derivatives on a suitable subset P of Ω (or of $\tilde{\Omega}$). More precisely, we start with P and some differential operators δ_1 = identity, $\delta_2, \cdots, \delta_\ell$. We then consider the map

(6) $$\alpha : f \to (\delta_1 f|_P, \delta_2 f|_P, \cdots, \delta_\ell f|_P) \,.$$

To parametrize solutions of $\vec{D}f = 0$ on Ω means to describe $\{\alpha f\}$ in some reasonable way.

Of course, we can always choose $P = \Omega$, $\ell = 1$ which gives no information. The success of this method depends on describing $\{\alpha f\}$ without using $\vec{D}$. For this we want to take P as small as possible. In particular, if the D_j have constant coefficients it is best to take P with real dim P = complex dim V. (In [5] we consider the case $\dim P < \dim V$ in which case $\ell = \infty$.)

Examples of parametrization problems are the Cauchy, Dirichlet, and usual boundary value problems.

To prove the extension result we need to know that the parametrization data for solutions on Ω and on $\tilde{\Omega}$ are the same. For the usual edge-of-the-wedge theorem we shall choose $P = R$.

2. *Strongly free systems*

The first problem in finding suitable parametrization sets is to find sets P which are "close to" uniqueness sets for $\vec{D}$. We shall begin with the study of some systems for which we know how to find sets P. These are called *strongly free.*

We consider systems $\vec{D}$ for which the polynomials $\hat{D}_j$ are real, homogeneous, and the sets $\hat{D}_j(\hat{x}) = c_j$ are complete intersections for all choices of $\{c_j\}$. Examples of such $\hat{D}$ are generators for the ring of invariants of a finite reflection group (in which case $\dim V = 0$) (see e.g. [6]). Another example is the generators of the ring of reflections for the orthogonal group $\mathcal{O}(k)$ acting on $R^n = R^k \oplus \cdots \oplus R^k$ where $n \leqq k^2$. Other examples related to groups are found in Weyl's book [9].

(These examples do not contain the Cauchy-Riemann system because we require $\hat{D}_j$ to be real; this point will be clarified later.)

In case $\hat{D}_j$ generate the ring of invariants for a finite reflection group, then it is known (see [4]) that the set P of the form

$$P = \{x \mid \hat{D}_j(x) = c_j\} \tag{7}$$

form a uniqueness set for polynomial solutions of $\vec{D}f = 0$ for arbitrary constants $\{c_j\}$. It is proved in [4] that this uniqueness result is true for any strongly free $\{\hat{D}_j\}$. In fact, more is true: Instead of choosing the c_j in (7) to be constants, they can be any polynomials with degree $c_j <$ degree $\hat{D}_j$.

Some care must be made in the definition of P. For, some components of the algebraic variety P defined by (7) may be multiple. For such $\{c_j\}$ we must consider P (or some components of it) as defined together with certain differential operators $\{\partial_k\}$ to be a multiplicity variety in the sense of [2]. These operators are part of the intrinsic nature of P and are what we mean when we speak of the restriction $f|_P$. Their role is different from that of the operators δ_j. The δ_j measure the preponderance of C^∞ solutions in a neighborhood of P over entire solutions (see the Conjecture below).

One might be somewhat puzzled at the lack of symmetry between $\vec{D}f = 0$ and $\hat{D}_j(x) = c_j$. In the latter case we drop the homogeneity, while in the former we cannot. This can be seen from the simplest example where $n = 1$ and $D = d^2/dx^2$. If we consider solutions of

$$\left(\frac{d^2}{dx^2} + 1\right) f = 0 \tag{8}$$

then f can vanish on $x^2 = \pi^2$. (f cannot vanish on $x^2 = 0$ since that would mean $f(0) = f'(0) = 0$.)

The symmetry is reestablished by

THEOREM 1. *Let us consider entire solutions of*

$$(D_j + \lambda_j) f = 0 . \tag{9}$$

Here λ_j *is a real constant coefficient operator of order* $<$ *order* D_j. *The coefficients of* λ_j *are all* $< \varepsilon$. *Then* f *cannot vanish on a set of the form*

(10) $$P = \{x | \hat{D}_j(x) = c_j\}$$

where c_j *are polynomials with degree* $c_j <$ *degree* $\hat{D}_j$ *unless some coefficient of some* c_j *is* $> \beta\varepsilon^{-t}$. *Here* β *and* t *depend only on* $\vec{D}$.

REMARK 1. It is easy, using the Fischer inner product (see [4], [6]) to prove that α is surjective, that is, any function h on P which is the restriction to P of an entire function is the restriction of an entire solution of (9). In fact, when all coefficients of the c_j are sufficiently small, then α is a topological isomorphism in the "natural" topologies.

REMARK 2. When the λ_j are constants our result gives estimates for the smallest eigenvalue of the Dirichlet Problem with data on P.

Note that this result is clear for $m = 1$, $n = 1$, $D = d^2/dx^2$.

Note that the assumption that f be entire is important. Thus the function

(11) $$f = \frac{x_1}{1+x_1^2} - \frac{x_2}{1+x_2^2}$$

satisfies

(12) $$\frac{\partial^2 f}{\partial x_1 \partial x_2} = 0 .$$

However, f vanishes on $x_1 x_2 = 1$.

f is, in fact, real analytic, so it is only the regularity in the entire complex domain that accounts for the uniqueness property.

Of course, Theorem 1 has no relevance for the extension problem since it relates to entire solutions. However, much evidence supports the following conjecture:

CONJECTURE 1. We can find ℓ operators $\delta_1 = \mathrm{id}, \delta_2, \cdots, \delta_\ell$ so that the map α defined using P as in (10) is one-one on solutions f of $\vec{D}f = 0$ with $f \in C^\infty$ on Ω. Here Ω is any "cone" over P, that is,

(13) $$\Omega = \{x | \hat{D}_j(x) = \gamma_j c_j\}$$

where each γ_j lies in an interval of the form

$$1 \leqq \gamma_j \leqq M \tag{14}$$

or

$$1 \geq \gamma_j \geq M^{-1} \tag{15}$$

[that is, some γ_j satisfy (14) the others satisfy (15)].

We shall see later that Conjecture 1 does *not* hold for C^r solutions.

Edge-of-the-Wedge Problem. Suppose we are given two regions Ω_1, Ω_2 of the form (13) and functions f_j which are C^∞ on interior Ω_j and satisfy $\vec{D}f_j = 0$ there and such that $\delta_k f_j$ is continuous on Ω_j for all k. Suppose $\alpha f_1 = \alpha f_2$. Can we find a larger region $\tilde{\Omega} \supset \Omega_1 \cup \Omega_2$ and a solution g of $\vec{D}g = 0$ on $\tilde{\Omega}$ such that $g = f_j$ on Ω_j ?

In case the edge-of-the-wedge problem has a positive solution for suitably chosen Ω_1, Ω_2 we say that the *edge-of-the-wedge property* holds.

In what follows we shall restrict our consideration to the case $c_j = \text{constant}$ and $\lambda_j = 0$.

3. *The compact case*

Let us study the case in which P is compact. We start with two examples.

EXAMPLE 1. *Laplacian in the plane.* The simples nontrivial example of the above is for $n = 2$, $\ell = 1$, $D = \Delta = \partial^2/\partial x_1^2 + \partial^2/\partial x_2^2$. Call P the unit circle and set $\delta_1 = \text{id}$, $\delta_2 = r\partial/\partial r$. Let f_1 be harmonic in $M^{-1} < r < 1$ and f_2 harmonic in $1 < r < M$. We assume that f_j and $r\partial f_j/\partial r$ are continuous and equal on the unit circle.

Using Fourier series we write f_j in the form

$$f_j(r, \theta) = \sum a_j^k r^k e^{ik\theta} + \sum b_j^k r^{-k} e^{ik\theta}\,. \tag{16}$$

The assumption that f_1 is harmonic in $M^{-1} < r < 1$ means that

$$a_1^k = \mathcal{O}(\tilde{M}^{-|k|}) \quad \text{for} \quad k < 0$$

(17)

$$b_1^k = \mathcal{O}(\tilde{M}^{-|k|}) \quad \text{for} \quad k > 0 \, .$$

$\tilde{M}$ is any number $< M$.

Similarly, using the fact that f_2 is harmonic in $1 < r < M$ we find

$$a_2^k = \mathcal{O}(\tilde{M}^{-|k|}) \quad \text{for} \quad k > 0$$

(18)

$$b_2^k = \mathcal{O}(\tilde{M}^{-|k|}) \quad \text{for} \quad k < 0 \, .$$

We now use the fact that $\alpha f_1 = \alpha f_2$.

$$\delta_1 f(\theta) = \sum (a_1^k + b_1^k)\, e^{ik\theta}$$

$$= \sum (a_2^k + b_2^k)\, e^{ik\theta}$$

(19)

$$\delta_2 f(\theta) = \sum [k(a_1^k - b_1^k)\, e^{ik\theta}$$

$$= \sum [k(a_2^k - b_2^k)\, e^{ik\theta} \, .$$

This leads to the equations (for $k \neq 0$)

$$a_1^k + b_1^k = a_2^k + b_2^k$$

(20)

$$a_1^k - b_1^k = a_2^k - b_2^k$$

that is,

$$a_1^k = a_2^k$$

(21)

$$b_1^k = b_2^k$$

for $k \neq 0$. Using (17) this yields

$$a_j^k = \mathcal{O}(\tilde{M}^{-|k|}) \quad \text{all } k$$

(22)

$$b_j^k = \mathcal{O}(\tilde{M}^{-|k|}) \quad \text{all } k .$$

This means that the series (16) converges for $M^{-1} < r < M$ and defines a harmonic function g there which equals f_1 in $M^{-1} < r < 1$ and equals f_2 in $1 < r < M$.

This example shows us the true workings of the parametrization method. For we have good descriptions of αf_j and also of αg. Hence we can check directly that $\alpha f_1 = \alpha f_2$ implies that $\alpha(f_j)$ is the parametrization data of a function harmonic on $M^{-1} < r < M$.

EXAMPLE 2. *Holomorphic functions on the bidisc.* The simplest analog of the actual edge-of-the-wedge theorem concerns holomorphic functions of two complex variables $z_1 = x_1 + ix_3$ and $z_2 = x_2 + ix_4$. (The case $n > 2$ presents no new difficulties.) The polynomials $x_j + ix_{j+2}\,(j = 1, 2)$ corresponding to the Cauchy-Riemann system are not real so the ideas centering around Theorem 1 do not apply. Indeed, setting $z_1 = c_1$, $z_2 = c_2$ gives us a point in $R^4 = C^2$ which is certainly not a uniqueness set for holomorphic functions on C^2. Note however that if uniqueness holds on a set A then it holds for any set $\supset A$. We can thus replace complex polynomials $\hat{D}_j$ by $\hat{D}_j \overline{\hat{D}}_j$ in the definition of P and we could expect uniqueness (if the c_j are also real). Indeed this is the case. But the result is weaker than for real polynomials where we have existence and uniqueness. However, for the purpose of the parametrization problem we are most interested in the uniqueness property, so the non-existence does not bother us.

It is, therefore, reasonable to consider the parametrization problem for the Cauchy-Riemann system with data on the bicylinder $r_1 = r_2 = 1$. Here $r_j^2 = x_j^2 + x_{j+2}^2 = |z_j|^2$. For simplicity we choose $M = \infty$. Let f_1 be holomorphic in $\Omega_1 : r_1 < 1,\ r_2 < 1$ and f_2 holomorphic in $\Omega_2 : r_1 > 1,\ r_2 > 1$. $\alpha(f)$ is the restriction of f to the bicylinder. We assume f_1 and f_2 are continuous at $r_1 = r_2 = 1$ and that $\alpha f_1 = \alpha f_2$.

Edge-of-the-wedge theorem. There is a function g holomorphic for all z except $z_1 = 0$ or $z_2 = 0$ such that $g = f_j$ on Ω_j.

Let us analyze this theorem by means of the parametrization data. Any function h on $r_1 = r_2 = 1$ has a Fourier series

$$h(\theta_1, \theta_2) = \sum a^{k_1 k_2} e^{ik_1\theta_1 + ik_2\theta_2} . \tag{23}$$

The only possible analytic extension of h is

$$\tilde{h}(z_1, z_2) = \sum a^{k_1 k_2} r_1^{k_1} r_2^{k_2} e^{ik_1\theta_1 + ik_2\theta_2} . \tag{24}$$

For $\tilde{h}$ to be analytic in Ω_1, all the $a^{k_1 k_2}$ must be small when at least one of k_1, k_2 is negative. We see easily that

$$a_1^{k_1 k_2} = \mathcal{O}(\tilde{M}^{-|k_1|-|k_2|}) \tag{25}$$

for all $\tilde{M} > 0$ except on the set of k_1, k_2 satisfying

$$\min(k_1, k_2) \geq -\varepsilon \max(|k_1|, |k_2|) \tag{26}$$

that is, except in a cone strictly containing the first quadrant.

Similarly

$$a_2^{k_1 k_2} = \mathcal{O}(\tilde{M}^{-|k_1|-|k_2|}) \tag{27}$$

except in cones containing the third quadrant.

Now setting $\alpha f_1 = \alpha f_2$ we see that

$$a_1^{k_1 k_2} = a_2^{k_1 k_2} = \mathcal{O}(\tilde{M}^{-|k_1|-|k_2|}) \tag{28}$$

for all k_1, k_2. Thus

$$g = \sum a_j^{k_1 k_2} z_1^{k_1} z_2^{k_2} \tag{29}$$

is holomorphic off $\{z_1 = 0\} \cup \{z_2 = 0\}$ and $g = f_j$ on Ω_j.

The same method applies to establish an analogous result for pleuriharmonic functions on the bidisc (even though the equations for pleuriharmonic functions are not strongly free). For pleuriharmonic functions we can use $\alpha f = (f|_P, r_1\, \partial f/\partial r_1|_P)$.

We can also treat the equations

$$\frac{\partial^2 f}{\partial z_1 \partial \bar{z}_1} = \frac{\partial^2 f}{\partial z_2 \partial \bar{z}_2} = 0$$

on the bidisc. In this case

$$\alpha f = \left(f|_P, r_1 \frac{\partial f}{\partial r_1}\bigg|_P, r_2 \frac{\partial f}{\partial r_2}\bigg|_P, r_1 r_2 \frac{\partial^2 f}{\partial r_1 \partial r_2}\bigg|_P\right).$$

This, and other results we have checked for the compact orthogonal groups, leads us to make

CONJECTURE 2. If $\vec{D}$ is strongly free and if the kernel of $\vec{D}$ is invariant under a compact linear group G and P is an orbit of G then the edge-of-the-wedge property holds.

4. *The non-compact case*

We now pass to the case in which P is not compact. The lesson we learned from the previous section is that the parametrization method works well when P is the orbit of a real Lie group G and the kernel of $\vec{D}$ is invariant under G. For we can then use harmonic analysis on G. In the classical edge-of-the-wedge theorem $P = R^n$ and the underlying space is R^{2n}. It is thus reasonable to study the edge-of-the-wedge theorem by ordinary Fourier analysis.

We should note that R^n is not of the form P according to our previous descriptions of P as in (10). For, if we call $z_j = x_j + ix_{n+j} = x_j + iy_j$ then R^n is defined by $y_j = 0$ for all j. This is not of the form described in (10) (which was meant to apply only to real $\hat{D}_j$) or of the modified form discussed in Section 3 above. The point is that we may regard $\hat{D}_j - c_j$ in

(10) as a deformation of $\hat{D}_j$. This makes good algebraic sense because degree $c_j <$ degree $\hat{D}_j$. However, we can still have certain "deformations" in which degree $c_j =$ degree $\hat{D}_j$ but Theorem 1 still holds. This is because $\hat{D}_j$ is larger at infinity than c_j in a "sufficiently large" part of the infinite points. It would take us too far afield to discuss this here. In any case, it is clear that we have uniqueness on $y_1 = y_2 = \cdots = y_n = 0$ for the case of the Cauchy-Riemann system.

Similar remarks apply to the Cauchy Problem for any strongly free $\hat{D}$. (For example, we deform $x_1^2 + x_2^2$ into x_1^2. Since $x_1^2 = 0$ is the x_2 axis with multiplicity 2, this yields the usual Cauchy Problem on the x_2 axis.)

Let us start with the simplest form of the edge-of-the-wedge theorem, namely when $\Omega_j = \{x+iy\}$ where y is in a proper cone Γ_j with $\Gamma_1 = -\Gamma_2$. (Simple modifications apply to the general case.) Of course, we cannot take the Fourier transform of $h = \alpha(f_j)$ in the usual sense; we must use the generalized Fourier transform of [2]. In any case, the formulas (23)-(27) suggest that we attempt to represent h (thought of as αf_j) as Fourier transforms of $\hat{h}_j$ say where

(30) $$\hat{h}_j = \hat{u}_j + \hat{v}_j .$$

Here $\hat{u}_j$ is exponentially decreasing everywhere, so u_j is entire and

(31) $$\text{support } \hat{v}_j \subset \tilde{\Gamma}_j' .$$

Here $\tilde{\Gamma}_j'$ is a slightly larger cone than Γ_j' which is the dual of Γ_j. The cones Γ_j' are considered to be in the real part of the dual of R^n.

Assuming we knew this then we could proceed as follows: Since $\alpha f_1 = \alpha f_2$

(32) $$u_1 + v_1 = u_2 + v_2$$

that is, $v_1 - v_2$ is entire. We want to conclude that both v_1 and v_2 are entire for then $h = u_j + v_j$ is the Cauchy data of an entire function g whose restriction to Ω_j is clearly f_j.

THEOREM 2. *Let* Δ_1 *and* Δ_2 *be two proper cones in* R^n *with* $\Delta_1 \cap \Delta_2 = \{0\}$. *Let support* $\hat{v}_j \subset \Delta_j$ *and suppose that* $v_1 - v_2$ *is entire. Then both* v_1 *and* v_2 *are entire.*

The proof of Theorem 2 uses some simple properties of the operator of convolution by the Fourier transform of the characteristic function of Δ_j. For $n = 1$ this is just the Hilbert transform.

The proof of the edge-of-the-wedge theorem is completed by

THEOREM 3. *If* h_j *is the Cauchy data of a function analytic in* Ω_j *then we can write* h_j *as the Fourier transform of an* $\hat{h}_j$ *of the form* (30) *where* u_j *is entire and support* $v_j \subset \tilde{\Gamma}_j'$.

The proof of Theorem 2 relies heavily on the theory of sufficient sets studied in [2].

REMARK. At infinity Δ_1 and Δ_2 are very far apart, namely there is a positive angle between them. If Δ_1 and Δ_2 intersected at infinity (e.g. Δ_1, Δ_2 are complementary half-spaces) then no analog of Theorem 2 could hold. We are thus led to

PROBLEM. How far apart do sets Δ_1, Δ_2 (not necessarily cones) have to be at infinity in order that the conclusion of Theorem 3 hold?

We suspect that angular distance is not quite the right answer but cannot be modified too much.

As in the compact case we can use our method to handle pleuriharmonic functions and solutions of

$$\frac{\partial^2 f}{\partial x_1^2} + \frac{\partial^2 f}{\partial y_1^2} = \frac{\partial^2 f}{\partial x_2^2} + \frac{\partial^2 f}{\partial y_2^2} = 0 .$$

So much for the case $P = R^n$ and the group R^n. How about other groups? The simplest case where new phenomena occur is $G = S\mathcal{O}(1,2)$ acting on R^3. Using coordinates (t, x, y) in R^3, the orbits of G are:

(a) The origin.

(b) The positive and negative light cones $t^2 = x^2 + y^2$, $t > 0$ and $t < 0$.

(c) Half of hyperboloid of two sheets $t^2 - x^2 - y^2 = a^2$, $t > 0$ and $t < 0$.

(d) Hyperboloid of one sheet $t - x^2 - y^2 = -a^2$.

The G invariant operator is the wave operator

$$\square = \frac{\partial^2}{\partial t^2} - \frac{\partial^2}{\partial x^2} - \frac{\partial^2}{\partial y^2} .$$

As far as uniqueness goes, we can find C^k solutions of $\square f = 0$ which vanish together with $k-3$ derivatives on the light cone. k may be taken arbitrarily large. Thus Conjecture 1 is not true for C^k solutions; we conjecture it for C^∞ solutions.

Calling r the Minkowski distance, it is natural to call α the map

$$\alpha f = \left(f\,|_P,\, r \frac{\partial f}{\partial r}\Big|_P \right) .$$

For P one of the orbits of type (c) or (d) there is uniqueness. Thus we can pose an edge-of-the-wedge problem for such P.

We note that P of type (c) is space-like for the hyperbolic operator $\square$. Thus the edge-of-the-wedge property is trivial since any pair of C^∞ functions on P is the Cauchy data for a C^∞ solution in the whole interior of the light cone.

P of type (d) is not space-like for $\square$. In order to prove an edge-of-the-wedge property we must study Fourier analysis on P which is of the form G/A, A being the noncompact Cartan subgroup of G. If we proceed as in our treatment of the classical edge-of-the-wedge theorem, we find that the crucial aspect of Fourier analysis on G/A that enters is the discrete series of representations of G. In fact, it is the existence of the discrete series in G/A which accounts for the nonspace-like nature of P.

If we break the Fourier analysis on G/A into continuous and discrete parts then we can proceed as before and deduce the edge-of-the-wedge property.

This same type of analysis works for some finite dimensional representations of $S\mathcal{O}(1,n)$ in place of the standard representation of $S\mathcal{O}(1,2)$. A general theory is still lacking.

BIBLIOGRAPHY

[1] S. Bochner, Partial differential equations and analytic continuation, *Proc. Natl. Acad. Sci. 38* (1952), 227-230.

[2] L. Ehrenpreis, *Fourier Analysis in Several Complex Variables*, New York (Interscience), 1970.

[3] ———, Reflection, removable singularities, and approximation for partial differential equations, Part I, to appear in *Annals of Math.*

[4] ———, Harmonic functions and hyperbolic equations, to appear.

[5] ———, Some non-standard Cauchy Problems, to appear.

[6] L. Flatto, Finite reflection groups, to appear.

[7] T. Kawai, Extension of solutions of systems of linear differential equations, *Research Inst. Kyoto, 12* (1976), 215-227.

[8] A. Martineau, Théorèmes sur le prolongement analytique du type "Edge of the Wedge Theorem," Séminaire Bourbaki, 20-ième année, No. 340, 1967/68.

[9] H. Weyl, *The Classical Groups*, Princeton, 1939.

THE RADIAL DERIVATIVE, FRACTIONAL INTEGRALS, AND THE COMPARATIVE GROWTH OF MEANS OF HOLOMORPHIC FUNCTIONS ON THE UNIT BALL IN $\mathbb{C}^n$

Ian Graham*

A classical theorem of Hardy and Littlewood ([1, Theorem 5.12] or [5, Theorem 33]) states that if f is a holomorphic function on the unit disk D such that $f' \in H^p(D)$ for some $p < 1$ then

$$f \in H^{\frac{p}{1-p}}(D) .$$

The original proof involved various auxiliary results on the comparative growth of means of holomorphic functions (which have other applications) and relied heavily on Blaschke products. In [4] we considered an analogue of this theorem on the unit ball B_n in $\mathbb{C}^n$ in which the radial derivative

$$Rf = \sum_{j=1}^{n} z_j \frac{\partial f}{\partial z_j}$$

introduced by Rudin [8] played the role of the ordinary derivative in one variable. Results on the comparative growth of means of holomorphic functions were obtained using the fact that a continuous function on $\overline{B_n}$

*Research partially supported by the National Research Council of Canada.

Recent Developments in Several Complex Variables
0-691-08285-5/81/000171-08 $00.50/0 (cloth)
0-691-08281-2/81/000171-08 $00.50/0 (paperback)
For copying information, see copyright page.

which is plurisubharmonic on B_n is majorized by the Poisson-Szegö integral of its boundary values. In some cases sharp results were not obtained and one purpose of this paper is to note that these methods do yield sharp results if an argument of Flett [3] is incorporated.

The Hardy-Littlewood theorem can be stated more generally in terms of the fractional integrals of a holomorphic function and in fact this is done in [5]. The (Riemann-Liouville) fractional integral of order $\mu > 0$ of a holomorphic function on D may be defined by [5, p. 409]

$$f_\mu(z) = \frac{1}{\Gamma(\mu)} \int_0^z (z-u)^{\mu-1} f(u)\, du$$

where the integration is along the straight line from 0 to z and $(z-u)^{\mu-1}$ has its principal value when $u = 0$. The function f_μ is discontinuous on the negative real axis unless μ is an integer but is holomorphic on the remainder of the unit disk. If f has a radial boundary value at a boundary point $\zeta = e^{i\theta}$ then so does f_μ and it is given by

$$\begin{aligned} f_\mu(\zeta) &= \frac{1}{\Gamma(\mu)} \int_0^\zeta (\zeta-u)^{\mu-1} f(u)\, du \\ &= \frac{\zeta^\mu}{\Gamma(\mu)} \int_0^1 (1-r)^{\mu-1} f(r\zeta)\, dr\,. \end{aligned} \tag{1}$$

If we delete the factor ζ^μ which does not affect the size of f_μ and express $f(r\zeta)$ in terms of the Poisson(-Szegö) integral of its boundary values, which we may do if $f \in H^p(D)$ for some $p \geq 1$, we obtain the expression

$$\frac{1}{\Gamma(\mu)} \int_0^1 (1-r)^{\mu-1} \int_T \mathcal{P}(r\zeta, \eta) f(\eta)\, d\sigma(\eta)\, dr \tag{2}$$

using notation which generalizes readily to several variables. Here $\mathcal{P}(z, \eta) = \frac{1}{2\pi}\frac{1-|z|^2}{|1-z\bar{\eta}|^2}$ denotes the Poisson-Szegö kernel which coincides with the Poisson kernel on the disk except for a classical factor of π. If we change the order of integration in (2) we are led to formulate fractional integration theorems in terms of the integral operator on T with kernel $\int_0^1 (1-r)^{\mu-1}\mathcal{P}(r\zeta, \eta)\,dr$. Actually this approach seems to be of little use in one variable but in several variables it is very fruitful. Its relevance in the present context to proving the sharp form of Theorem 2 below was pointed out by Professor E. M. Stein [12].

The theorem on fractional integration has been generalized to various contexts. In R^n the analogous theorem is due to Sobolev (see [9, p. 119]). On the Heisenberg group it is due to Folland-Stein [2] for the case $p > 1$ and Krantz [7] for the case $0 < p \leq 1$. The results of Krantz of which I learned at this conference give an alternative proof of Theorem 2 and answer some of the questions posed in [4].

We shall write c_n for the numerical factor $\frac{(n-1)!}{2\pi^n}$. If f is holomorphic on the unit ball B_n in C^n then we define

$$M_p(r, f) = \left(c_n \int_{\partial B_n} |f(r\zeta)|^p d\sigma(\zeta)\right)^{\frac{1}{p}}$$

for $0 < r < 1$, $0 < p < \infty$, and

$$M_\infty(r, f) = \sup_{\|\zeta\|=1} |f(r\zeta)| .$$

For $0 < p \leq \infty$ we define

$$H^p(B_n) = [f \text{ holomorphic on } B_n | \sup_{r<1} M_p(r, f) < \infty\}$$

and denote the supremum in question by $\|f\|_p$. We shall sometimes apply the preceding definitions to functions which are not actually holomorphic.

The Poisson-Szegö kernel is denoted by

$$\mathcal{P}(z,\zeta) = c_n \cdot \frac{(1-\|z\|^2)^n}{|1-<z,\zeta>|^{2n}} .$$

If f is holomorphic in B_n then $A_\alpha f$ denotes the admissible maximal function of Koranyi [6].

We first collect the sharp results on the comparative growth of means.

THEOREM 1. *Suppose that* g *is a continuous complex-valued function on* B_n *such that* $|g|^p$ *is plurisubharmonic for all* p *which satisfy* $0 < p_0 < p < \infty$. *Suppose that* $M_p(r,g) \le \frac{C}{(1-r)^\beta}$ *where* C *is a positive constant and* β *is a nonnegative constant, and* p *lies in the indicated range. Then if* $p < q \le \infty$ *and* $\alpha = \frac{1}{p} - \frac{1}{q}$ *we have*

(i) $M_q(r,g) \le \frac{KC}{(1-r)^{\beta+n\alpha}}$.

(K *is a constant depending on* n, p, *and* β.)

(ii) *If* $\beta = 0$ *and* $\lambda \ge p$ *then*

$$\left(\int_0^1 (1-r)^{\lambda n\alpha-1} M_q(r,f)^\lambda dr\right)^{1/\lambda} \le D\|f\|_p .$$

(D *is a constant depending on* n, p, q, *and* λ.)

Remarks on the proof. Part (i) is proved in [4] using the fact that a continuous function on $\overline{B_n}$ which is plurisubharmonic on B_n is majorized by the Poisson-Szegö integral of its boundary values. Part (ii) follows from part (i) as indicated in [3, Theorem 1]. The argument which yields (ii) given in [3] uses the Marcienkiewicz interpolation theorem. This replaces the use of Blaschke products in the classical one-variable proof. A reduction to the case $\lambda = p > 1$ is made, and for this reason it is necessary to state the theorem in greater generality than simply for holomorphic functions. (If $f \in H^p(B_n)$ where $p < 1$ we need to consider $g = |f|^a$ where $a < p$.)

REMARK. Part (ii) of Theorem 1 together with the methods of [4] yield the following sharp result. However, we shall give a variation on the proof which was suggested by Stein [12].

THEOREM 2. *Suppose* f *is holomorphic in* B_n. *Suppose* $Rf \in H^p(B_n)$ *for some* $p < n$. *Then* $f \in H^{\frac{np}{n-p}}(B_n)$.

Proof. Fix r_0 such that $0 < r_0 < 1$. For $r_0 < r < 1$ we have

$$f(r\zeta) - f(r_0\zeta) = \int_{r_0}^{r} \frac{d}{ds} f(s\zeta)\, ds = \int_{r_0}^{r} \frac{1}{s} Rf(s\zeta)\, ds\ ,$$

hence

$$|f(r\zeta)| \leq |f(r_0\zeta)| + \frac{1}{r_0} \int_{r_0}^{r} |Rf(s\zeta)|\, ds$$

$$\leq |f(r_0\zeta)| + \frac{1}{r_0} \int_{0}^{1} |Rf(s\zeta)|\, ds\ .$$

Choosing $t < p$ we write

$$\int_0^1 |Rf(s\zeta)|\, ds = \int_0^1 |Rf(s\zeta)|^{1-t} |Rf(s\zeta)|^t\, ds$$

$$\leq (A_\alpha(Rf)(\zeta))^{1-t} \int_0^1 |Rf(s\zeta)|^t\, ds = \phi(\zeta)\psi(\zeta)$$

with an obvious notation. Thus $\phi \in L^{\frac{p}{1-t}}(\partial B_n)$ [10, Corollary to Theorem 10]. Writing $g(\zeta) = |Rf(\zeta)|^t$, we have

$$\int_0^1 |Rf(s\zeta)|^t ds \le \int_0^1 \int_{\partial B_n} \mathcal{P}(s\zeta, \eta)\, g(\eta)\, d\sigma(\eta)\, ds = \int_{\partial B_n} K(\zeta, \eta)\, g(\eta)\, d\sigma(\eta)$$

where $K(\zeta, \eta) = \int_0^1 \mathcal{P}(s\zeta, \eta)\, ds \simeq \dfrac{1}{|1-<\zeta, \eta>|^{n-1}}$. We now apply Theorem 15.11 of [2] (note that the ambient space has dimension $n+1$ in that paper) to conclude that the integral operator with kernel K maps $L^p(\partial B_n)$ to $L^q(\partial B_n)$ where $\frac{1}{q} = \frac{1}{p} - \frac{1}{n}$ and $1 < p < q < \infty$. We apply this theorem with p replaced by p/t and conclude that $\psi \in L^{q_1}(\partial B_n)$ where $\frac{1}{q_1} = \frac{t}{p} - \frac{1}{n}$. (If in the hypotheses of the theorem we have $Rf \in H^p(B_n)$ with $p > 1$ then we may take $t = 1$ and apply Theorem 15.11 of [2] directly.) Now from Hölder's inequality we obtain

$$\int_{\partial B_n} (\phi(\zeta)\psi(\zeta))^q\, d\sigma(\zeta) \le \left(\int_{\partial B_n} \phi(\zeta)^{qP} d\sigma(\zeta)\right)^{1/P} \left(\int_{\partial B_n} \psi(\zeta)^{qQ}\, d\sigma(\zeta)\right)^{1/Q}$$

with $q = \dfrac{np}{n-p}$, $P = \dfrac{n-p}{n(1-t)}$, $Q = \dfrac{n-p}{nt-p}$. (P and Q are conjugate exponents.) It follows that $f \in H^{\frac{np}{n-p}}(B_n)$.

The foregoing methods yield the following analogue of the fractional integration theorem on the ball:

THEOREM 3. *For* $0 < \mu < n$ *let* K_μ *be the kernel on* $\partial B_n \times \partial B_n$ *given by* $K_\mu(\zeta, \eta) = \int_0^1 (1-r)^{\mu-1}\, \mathcal{P}(r\zeta, \eta)\, dr$. *Then the mapping*

$$f \mapsto \int_{\partial B_n} K_\mu(\cdot, \eta)\, f(\eta)\, d\sigma(\eta)$$

sends $L^p(\partial B_n)$ *to* $L^q(\partial B_n)$ *where* $\frac{1}{q} = \frac{1}{p} - \frac{\mu}{n}$ *and* $1 < p < q < \infty$.

Proof. We observe that $K_\mu(\zeta, \eta) \simeq \dfrac{1}{|1-\langle\zeta, \eta\rangle|^{n-\mu}}$ and apply Theorem 1.5.11 of [2].

To complement Theorem 2 we have

PROPOSITION. *If* f *is holomorphic on* B_n *and* $Rf \in H^p(B_n)$ *for some* p, $n < p < \infty$ *then* $f \in \Gamma_{1-n/p}(B_n)$. (*The Lipschitz space* $\Gamma_\alpha(B_n)$ *is defined in* [11].)

Proof. We have

$$|Rf(z)|^p \le \int_{\partial B_n} \mathcal{P}(z,\zeta)|Rf(\zeta)|^p \, d\sigma(\zeta) \le \frac{C}{(1-\|z\|)^n} \|Rf\|_p{}^p .$$

Thus

$$|Rf(z)| \le \frac{C^{1/p}\|Rf\|_p}{(1-\|z\|)^{n/p}} .$$

It follows that each slice function f_ζ defined by $f_\zeta(\tau) = f(\tau\zeta)$ where $\zeta \in \partial B_n$ belongs to $\Lambda_{1-n/p}(D)$ with a uniform Lipschitz constant. This is obtained by combining the observation of Rudin [8] that $Rf(\tau\zeta) = \tau f_\zeta'(\tau)$ with a theorem of Hardy and Littlewood [1, Theorem 5.1]. By the main result of [8] it follows that $f \in \Gamma_{1-n/p}(B_n)$.

REMARKS. This result has also been obtained by Krantz [7]. Krantz likewise shows that $Rf \in H^n(B_n)$ implies $f \in$ BMOA, thus answering a question posed in [4].

PROBLEM. Give a counterexample in dimension $n > 2$ to show that $Rf \in H^n(B_n)$ does not imply $f \in H^\infty(B_n)$. (A counterexample is known for dimension $n = 2$ only [4].)

Note added November 1979:

Professor Joseph A. Cima has drawn my attention to some work of K. T. Hahn and J. Mitchell which is described in J. Mitchell, Representations of linear functionals on H^p spaces over bounded homogeneous domains in $\mathbb{C}^n$, P.S.P.M. *27*(1975), part 2, 363-372. This work, which deserves to be better known, contains results which are similar to the theorem in reference [4] below and to Theorem 1 of the present paper. It also contains an identification of the dual space of $H^p(B_n)$ for $0 < p < 1$, a result which was obtained by a different method in J. B. Garnett and R. H. Latter, The atomic decomposition for Hardy spaces in several variables, Duke Math. J. *45*(1978), 815-845. See also [13].

REFERENCES

[1] P. Duren, Theory of H^p Spaces. Academic Press, New York, 1970.

[2] G. B. Folland and E. M. Stein, Estimates for the $\bar{\partial}_b$ complex and analysis on the Heisenberg group. Comm. Pure Appl. Math. *27*(1974), 429-522.

[3] T. M. Flett, On the rate of growth of mean values of holomorphic and harmonic functions. Proc. London Math. Soc. *20*(1970), 749-768.

[4] I. Graham, An H^p-space theorem for the radial derivative of holomorphic functions on the unit ball in $\mathbb{C}^n$. University of Toronto preprint.

[5] G. H. Hardy and J. E. Littlewood, Some properties of fractional integrals, II. Math. Z. *34*(1932), 403-439.

[6] A. Koranyi, Harmonic functions on hermitian hyperbolic space. Trans. Amer. Math. Soc. *135*(1969), 507-516.

[7] S. G. Krantz, Analysis on the Heisenberg group and estimates for functions in Hardy classes of several complex variables. To appear.

[8] W. Rudin, Holomorphic Lipschitz functions in balls. Comment. Math. Helv. *53*(1978), 143-147.

[9] E. M. Stein, Singular integrals and differentiability properties of functions. Princeton University Press, Princeton, N. J. 1970.

[10] ————, Boundary behavior of holomorphic functions of several complex variables. Princeton University Press, Princeton, N. J. 1972.

[11] ————, Singular integrals and estimates for the Cauchy-Riemann equations. Bull. Amer. Math. Soc. *79*(1973), 440-445.

[12] ————, private communication.

[13] K. T. Hahn and J. Mitchell, Representation of linear functionals in H^p spaces over bounded symmetric domains in $\mathbb{C}^n$. J. Math. Anal. Appl. *56*(1976), 379-396.

STABILITY PROPERTIES OF THE BERGMAN KERNEL AND CURVATURE PROPERTIES OF BOUNDED DOMAINS

R. E. Greene* and Steven G. Krantz*

Little information about the Bergman kernel of a bounded domain in $\mathbb{C}^n$ or about the associated metric is deducible directly from their definitions. However, in recent years considerable information has nevertheless been obtained about their behavior near the boundary of C^∞ strictly pseudoconvex domains (e.g. [9], [11], and [5]; cf. also [3]). In view of the boundary behavior's being well understood, it is natural to inquire about the interior behavior as well. To try to treat such inquiries in entire generality seems to pose great difficulties. The purpose of the present paper is to present some results on interior behavior having to do with the restricted issue of stability of the kernel and metric behavior under (small) perturbations of the boundary of the domain.

The perturbations to be considered will be small in the C^∞ or C^k topology: If D is a domain in $\mathbb{C}^n$ with $C^\infty(C^k)$ boundary and $\mathfrak{U}$ a neighborhood of the injection $\iota : \partial D \to \mathbb{C}^n$ in the $C^\infty(C^k)$ topology of maps of ∂D to $\mathbb{C}^n$, then by definition a $C^\infty(C^k)$ neighborhood of D in the set of domains in $\mathbb{C}^n$ will be $\{D' \subset \mathbb{C}^n | D'$ is $C^\infty(C^k)$ and $\partial D' = h(\partial D)$ for some $h \in \mathfrak{U}\}$. An equivalent idea would be obtained by considering the

*Research supported by the National Science Foundation (both authors) and a Sloan Foundation Fellowship (Greene).

Recent Developments in Several Complex Variables
0-691-08285-5/81/000179-20$01.00/0 (cloth)
0-691-08281-2/81/000179-20$01.00/0 (paperback)

C^∞ or C^k topology on defining functions of the domains. The (finite) number of derivatives needed to define the appropriate neighborhood $\mathcal{U}$ can, in all the cases to be discussed, be determined explicitly; but for brevity most of the results are given in the unspecified C^∞ form.

The Bergman kernel of the unit ball B in $\mathbb{C}^n$, $K(z,w) = c_n/(1-z\cdot\overline{w})^{n+1}$, $c_n = n!\pi^{-n}$, has no zeroes on $B \times B$, and indeed is bounded away from zero. In [14], the question was raised of the absence of zeroes of the kernel for a general domain. The first example of a domain for which the kernel has zeroes was given in [17], and in [18] it was shown that for all nonsimply connected C^∞ domains in $\mathbb{C}$ there are zeroes. Examples of C^∞ strictly pseudoconvex domains in $\mathbb{C}^n$ for all n for which the kernel has zeroes can be obtained by considering smooth interior approximations of $D \times \Delta^{n-1}$, D being any domain in $\mathbb{C}$ the kernel of which has zeroes and Δ being the unit disc in $\mathbb{C}$. A suitable set of approximations can be obtained for instance as sublevel level sets of a strictly plurisubharmonic exhaustion of $D \times \Delta^{n-1}$. If all the approximations had a kernel without zeroes then so would the kernel of $D \times \Delta^{n-1}$ on $(D\times\Delta^{n-1}) \times (D\times\Delta^{n-1})$ by Hurwitz' Theorem (see [17, Theorem 3] and [8, p. 178]). However, no example of a simply connected domain for which the kernel has zeroes seems to be known. General results on nonvanishing would be of interest in terms of function theory on manifolds ([8, p. 177 ff.]). The following theorem implies in particular nonvanishing for perturbations of the unit ball.

THEOREM 1. *If* D_0 *is a* C^∞ *strictly pseudoconvex domain such that, for some positive number* c, $|K_{D_0}(z,w)| \geq c$ *for all* $z, w \in D_0$ *and if* $\lambda \in [0,1)$, *then there is a* C^∞ *neighborhood* $\mathcal{O}$ *of* D_0 *such that for any* $D \in \mathcal{O}$

$$|K_D(z,w)| \geq \lambda c \quad \textit{for all} \quad z, w \in D .$$

The outline of the proof of Theorem 1 is first to note that if z and w are simultaneously near each other and near the boundary of D_0 then $|K_{D_0}(z,w)|$ is large by the expansion of K given in [5]. On the other hand, the kernel has a smooth extension ([11]) to $\bar{D}_0 \times \bar{D}_0 - \{(z,z) | z \in \partial D_0\}$ ($\bar{\ }$ denotes closure). It is reasonable to expect this extension to be stable under perturbation of D_0 away from the diagonal, e.g., on $\{(z,w) | |z-w| \geq \xi > 0\}$. If this expectation is verified and if the size of the neighborhood of the boundary diagonal on which $|K(z,w)|$ is large is also stable under perturbation, then the proof is complete. The detailed verification is carried out in a later section.

The asymptotic expansion of the Bergman kernel ([5]) implies that the Riemannian sectional curvature of the Bergman metric of a C^∞ strictly pseudoconvex domain is negative near the boundary: in fact, the curvature tensor of the metric converges as the boundary is approached to the curvature tensor of the Bergman metric of the ball, i.e., to the curvature tensor

$$R(X,Y,Z,W) = -\frac{1}{n+1}\{g(X,Z)\,g(Y,W) - g(X,W)\,g(Y,Z) + g(X,JZ)\,g(Y,JW) - g(X,JW)\,g(Y,JZ) + 2g(X,JY)\,g(Z,JW)\}$$

of the Kähler metric g of constant negative holomorphic sectional curvature $-4/(n+1)$ ([12]). Thus for any $\varepsilon > 0$ the Riemannian sectional curvature lies between $-\frac{4}{n+1} - \varepsilon$ and $\frac{-1}{n+1} + \varepsilon$ at points sufficiently near the boundary. Although on any C^∞ strictly pseudoconvex domain there is a complete Kähler metric of negative, bounded from zero, holomorphic sectional curvature [12], there are C^∞ strictly pseudoconvex domains on which there is no complete Kähler metric of nonpositive Riemannian sectional curvature. A domain with such a metric would necessarily have universal cover (real) diffeomorphic to R^{2n} by the Cartan-Hadamard Theorem ([13, p. 102]). A tubular neighborhood of a totally real compact submanifold of dimension greater than one is an example of a C^∞ strictly pseudoconvex domain that does not have this covering property and hence has no such metric. The following theorem yields large families of examples of complete Kähler manifolds with negative Riemannian sectional curvature, bounded in fact away from zero (for a discussion of function theory on such manifolds, see [8]). The families, all sufficiently small deformations of the ball in C^n, are in fact, in the sense of biholomorphic inequivalence, infinite dimensional ([4]). Previously, the only known example not biholomorphic to the ball of a complete simply-connected Kähler manifold with Riemannian sectional curvature negative, bounded away from zero, was the example in [15]. (It is not known at present whether this example is biholomorphic to a domain in C^n.)

THEOREM 2. *For each* $n = 1, 2, \cdots,$ *there is a* C^∞ *neighborhood* $\mathcal{O}_n$ *of the unit ball* B *in* C^n *such that on each domain* $D \in \mathcal{O}_n$ *there is a complete Kähler metric with Riemannian sectional curvature* $\leq -1+\varepsilon$ *and* $\geq -4-\varepsilon$ *everywhere on* D.

For $n \geq 2$, the curvature restrictions are the strongest constant bounds possible: if the Riemannian sectional curvature is ≥ -4 and ≤ -1 then the holomorphic sectional curvature is constant (e.g. [13, p. 167]) and the (simply connected) domain would be necessarily biholomorphic to the ball. This remark is a special case of the fact that if the Riemannian sectional curvature of a Kähler metric is between $-4-\varepsilon$ and $-1+\varepsilon$ and if ε is small, then the curvature tensor is nearly that of constant holomorphic sectional curvature -4. Specifically, a Kähler manifold's having δ Riemannian sectional curvature pinching implies that $\delta < 1/4$ and that the holomorphic sectional curvature is $\delta(8\delta+1)(1-\delta)^{-1}$ pinched. Then the formula for Riemannian sectional curvature in terms of holomorphic sectional curvature and the corresponding formula for the Riemann curvature tensor in terms of sectional curvature imply, since $\delta(8\delta+1)(1-\delta)^{-1}$ is near 1 if δ is near $1/4$, that the curvature tensor is near the tensor of constant holomorphic sectional curvature. This argument shows also that if a Kähler metric has nearly constant (nonzero) holomorphic sectional curvature then its curvature tensor is close to the curvature of constant holomorphic sectional curvature (see [1]; also [12; note 23] for further remarks). The proof of Theorem 2 actually shows essentially directly that the whole curvature tensor of the metric constructed on D is close to the curvature tensor of constant negative holomorphic sectional curvature -4

(in particular the Proposition (*) in the proof could be phrased in terms of the behavior of the whole curvature tensor).

Proof of Theorem 2. Only the cases $n \geq 2$ need be considered. Assume for the moment the following auxiliary Proposition (*). (In this proposition, the term $-\log(2-\text{dis}(z,\partial D))$ is present for technical convenience. Since it and its derivatives are bounded near ∂D, it is easy to check that the proposition would still be true were this term omitted.)

(*) *There is a* C^∞ *neighborhood* $\mathcal{W}_n$ *of* B *and a positive number* δ *such that if* $D \in \mathcal{W}_n$ *and if* D_δ *is defined to be* $\{z \in D \mid \text{dis}(z,\partial D) < \delta\}$, *then: the function* $z \to \text{dis}(z,\partial D)$, *where* dis = *Euclidean distance, is* C^∞ *on* D_δ; *the form*

$$\sum_{i,j} \frac{\partial^2}{\partial z_i \partial \bar{z}_j} [-\log \text{dis}(z,\partial D) - \log(2-\text{dis}(z,\partial D))]\, dz_i \otimes d\bar{z}_j$$

is positive definite on D_δ; *and the Riemannian sectional curvatures of the metric determined by this form are* $\geq -4-\varepsilon$ *and* $\leq -1+\varepsilon$.

Now let $\rho: \mathbf{R} \to \mathbf{R}$ be a C^∞ function such that $\rho(x) = 0$ if $x \leq \delta/2$, $\rho(x) = 1$ if $x \geq \delta$ and $0 \leq \rho(x) \leq 1$ for all x. Set $\rho_D(z) = \rho(\text{dis}(z,\partial D))$, $z \in D$. Then ρ_D is C^∞ on D if $D \in \mathcal{W}_n$. If D is sufficiently C^∞-close to B then on $\{z \in D \mid \delta/2 \leq \text{dis}(z,\partial D) \leq \delta\}$ the function $1-|z|$ is positive and C^∞ and is C^∞ close to the function $z \to \text{dis}(z,\partial D)$.

Define $h: D \to \mathbf{R}$ by

$$\begin{aligned} h(z) = {} & (1-\rho_D(z))[-\log \text{dis}(z,\partial D) - \log(2-\text{dis}(z,\partial D))] \\ & + \rho_D(z)[-\log(1-|z|^2)]\,. \end{aligned}$$

Note that the metric $\sum_{i,j} \partial^2/\partial z_i \partial \bar{z}_j [-\log(1-|z|^2)]\, dz_i \otimes d\bar{z}_j$ on the unit ball has constant holomorphic sectional curvature -4 and hence Riemannian sectional curvature between -4 and -1. The metric $\sum_{i,j} (\partial^2 h/\partial z_i \partial \bar{z}_j)\, dz_i \otimes d\bar{z}_j$ is a metric on D with the required properties.

It satisfies the curvature restrictions on $\{z \in D \mid \mathrm{dis}(z, \partial D) < \delta/2\}$ because of (*) and the fact that h agrees with $-\log \mathrm{dis}(z, \partial D) - \log(2 - \mathrm{dis}(z, \partial D))$ there; on $\{z \in D \mid \mathrm{dis}(z, \partial D) > \delta\}$ because $\rho \equiv 1$ there so that h is $-\log(1 - |z|^2)$ there; and on $\{z \in D \mid \delta/2 \leq \mathrm{dis}(z, \partial D) \leq \delta\}$ because there $z \to (2 - \mathrm{dis}(z, \partial D)) \cdot \mathrm{dis}(z, \partial D)$ is C^∞ close to $1 - |z|^2$ so that h is C^∞ close to $-\log(1 - |z|^2)$ there.

The verification of proposition (*) is essentially a direct calculation. The calculation is closely related to that in [12], which shows that the curvature tensor of the metric $\sum \frac{\partial^2(\log \psi)}{\partial z_i \partial \bar{z}_j} dz_i \otimes d\bar{z}_j$ converges to the tensor of constant holomorphic sectional curvature -4 for any strictly plurisubharmonic defining function ψ for a C^∞ strictly pseudoconvex domain D. But it is easier to approach the present case directly than to check that the rate of convergence deduced in [12] has the required stability probability.

Let p_0 be a point of ∂D (where D is C^∞-close to B) and p be a point close to p_0 and on the (real) interior normal N_{p_0} to ∂D at p_0. Choose an orthonormal linear coordinate system $(z_1, \cdots, z_n)$ such that $p_0 = (1, 0, \cdots, 0)$ and $\frac{\partial}{\partial x_1} = N_{p_0}$. Set $\phi(z) = (\mathrm{dis}(z, \partial D)) \cdot (2 - \mathrm{dis}(z, \partial D))$. This function is C^∞ on a neighborhood in $\bar{D}$ of ∂D. Set $g_{i\bar{j}} = \frac{\partial^2}{\partial z_i \partial \bar{z}_j}(-\log \phi(z)) = \phi_i \phi_{\bar{j}} \phi^{-2} - \phi_{i\bar{j}} \phi^{-1}$. For D = B so that $\phi(z) = 1 - |z|^2$, $g_{i\bar{j}} = \bar{z}_i z_j (1 - |z|^2)^{-2} + \delta_{ij}(1 - |z|^2)^{-1}$ so at p the matrix $(\phi^2 g_{i\bar{j}})$, $i, j = 1, \cdots, n$, is diagonal with diagonal entries $x_1^2 + \phi, \phi, \cdots, \phi$. If D is C^∞-close to B then ϕ is C^∞-close to the function $1 - |z|^2$. By rotation in the $z_2, \cdots, z_n$ coordinates, $(\phi_{i\bar{j}})$, $i, j = 2, \cdots, n$, can be taken to remain diagonal. Then, $(\phi^2 g_{i\bar{j}})$, $i, j = 1, \cdots, n$, for D has the form

$$\begin{bmatrix} (\sim 1) & (\sim 0)\phi & \cdots & (\sim 0)\phi \\ (\sim 0)\phi & (\sim 1)\phi & & 0 \\ \vdots & \ddots & 0 & \\ \vdots & 0 & \ddots & \vdots \\ (\sim 0)\phi & 0 & & (\sim 1)\phi \end{bmatrix}$$

where: (a) the notation (~ 0) or (~ 1) means a quantity which is uniformly close (in the C^0 sense) near ∂D to 0 (or 1, respectively), the difference from 0 or 1 being an amount which can be made arbitrarily small if D is sufficiently C^∞-close to B and if p is sufficiently close to p_0 and (b) only the terms of lowest order in ϕ are considered, so that for instance $x_1^2 + \phi$ for B becomes (~ 1) for D. The metric $G = \sum \frac{\partial^2(-\log \phi)}{\partial z_i \partial \bar{z}_j} dz_i \otimes d\bar{z}_j$ thus has the property $G \geq \left(\frac{\sim 1}{\phi}\right) E$, if E = the Euclidean metric of C^n. Furthermore,

$$(\phi^2 g_{i\bar{j}})^{-1} = \begin{bmatrix} (\sim 1) & (\sim 0) & \cdots & (\sim 0) \\ (\sim 0) & (\sim 1)\frac{1}{\phi} & & \\ \vdots & & \ddots & (\sim 0) \\ \vdots & (\sim 0) & & \ddots \\ (\sim 0) & & & (\sim 1)\frac{1}{\phi} \end{bmatrix}$$

The curvature tensor of the metric $2G$ (the computation is done for $2G$ for consistency with the standard notations of [13]) is given by

$$R_{a\bar{b}c\bar{d}} = g_{a\bar{b}}g_{c\bar{d}} + g_{a\bar{d}}g_{c\bar{b}} - \frac{1}{\phi^2}(\phi\phi_{a\bar{b}c\bar{d}} - \phi_{ac}\phi_{\bar{b}\bar{d}})$$
$$- \frac{1}{\phi^4}\sum_{e,f} g^{f\bar{e}}(\phi\phi_{ac\bar{e}} - \phi_{ac}\phi_{\bar{e}})(\phi\phi_{\bar{b}\bar{d}f} - \phi_{\bar{b}\bar{d}}\phi_f) .$$

Now if X has unit 2G-length then $E(X, X)$ and hence also $E(JX, JX)$ are $\leq (\sim 2)\phi$. Thus for any 4-tensor T, $T(X, JX, X, JX)$ is bounded by $(2n)^4\phi^2 \cdot$ (maximum coefficient of T in Euclidean coordinates). In particular, every term in the Euclidean coordinate expression for R just given which has order of magnitude less than equal to $\frac{1}{\phi}$ as $\phi \to 0$ makes a contribution to sectional curvature of order ϕ or higher order in ϕ as $p \to \partial D$ (i.e., as $\phi \to 0^+$). Combining the formula for R and the information given about $(g^{i\bar{j}})$ yields that

$$R_{a\bar{b}c\bar{d}} = g_{a\bar{b}}g_{c\bar{d}} + g_{a\bar{d}}g_{c\bar{b}}$$
$$+ \left(\frac{\sim 1}{\phi^2}\right)\phi_{ac}\phi_{\bar{b}\bar{d}} - \left(\frac{\sim 1}{\phi^2}\right)\phi_{ac}\phi_{\bar{b}\bar{d}}$$
$$+ (\sim 0)\begin{pmatrix}\text{polynomial expressions in} \\ \phi \text{ and its derivatives}\end{pmatrix}$$
$$+ \text{terms of order } \leq (\sim 0)\phi^{-2} .$$

Since $g_{a\bar{b}}g_{c\bar{d}} + g_{a\bar{d}}g_{c\bar{b}}$ is the curvature tensor of constant holomorphic section curvature -2 ([13; p. 169]), the curvature tensor of $G = \frac{1}{2}(2G)$ is C^0 close to the curvature tensor of holomorphic sectional curvature $2(-2) = -4$ near ∂D. All estimates have the required uniformity in variation of D near B and Proposition (*) follows.

If care had been taken to distinguish those terms which were small because D is close to B from those which were small because p is close to ∂D then the result of [12] that R converges as $p \to \partial D$ to the tensor of constant holomorphic sectional curvature would have been recovered. This result is not actually needed here. However, the argument just given offers a slightly variant approach to the convergence result of [12] for arbitrary D: namely, to prove the result directly for domains with boundary given by quadratic polynomials and then deduce the general case by approximation.

In Theorem 2, the choice of the neighborhood $\mathcal{O}_n$ depends on ε. This dependence is necessary:

THEOREM 3. *If* D *is a simply connected complex manifold on which for every* $\varepsilon > 0$ *there is a complete Kähler metric with Riemannian sectional curvature* $\geq -4-\varepsilon$ *and* $\leq -1+\varepsilon$, *then* D *is biholomorphic to the unit ball* B.

To prove this result, suppose that $\{G_\ell | \ell = 1, 2, \cdots\}$ is a sequence of complete Kähler metrics on D with the Riemannian sectional curvature of G_ℓ greater than or equal to $-4-\frac{1}{\ell}$ and less than or equal to $-1+\frac{1}{\ell}$ for each ℓ. Pick a point O_D in D and let O_B = the origin in $\mathbb{C}^n$. For each ℓ, choose an isometry I_ℓ of the real tangent space at O_B with the metric $G = \sum_{i,j} -\frac{\partial^2(\log(1-|z|^2))}{\partial z_i \partial \bar{z}_j} dz_i \otimes d\bar{z}_j$ to the real tangent space at O_D with the metric G_ℓ such that $I_\ell \circ J = J \circ I_\ell$, where the J's are the complex structure tensors. Define maps $h_\ell : B \to D$ by $h_\ell(z) = \exp_{O_D,\ell}(I_\ell(\exp_{O_B}^{-1}(z)))$ where $\exp_{O_B}$ is the G-exponential map at O_B in B and $\exp_{O_D,\ell}$ is the G_ℓ-exponential map at O_D in D. By the Cartan-Hadamard Theorem, each h_ℓ is a diffeomorphism.

The generalized Schwarz Lemma of [20] shows that there is a constant C such that $G_k \leq CG_m$ for all k, m (C independent of k, m). Also, because for all ℓ the curvature of G_ℓ is ≥ -5 and ≤ 0, by the usual comparison theorems there is for each $r > 0$ a constant C_r such that h_ℓ and h_ℓ^{-1} on the G-ball of radius r are Lipschitz continuous with Lipschitz constant $\leq C_r$ relative to the G-metric on B and the G_ℓ-metric on D (C_r is independent of ℓ). Consequently, the families $\{h_\ell\}$ and

$\{h_\ell^{-1}\}$ are Lipschitz continuous on B_r with Lipschitz constant $< CC_r$ relative to the G-metric on B and the G-metric on D. Thus there is a subsequence $\{h_{\ell_k} | h = 1, 2, \cdots\}$ which converges uniformly on compact subsets of B to an invertible mapping h which is Lipschitz continuous on compact sets with an inverse that is also Lipschitz continuous on compact sets.

It is now to be shown that h is holomorphic. For this purpose, note that the differential of h_ℓ at a point $p \in B$ can be obtained as follows: If $c : [0, r] \to B$ is the unique arc length parameter geodesic from O_B to p, then $dh_\ell(c'(r))$ = the (unit) tangent of the geodesic from O_D to $h_\ell(p)$. If X is a tangent vector to B at p with $G(X, c') = 0$, then there is a unique Jacobi J field along c such that $J(0) = 0$ and $J(r) = X$. J is induced by a family $C_t : [0, r] \to B$ of arc-length parameter geodesics with $c_t(0) = 0$ and $\frac{d}{dt} c_t(r) = X$; $dh_\ell(X)$ is the end-value of the Jacobi field along the geodesic $s \to \exp_{O_D, \ell} s(I_n c')$ induced by the variation $\tilde{c}_t(s) = \exp_{O_D, \ell} s(I_n c'_t)$. Thus dh_ℓ is determined by the solutions of the Jacobi equation for G_ℓ. If G_ℓ had constant holomorphic sectional curvature -4 then $dh_\ell \circ J - J \circ dh_\ell$ would be zero. By the continuous dependence (in the C^0 topology on solutions and on curvature tensors) of the solutions of the Jacobi equation on the curvature tensor and by the observations following Theorem 2 on the Kähler curvature tensors satisfying Riemannian sectional curvature bounds, $\sup_{\{X : G(X,X)=1\}} \|(dh_\ell \circ J) X - J \circ dh_\ell X\|$ goes to 0 uniformly on compact subsets of B. It follows that $\bar{\partial} h$ is zero in the distribution sense on B so that h is a holomorphic, invertible map from B to D.

In view of Theorem 2, it is natural to ask whether the Bergman metrics of perturbations of the unit ball have the property of having nearly the same curvature behavior as the Bergman metric of the ball. The following theorem asserts that this is the case, but it does not supplant Theorem 2. Its proof is not elementary; and its proof requires, at least at the present stage of knowledge of the Bergman kernel, closeness of D to the ball in a much larger number of derivatives than the only four derivatives required in Theorem 2. (The proof as given of Theorem 2 shows that in fact C^4 closeness of D to B is sufficient for the conclusion.)

THEOREM 4. *For each* $\varepsilon > 0$ *and each* $n = 1, 2, \cdots,$ *there is a* C^∞ *neighborhood* $\mathcal{O}_n$ *of the unit ball* B *in* C^n *such that if* $D \in \mathcal{O}_n$ *then the Bergman metric of* D *is a complete metric with Riemannian sectional curvature* $\geq -\frac{4}{n+1} - \varepsilon$ *and* $\leq -\frac{1}{n+1} + \varepsilon$.

COROLLARY. *If* D *is sufficiently* C^∞ *close to* B, *then any compact subgroup of the automorphism group of* D *has a fixed point. In particular, the automorphism group of every* C^∞ *strictly pseudoconvex domain which is sufficiently* C^∞ *close to the ball but not biholomorphic to the ball has a fixed point.*

Not every action of a compact Lie group on a manifold diffeomorphic to Euclidean space has a fixed point [6]; thus the corollary is not a consequence of general group action properties. The first statement of the corollary follows from the fact that a compact group of isometries of a complete simply-connected Riemannian manifold of nonpositive curvature has a fixed point ([13; p. 111]). The second statement of the corollary follows from the first together with the fact that every C^∞ strictly pseudoconvex domain with a noncompact automorphism group is biholomorphic to the ball ([19]; cf. [12]). The corollary implies that if two Reinhardt domains are C^∞ close to the ball but not biholomorphic to it, then any biholomorphic map between them is a linear map: such a map must preserve the origin, the origin being the unique fixed point of the automorphism group of such domains, and is hence linear by a well-known theorem of H. Cartan. (See [7] for further details and for references to related general results on Reinhardt domains.)

Theorem 4 is a consequence of the stability near the boundary of the asymptotic expansion of the kernel function as given in Theorem 5.

Specifically, Theorem 4 and the proof of Proposition (*) together imply that there is a C^∞ neighborhood $\mathcal{O}_n$ of B and a positive number $\delta < 1$ such that the curvature of the Bergman metric is $\geq -\frac{4}{n+1} - \varepsilon$ and $\leq -\frac{1}{n+1} + \varepsilon$ on $\{z \in D \mid \mathrm{dis}(z, \partial D) < \delta\}$ for $D \in \mathcal{O}_n$. (The existence of stable curvature estimates of this type can also be used to prove that the set of C^∞ strictly pseudoconvex domains with no nontrivial automorphisms is C^∞ open: see [7].) On the other hand, by choosing a smaller $\mathcal{O}_n$ if necessary (with δ fixed) it can be arranged that the compact set $\left\{z \in B \mid \mathrm{dis}(z, \partial B) \geq \frac{\delta}{8}\right\}$ is contained in every D in $\mathcal{O}_n$ and that, for every $D \in \mathcal{O}_n$, $\{z \in D \mid \mathrm{dis}(z, \partial D) \geq \delta/2\} \subset \{z \in B \mid \mathrm{dis}(z, \partial B) \geq \delta/8\}$. It is elementary ([16]) to see that $K_D(z, z)$ is C^∞ close to $K_B(z, z)$ on $\{z \in B \mid \mathrm{dis}(z, \partial B) \geq \delta/8\}$. This fact can be established directly: The appropriate argument is given later as a portion of the proof of Theorem 1. It follows that on $\{z \in D \mid \mathrm{dis}(z, \partial D) \geq \delta/2\}$ the curvature of the Bergman metric is C^0 close (actually, C^∞ close) to that of the ball, and the theorem is proved. (Some further details concerning the calculations to establish the required analogue of Proposition (*) will be discussed after the statement of Theorem 5.)

To state Theorem 5, some notation following [5] will be used: Suppose D_0 is a C^∞ strictly pseudoconvex domain and ψ_0 is a C^∞ defining function for D_0, i.e., a C^∞ function with $D_0 = \{z \mid \psi_0(z) > 0\}$, $\partial D_0 = \{z \mid \psi_0(z) = 0\}$, with grad $\psi_0(p) \neq 0$ for each point $p \in \partial D$, and with

$$-\sum_1^n \frac{\partial^2 \psi_0}{\partial z_i \partial \bar{z}_j} dz_i \otimes d\bar{z}_j > 0 \text{ at each } p \in \partial D \text{ (this last property is possible}$$

by strict pseudoconvexity). Set

$$X(z, w) = \psi_0(w) + \sum_1 \frac{\partial \psi_0}{\partial w_j}\bigg|_w (z_j - w_j) + \frac{1}{2} \sum_{j,k} \frac{\partial^2 \psi_0}{\partial w_j \partial w_k}\bigg|_w (z_j - w_j)(z_k - w_k).$$

Then according to [5; p. 9] there exist C^∞ functions $\phi_0, \tilde{\phi}_0$ such that

$$K(z, w) = \phi_0(z, w) X^{-(n+1)}(z, w) + \tilde{\phi}_0(z, w) \log X(z, w)$$

for z, w close together and near the boundary (log here denotes the principal branch on C^1 with the negative real axis removed). The stability property of this expansion is essentially that ϕ_0 and $\tilde{\phi}_0$ are stable in the C^∞ topology. It is convenient for the precise statement to use the norm defined on $C^\infty(C)$, where C is a compact set in C^n: if $f \in C^\infty(C)$ then by definition

$$\|f\|_{C,\infty} = \sup_C |f| + \sum_{\ell=1}^{\infty} \left[(4n)^{-\ell} \sum_{|\alpha|=\ell} \max\left(1, \sup_C |D_1^{\alpha_1} \cdots D_{2n}^{\alpha_{2n}} f|\right)\right]$$

where α is a $2n$ multi-index with nonnegative entries and $D_1, \cdots, D_{2n}$ are $\frac{\partial}{\partial x_1}, \frac{\partial}{\partial y_1}, \cdots, \frac{\partial}{\partial x_n}, \frac{\partial}{\partial y_n}$. The topology on $C^\infty(C)$ determined by this (finite-valued) norm is of course equivalent to the C^∞ topology in $C^\infty(C)$.

THEOREM 5. *If D_0 is a C^∞ strictly pseudoconvex domain in C^n and ψ_0 is a defining function for D_0, if $\phi_0, \tilde{\phi}_0$ are the functions in the expansion of K_{D_0}, and if ε is a positive number, then there is a C^∞ neighborhood $\mathcal{O}_n$ of D_0 and a positive number δ such that for each $D \in \mathcal{O}_n$ there is a choice of ψ, ϕ, and $\tilde{\phi}$ for the expansion of K_D and a diffeomorphism $F : \bar{D} \to \bar{D}_0$ with*

$$\|F - I\|_{\bar{D},\infty} < \varepsilon$$

$$\|\psi - \psi_0\|_{C,\infty} < \varepsilon$$

$$\|\phi - (\phi_0 \circ F)\|_{E,\infty} < \varepsilon$$

$$\|\tilde{\phi} - (\tilde{\phi}_0 \circ F)\|_{E,\infty} < \varepsilon$$

where

$$C = \{z \in C^n \,|\, \mathrm{dis}\,(x, \partial D_0) \le \delta\}$$

$$E = \{(z, w) \in \bar{D} \times \bar{D} \,|\, |z-w| + \mathrm{dis}\,(z, \partial D) + \mathrm{dis}\,(w, \partial D) \le \delta\}$$

and I is the identity map of D to itself.

The proof of Theorem 5 will be given in [7]. The restriction of the estimates on $\psi-\psi_0$, $\phi-\phi_0$, and $\tilde{\phi}-\tilde{\phi}_0$ to points near the boundary of D_0 is of course not essential; but the statement as given carries all essential information, since the C^∞ interior stability of the kernel is clear, as remarked earlier.

In the calculation for the Bergman metric analogue of Proposition (*), the formula for the curvature tensor of $2\sum \frac{\partial^2 \log K(z,z)}{\partial z_i \partial \bar{z}_j} dz_i \otimes d\bar{z}_j$ is

$$R_{a\bar{b}c\bar{d}} = g_{a\bar{b}} g_{c\bar{d}} + g_{a\bar{d}} g_{c\bar{b}} - \frac{1}{K^2}(KK_{a\bar{b}c\bar{d}} - K_{ac}K_{\bar{b}\bar{d}})$$
$$+ \frac{1}{K^4} \sum_{\bar{e},f} g^{f\bar{e}}(KK_{ac\bar{e}} - K_{ac}K_{\bar{e}})(KK_{\bar{b}\bar{d}f} - K_{\bar{b}\bar{d}}K_f).$$

Since $X(z,z) = \psi(z)$, direct calculation of the metric coefficients $g_{a\bar{b}} = \frac{\partial^2 \log K}{\partial z_i \partial \bar{z}_j}$ shows that the $g_{a\bar{b}}$ have order $\frac{1}{\psi}$ and as before terms in R of order less than $\frac{1}{\psi^2}$ can be neglected. The reasoning then continues as in Proposition (*), uniformity of the estimates following from Theorem 4 (applied to $D_0 = B$, $\psi_0(z) = -1 + |z|^2$, $\phi_0 = \frac{n!}{\pi^n}$, $\tilde{\phi}_0 \equiv 0$).

§2. *The proof of Theorem 1*

The bound of $K(z,w)$ away from zero, the existence of which is the content of Theorem 1, is to be established in three parts:

Case I: z, w in D, both bounded away from ∂D.

Case II: z, w bounded away from each other.

Case III: z, w close to ∂D and to each other.

Of course, Cases I and II are not disjoint; but this overlapping is irrelevant and the establishing of Case I in the form indicated is of independent interest. As noted, Case III is disposed of by the stability of the asymptotic expansion demonstrated in [7]. Arguments will now be given for the first two cases.

Case I is comparatively elementary. The precise result that will be shown is:

(*) If ε is a positive number and C is a compact subset of D_0, then there is a C^∞ neighborhood $\mathcal{O}$ of D_0 with the property that if $D \in \mathcal{O}$ then $\sup_{z,w \in C \cap D} |K_D(z,w) - K_{D_0}(z,w)| < \varepsilon$.

For proving Theorem 1, this statement will be applied in the case where $\mathcal{O}$ is chosen so small that $C \subset D$ for every $D \in \mathcal{O}$. The proof of (*) will use the standard relationship between the Bergman kernel and the extremizing functions: For any bounded domain D, let $A^2(D)$ = the L^2 holomorphic functions on D, the L^2 norm being denoted by $\| \;\|_D$. For any $p \in D$, there is a unique L^2 holomorphic function $f_{p,D}$ such that $f_{p,D}(p) = 1$ and $\|f_{p,D}\|_D \le \|g\|_D$ for any $g \in A^2(D)$ with $g(p) = 1$. Also, $f_{p,D}(z) = K(z,p)/K(p,p)$, $z \in D$ and $\|f_{p,D}\|^{-1} = K(p,p)$. Now suppose for the moment that:

(**) Given $\eta > 0$, there exist domains D_1, D_2 with $C \subset D_1 \subset \bar{D}_1 \subset D_0 \subset \bar{D}_0 \subset D_2$ such that for each $p \in C$, $\|f_{p,D_1} - f_{p,D_2}\|_{D_1} < \eta$ and $\|f_{p,D_2}\|_{D_2 - D_1} < \eta$.

Then for any $p \in C$ and any D with $D_1 \subset D \subset D_2$, $\|f_{p,D_1}\|_{D_1} \le \|f_{p,D}\|_{D_1} \le \|f_{p,D}\|_D \le \|f_{p,D_2}\|_D \le \|f_{p,D_2}\|_{D_2}$. Since

$$\|f_{p,D_2}\|_{D_2} = \|f_{p,D_2}\|_{D_1} + \|f_{p,D_2}\|_{D_2 - D_1}$$

$$\le \|f_{p,D_1}\|_{D_1} + 2\eta ,$$

$$0 \le \|f_{p,D}\|_{D_1} - \|f_{p,D_1}\|_{D_1} < 2\eta .$$

Thus $|K_{D_1}(p,p) - K_D(p,p)|$ is small uniformly for $p \in C$ if η is small. The function f_{p,D_1} is the minimum norm element in the affine hyperplane $\{f \in A^2(D_1) | f(p) = 1\}$ in $A^2(D_1)$; so f_{p,D_1} is perpendicular to this hyperplane.

Thus that $\|f_{p,D}\|_{D_1} - \|f_{p,D_1}\|_{D_1} < 2\eta$ implies by the Pythagorean theorem that

$$\|f_{p,D} - f_{p,D_1}\|_{D_1} \leq (\|f_{p,D}\|^2_{D_1} - \|f_{p,D_1}\|^2_{D_1})^{1/2}$$

$$\leq (2\|f_{p,D_1}\| + 2\eta)^{1/2}(2\eta)^{1/2} .$$

Hence $f_{p,D_1}(z)$ and $f_{p,D}(z)$ are uniformly (in z and p) close together for $z, p \in C$ if η is small. The same conclusion consequently holds for $K_{D_1}(p,p)f_{p,D_1}(z) = K_{D_1}(z,p)$ and $K_D(p,p)f_{p,D} = K_D(z,p)$. Thus statement (**) implies statement (*): It is only necessary to pick $\mathcal{O}$ such that $D \in \mathcal{O}$ implies $D_1 \subset D \subset D_2$ for suitably chosen D_1, D_2; then both K_D and K_{D_0} will be close uniformly on $C \times C$ to K_{D_1} and hence to each other.

To prove statement (**), first note that it is a standard result (e.g., [16]) that there is a domain D_1 with $\bar{D}_1 \subset D_0$ such that $\|f_{p,D_1} - f_{p,D_0}\|_{D_1}$, $\|f_{p,D_0}\|_{D_1} - \|f_{p,D_1}\|_{D_1}$ and $\|f_{p,D_0}\|_{D_0 - D_1}$ are small (nonnegative) numbers. To find D_2, choose (e.g., by [10]; Theorem 1.4.1) a domain G with $\bar{D}_0 \subset G$ such that the holomorphic functions in $L^2(G)$ are $L^2_{D_0}$-dense in the holomorphic functions in $L^2(D_0)$. Since the set $\{f_{p,D_0} | p \in C\}$ is compact in $L^2(D_0)$, there is a compact set $\{\hat{f}_p | p \in C\}$ of holomorphic functions in $L^2(G)$ with $\hat{f}_p(p) = 1$ and $\|f_{p,D_0} - \hat{f}_p\|_{D_0}$ small. By choosing $D_2 \subset G$ and D_2 near to D_0 but with $\bar{D}_0 \subset D_2$, it can be arranged by the compactness of $\{\hat{f}_p | p \in C\}$ that $\|\hat{f}_p\|_{D_2 - D_0}$ is small for all $p \in C$. In particular $\|\hat{f}_p\|_{D_2}$ is near $\|f_{p,D_0}\|_{D_0}$. Since $\|f_{p,D_0}\|_{D_0} \leq \|f_{p,D_2}\|_{D_2} \leq \|\hat{f}_p\|_{D_2}$, $\|f_{p,D_2}\|_{D_2}$ is close to $\|\hat{f}_p\|_{D_2}$. So f_{p,D_2} is L^2 close to $\hat{f}_p$ on D_2 (again by the Pythagorean Theorem). Similarly, f_{p,D_2} is L^2 close to f_{p,D_0} on D_0. Thus $\|f_{p,D_1} - f_{p,D_2}\|_{D_1}$ is close to $\|f_{p,D_1} - \hat{f}_p\|_{D_1}$ and hence to $\|f_{p,D_0} - \hat{f}_p\|_{D_1}$, which is small, and $\|f_{p,D_2}\|_{D_2 - D_1}$ is close to the quantity $\|\hat{f}_p\|_{D_2 - D_0} + \|f_{p,D_0}\|_{D_0 - D_1}$, which is small. Thus statement (**) is established.

It is worth noting that the arguments just given hold in great generality: The construction of an exterior approximating domain D_2 used only the property of D_0 that for each element of $A^2(D_0)$ and $\varepsilon > 0$, there is a larger domain G an element f_1 of $A^2(G)$ with $\|f - f_1\|_{D_0} < \varepsilon$. The construction of an interior approximating domain D_1 is valid for any bounded domain. In particular, this second construction justifies the Hurwitz theorem argument summarized earlier in the discussion of the LuQu'eng conjecture. And the two constructions together justify the interior stability argument needed for Theorem 1.

To dispose of Case II, it is enough to show that:

(***) If τ is a positive number, then there are a C^∞ neighborhood $\mathcal{O}$ of D_0 and positive numbers η and M such that for any $D \in \mathcal{O}$ and any length 1 vector field X on D,

$$\sup \{ |XK_D(z, w)| : z, w \in D, |z-w| > \tau, \operatorname{dis}(z, \partial D) < \eta \} < M$$

(where the action of X is on the first variable).

(By conjugate symmetry of $K_D(z, w)$, a corresponding estimate holds for first derivatives in the second variable if $\operatorname{dis}(w, \partial D) < \eta$, $|z-w| > \tau$.) The required Case II bound of $K_D(z, w)$ from zero then follows easily from Case I: for a fixed small $\varepsilon > 0$, $\varepsilon < \eta/2$, and for a suitable choice of $\mathcal{O}$, $K_D(z, w)$ is by Case I bounded appropriately from zero for z, w in the set $\{z | \operatorname{dis}(z, \partial D_0) \geq \varepsilon\}$. Moving z and/or w out from this set to arbitrarily near ∂D will involve motions of length at most (say) 2ε if D is close to D_0, so that the resulting change in $K_D(z, w)$ is at most $2\varepsilon M + 2\varepsilon M$. Here ε is to have been chosen so small that if $D \in \mathcal{O}$ and if $z, w \in D$ with $|z-w| > 2\tau$ and with z, w within distance ε of ∂D, then there are points z_0, w_0 of $\{z | \operatorname{dis}(z, \partial D_0) = \varepsilon\}$ with: (1) $|z-z_0| < \varepsilon$, $|w-w_0| < \varepsilon$; (2) every point of the line segment from z_0 to z having distance at least τ from every point of the line segment from w_0 to w;

and (3) both these line segments lying in the set $\{z \in D | \mathrm{dis}(z, \partial D) < \eta\}$. Such a choice of $\mathcal{O}$ and ε is possible by elementary geometry. Then the Case II bound of $K_D(z, w)$ away from zero holds for $|z-w| > 2\tau$.

To check statement (***), the method introduced in [11] will be used. For the present purpose, the method of the main body of [11], rather than the alternative methods discussed at the conclusion of that paper, is more appropriate. First note that by an appropriate preliminary choice of $\mathcal{O}$ (with further restriction on $\mathcal{O}$ to be made later) and a choice of smaller τ if necessary, involving no loss of generality, it can and will be assumed that for any $z_0 \in \partial D$ $\{z : |z-z_0| = \tau\}$ meets ∂D transversally and the intersection $D \cap \bar{B}(z_0, \tau)$ of D with the closed ball around z_0 of radius τ is diffeomorphic to $\left\{x \in R^{2n} | x_{2n} \geq 0, \sum x_j^2 \leq 1\right\}$. Then by choosing a smaller $\mathcal{O}$ if necessary, it can be arranged that the constants in the Sobolev estimates can be chosen uniformly in z_0 and D. Precisely, there is a constant c such that for any $D \in \mathcal{O}$ and any $z_0 \in \partial D$

$$(\dagger) \qquad \sup_{B(z_0, \tau/8)} |Xg| \leq c\|g\|_{H_{n+2}(B(z_0,\tau/4))}$$

where X is any length one vector field and g is any element of the Sobolev space $H_{2n+1}(B(z_0, \tau/4))$.

Now following [11], let $F : C^n \to R$ be a nonnegative C^∞ function with $F(z) = 1$ if $|z| < \frac{1}{3}$, $F(z) = 0$ if $|z| > \frac{1}{2}$; and let $\tilde{F} : C^n \to R$ be a nonnegative C^∞ function with $\tilde{F}(z) = 0$ if $|z| < \frac{1}{4}$, $\tilde{F}(z) = 1$ if $\frac{1}{3} < |z| < \frac{1}{2}$ and $\tilde{F}(z) = 0$ if $|z| > \frac{3}{4}$. For fixed $w \in D$, set $\phi_w(z) = F((z-w)/\min(\frac{\tau}{4}, \mathrm{dis}(w, \partial D)))$ and $\psi_w(\tau) = \tilde{F}((z-w)/\min(\frac{\tau}{4}, \mathrm{dis}(w, \partial D)))$. Let $\Gamma_w(\tau) = c_n|z-w|^{2-2n}$ be the fundamental solution for the Laplacian on R^{2n}. Then

$$K_D(z, w) = P_D(\phi_w \Delta(\psi_w \Gamma_w)),$$

where P is the orthogonal projection $L^2(D) \to A^2(D)$.

To estimate $|XK_D(z, w)|$ on $B(z_0, \frac{\tau}{8})$ it is enough by the (uniform) Sobolev estimate already given to estimate $K_D(z, w)_{H_{n+2}(B(z_0, \frac{\tau}{4}))}$. For this purpose, recall that

$$P_D g = g - \mathfrak{D} N \bar{\partial} g$$

for any $g \in L^2(D)$, where N is the Neumann operator and $\mathfrak{D}$ is the formal adjoint of $\bar{\partial}$ on (0, 1) forms, i.e. $\sum_j f_j \, d\bar{z}_j \to -\sum_j \frac{\partial f_j}{\partial z_j}$. Thus on $B(z_0, \tau/4)$

$$K(z, w) = \mathfrak{D} N \bar{\partial}(\phi_w \Delta(\psi_w \Gamma_w)) ,$$

since $\phi_w \Delta(\psi_w \Gamma_w) = 0$ on $B(z_0, \tau/2)$. Now the pseudolocal property of the Neumann operator N implies that

$$(\dagger\dagger) \qquad \|K_D(z, w)\|_{H_{n+2}(B(z_0,\tau/4))} \le c_0 \|\phi_w \Delta(\psi_w \Gamma_w)\|_{-2n-2}$$

where $\| \ \|_{-2n-2}$ is the $H_{-2n-2}(D)$ Sobolev norm. Furthermore the constant C_0 can be chosen uniformly for all $z_0 \in \partial D$ and $D \in \mathcal{O}$ (for a sufficiently small $\mathcal{O}$). This somewhat delicate stability property for the Neumann operator is treated in detail in [7].

It is straightforward using the argument in [11] to check that the norm $\|\phi_w \Delta(\phi_w \Gamma_w)\|_{-2n-2}$ is uniformly bounded in $w \in D$ and $D \in \mathcal{O}$ for suitable choice of $\mathcal{O}$. This fact and the estimates (†) and (††) yield statement (***).

The argument just given for Case II in fact easily yields a more refined conclusion, namely that the Bergman kernel is stable away from the diagonal in the C^∞ topology under small (in the C^∞ sense) perturbations of the domain. The only additional information needed to draw this conclusion is the appropriate stability of the Neumann operator. Further details are given in [7].

Cases I and II having been established and Case III being a consequence of Theorem 5, the proof of Theorem 1 is complete.

REFERENCES

[1] Berger, M., Pincement riemannian et pincement holomorphe, Ann. Scuola Norm. Sup. Pisa 14(1960), 151-159.

[2] Bishop, R. L. and Goldberg, S., On the 2nd cohomology group of a Kaehler manifold of positive curvature, Proc. Amer. Math. Soc. 16 (1965), 119-122.

[3] Boutet de Monvel, L. and Sjostrand, J. Sur la singularité des noyaux de Bergman et de Szego, Soc. Mat. de France, Astérisque 34-35 (1976), 123-164.

[4] Burns, D., Shnider, S. and Wells, R. O., On deformations of strictly pseudoconvex domains, *Invent. Math.* 46 (1978), 237-253.

[5] Fefferman, C., The Bergman kernel and biholomorphic mappings of pseudoconvex domains, Invent. Math. 26 (1974), 1-65.

[6] Floyd, E. E. and Richardson, R. W. An action of a finite group on an n-cell without stationary points, Bull. Amer. Math. Soc. (1959), 73-76.

[7] Greene, R. E. and Krantz, S., Deformations of complex structures, estimates for the $\bar{\partial}$ equation, and stability of the Bergman kernel *Advances in Math.*, to appear.

[8] Greene, R. E. and Wu, H., Function Theory on Manifolds Which Possess a Pole, Lecture Notes in Math. 699, Springer-Verlag, 1979.

[9] Hörmander, L., L^2 estimates and existence theorems for the $\bar{\partial}$ operator. Acta Math. 113 (1965), 89-152.

[10] Kerzman, N., Hölder and L^p estimates for solutions of $\bar{\partial}u = f$ in strong pseudoconvex domains, Comm. Pure App. Math. XXIV (1971), 301-380.

[11] ———, The Bergman kernel function. Differentiability at the boundary. Math. Ann. 195 (1972), 149-158.

[12] Klembeck, P., Kähler metrics of negative curvature, the Bergman metric near the boundary and the Kobayashi metric on smooth bounded strictly pseudoconvex sets, Indiana Univ. Math. Jour. 27, (1978), No. 2, 275-282.

[13] Kobayashi, S. and Nomizu, K., *Foundations of Differential Geometry*, Vol. II, Interscience-Wiley, New York, 1969.

[14] Lu Qi-Keng (= K. H. Look), On Kähler manifolds with constant curvature, Acta. Math. Sinica 16 (1966), 269-281 (Chinese) (= Chinese Math. 9 (1966), 283-298).

[15] Mostow, G. D. and Siu, Y. T., A compact Kähler surface of negative curvature not covered by the ball, *Ann. Math.* 112 (1980), 321-360.

[16] Ramadanov, I., Sur une propriétie de la fonction de Bergman, C. R. Acad. Bulgare des Sci. 20 (1967), 759-762.

[17] Skwarczynski, M., The distance in the theory of pseudo-conformal transformations and the Lu Qi-Keng conjecture, Proc. Amer. Soc. 22 (1969), 305-310.

[18] Suita, N. and Yamada, A., On the Lu Qi-Keng conjecture, Proc. Amer. Math. Soc. 59 (1976), 222-224.

[19] Wong, B., Characterization of the Unit Ball in $\mathbb{C}^n$ by its Automorphism Group, Invent. Math. 41 (1977), No. 3, 253-257.

[20] Yau, S. T., A general Schwarz lemma for Kähler manifolds, Amer. J. Math. Vol. 100 (1978), No. 1, 197-204.

GLOBALE HOLOMORPHE KERNE ZUR LÖSUNG DER CAUCHY-RIEMANNSCHEN DIFFERENTIALGLEICHUNGEN

Michael Hortmann

Abstract

On a strictly pseudoconvex subset of a Stein-Manifold we construct a global holomorphic integral-kernel for solving the $\bar{\partial}$-equation.

§0. *Einleitung*

(0.1) Ist $K(\zeta, z) = \frac{1}{2\pi i} \frac{d\zeta - dz}{\zeta - z}$ der Cauchykern in $\mathbb{C}$, so hat man für Gebiete $G \subset\subset \mathbb{C}$ mit glattem Rand die Repräsentationsformel

(0.1.1) $$f(z) = \int_{\zeta \in \partial G} K(\zeta, z) \wedge f(\zeta) \qquad (z \in G)$$

für holomorphe Funktionen oder Differentialformen, die auf $\bar{G}$ definiert sind. Ausserdem ist

(0.1.2) $$u(z) = \int_{\zeta \in G} K(\zeta, z) \wedge f(\zeta) \qquad (z \in G)$$

für integrable (p,q)-Formen f mit $\bar{\partial} f = 0$ eine Lösung der Gleichung $\bar{\partial} u = f$, [6].

(0.2) Integralkerne zur Repräsentation holomorpher Differentialformen und zur Lösung der $\bar{\partial}$-Gleichung sind auch im $\mathbb{C}^n$, $n > 1$, bekannt. Zum

Recent Developments in Several Complex Variables
0-691-08285-5/81/000199-28$01.40/0 (cloth)
0-691-08281-2/81/000199-28$01.40/0 (paperback)

einen hat man als direkte Verallgemeinerung des Cauchy-Kerns den Bochner-Martinelli-Kern

$$K(\zeta, z) = \frac{(n-1)!}{(2\pi i)^n} \frac{1}{\|\zeta - z\|^{2n}} \sum_{i=1}^{n} (-1)^{i-1}(\overline{\zeta}_i - \overline{z}_i) \bigwedge_{j\neq i} (d\overline{\zeta}_j - d\overline{z}_j) \wedge \omega(\zeta - z)$$

wobei $\omega(\zeta - z) = (d\zeta_1 - dz_1) \wedge \cdots \wedge (d\zeta_n - dz_n)$. Allerdings ist dieser Kern nicht mehr holomorph in z, und für die Gültigkeit von (0.1.2) muss man sich auf Formen mit kompaktem Träger beschränken.

Ist jedoch $G \subset\subset \mathbf{C}^n$ streng pseudokonvex, so lassen sich Kerne $K(\zeta, z)$ konstruieren, die für $\zeta \in \partial G$ holomorph in $z \in G$ sind und für die (0.1.1) und (0.1.2) voll gültig bleiben, vgl. [4], [10], [11]. Für $q > 1$ kann man sogar auf die Integrabilität von f verzichten, siehe [7]. Erst mit Hilfe solcher Kerne werden auch genauere Abschätzungen für die Lösungen von $\overline{\partial} u = f$ möglich.

(0.3) Es stellt sich nun die Frage nach der Verallgemeinerbarkeit dieser Ergebnisse auf Steinsche Mannigfaltigkeiten. Eine erste Antwort stammt von A. Palm [13]; jedoch braucht er zur Durchführung seiner Konstruktion ein "trivial-stabiles" Tangentialbündel.

Nun haben Toledo und Tong in [15] ein Verfahren angegeben, nach dem sich die durch Koordinaten gegebenen lokalen Bochner-Martinelli-Kerne zu einem globalen Kern zusammenkleben lassen. Dies schliesst auch den eindimensionalen Fall ein, der auch bereits von Behnke-Stein [1] behandelt wurde: auf offenen Riemannschen Flächen existiert ein globaler "Cauchy-Kern."

In der vorliegenden Arbeit wird gezeigt, dass für ein streng pseudokonvexes Gebiet G in einer Steinschen Mannigfaltigkeit X der Toledo-Tongsche Prozess so modifiziert werden kann, dass sich auch die lokalen Henkin-Ramirezschen Kerne verkleben lassen. Als Ergebnis erhalten wir

(0.3.1) SATZ. *Sei* G *ein streng pseudokonvexes Gebiet mit glattem Rand in einer Steinschen Mannigfaltigkeit. Dann gibt es eine ausserhalb der*

Diagonalen auf $\overline{G} \times G$ *erklärte glatte* $(n, n-1)$*-Form* $\Omega(\zeta, z)$ *mit integrabler Singularität und folgenden Eigenschaften:*

1. *Ist* $u \in \mathcal{C}^{\infty}_{p,q}(G)$, $0 \leq p,q \leq n$, *so konvergiert das Integral*

$$Tu(z) = \int_{\zeta \in G} \Omega(\zeta, z) \wedge u(\zeta) \qquad (z \in G),$$

falls $q \neq 1$. *Ist* $q = 1$, *so muss* u *zusätzlich integrabel sein. In beiden Fällen ist Tu eine glatte* $(p, q-1)$*-Form auf* G.

2. *Der Operator* T *löst die* $\overline{\partial}$*-Gleichung. Ist* $u \in \mathcal{C}^{\infty}_{p,q}(G)$ *(falls* $q = 1$, *muss* u *zusätzlich integrabel sein), so gilt*

$$u = \overline{\partial}Tu + T\overline{\partial}u .$$

3. *Der Kern des Randoperators*

$$Su(z) = \int_{\zeta \in \partial G} \Omega(\zeta, z) \wedge u(\zeta) \qquad (u \in C^0_*(\partial G), z \in G)$$

ist holomorph in der freien Variablen. S *reproduziert holomorphe Differentialformen.*

(0.4) Die Arbeit gliedert sich wie folgt: In §1 reduzieren wir das Problem der Konstruktion globaler $\overline{\partial}$-lösender Kerne im wesentlichen auf die Lösung einer Folge von $\overline{\partial}$-Gleichungen.

In §2 entwickeln wir eine vereinfachte und polierte Version des Toledo-Tongschen Apparates, der diese Gleichungen zu lösen erlaubt.

In §3 werden dann die Besonderheiten bei der Konstruktion des globalen Bochner-Martinelli-Kerns B (3.1), des globalen Henkin-Ramirez-Kerns H (3.2) sowie einer globalen Lösung $\overline{\partial}L = B - H$ behandelt. Schliesslich werden in (3.4) B, H und L zu einem globalen Kern verklebt, für den dann die Aussagen von (0.3.1) gelten.

§1. *Fundamentallösungen für* $\bar\partial$

(1.1) *Fundamentallösungen und Integralformeln.* Sei G ein Teilgebiet einer Steinschen Mannigfaltigkeit X, dim X = n. Wir suchen auf $\bar{G} \times G$ "Fundamentallösungen für den $\bar\partial$-Operator," d.h. (n, n−1)-Courants $\Omega \in \mathfrak{D}'_{n,n-1}(\bar{G} \times G)$ mit

$$\bar\partial\Omega = \delta_\Delta \tag{1.1.1}$$

Dabei ist $\delta_\Delta \in \mathfrak{D}'_{n,n}(\bar{G} \times G)$ gegeben als lineares Funktional $\mathfrak{D}_{n,n}(\bar{G} \times G) \ni \phi \to \int_\Delta \phi$, d.h. glatte (n,n)-Formen mit kompaktem Träger in $\bar{G} \times G$ werden über der Diagonalen Δ integriert. δ_Δ hat als Träger die Diagonale. Auf Grund der Regularitätssätze für den $\bar\partial$-Operator [9a] muss dann der singuläre Träger von Ω ebenfalls die Diagonale, Ω ausserhalb Δ also glatt sein.

Wir beschränken unsere Untersuchung auf die Fälle G = X bzw. G streng pseudokonvex; in beiden Fällen hat G eine Steinsche Umgebungsbasis in X, und aus Kohomologiegründen ist die Existenz einer Fundamentallösung Ω sichergestellt.

Nehmen wir an, es gäbe eine Fundamentallösung mit integrabler Singularität. Für $\phi \in \mathfrak{D}_{p,q}(\bar{G})$ ist dann der Integraloperator

$$T\phi(z) = \int_{\zeta \in G} \Omega(\zeta, z) \wedge \phi(\zeta)$$

für fast alle $z \in G$ definiert (Fubini) und man überlegt sich leicht, dass $T\phi$ eine lokal-integrable (p, q−1)-Form sein muss.

Wenn wir für $\phi \in \mathfrak{D}_{p,q}(\bar{G})$ das Randintegral

$$\int_{\zeta \in \partial G} \Omega(\zeta, z) \wedge \phi(\zeta) = \psi(z) \qquad (z \in G)$$

betrachten (G habe glatten Rand), so stellen wir fest, dass ψ in Komponenten $\psi_1 + \psi_2$ zerfällt, wobei $\psi_1 \in \mathcal{C}^\infty_{p,q}(G)$ und $\psi_2 \in \mathcal{C}^\infty_{p+1,q-1}(G)$.

Mit $S_{p,q}$ wollen wir den Operator $\phi \to \psi_1$ bezeichnen, und wir setzen
$S := \bigoplus_{0 \leq p,q \leq n} S_{p,q}$.

Aus $\bar{\partial}\Omega = \delta_\Delta$ ergibt sich für Differentialformen $\phi \in \mathfrak{D}_{**}(\overline{G})$ auf $\overline{G}$ die folgende Identität (Koppelman-Formel):

$$\phi = S\phi + (-1)^n(\bar{\partial}T\phi + T\bar{\partial}\phi) . \tag{1.1.2}$$

Dies erhält man durch eine einfache Anwendung des Stokesschen Satzes, vgl. [11], [12].

Man sieht dass für Formen mit kompaktem Träger in G $S\phi$ verschwindet; ist darüberhinaus $\bar{\partial}\phi = 0$, so erhalten wir eine Lösung der $\bar{\partial}$-Gleichung.

Für holomorphe Differentialformen in $\mathfrak{D}_{**}(\overline{G})$ ist $\bar{\partial}\phi = 0$; ausserdem verschwindet $T\phi$, da sonst $\Omega(\zeta, z)$ eine Komponente vom Typ $(*, n)$ in ζ haben müsste. Übrig bleibt die Repräsentationsformel $\phi = S\phi$.

Offenbar ist unsere Fundamentallösung Ω aber noch verbesserungswürdig. Zum einen sollte natürlich für glattes ϕ auch $T\phi$ glatt sein. Ist dies immer der Fall, so wollen wir den Operator T wie auch Ω "regulär" nennen. Die Regularität ist eine Symmetriebedingung an die Singularität von Ω, vgl. [15].

Für den $\mathbf{C}^n$ ist bekanntlich der Bochner-Martinelli-Kern eine reguläre Fundamentallösung, und in (3.1) konstruieren wir eine solche für $X \times X$.

Zum anderen stört die Beschränkung auf Formen mit kompaktem Träger in G beim Lösen der $\bar{\partial}$-Gleichung. Hier ist Abhilfe zu schaffen, wenn Ω die folgende Holomorphieeigenschaft hat:

(H) Zu kompaktem $K \subset G$ gibt es ein Kompaktum $L \subset G$, so dass für alle $\zeta \in \overline{G} \setminus L$ $\Omega(\zeta, z)$ in einer Umgebung von K eine holomorphe Differentialform bzgl. z ist.

(1.1.3) SATZ. $G \subset\subset X$ *habe glatten Rand, und* Ω *sei eine reguläre Fundamentallösung für* $\bar{\partial}$ *auf* $\overline{G} \times G$ *mit der Eigenschaft (H). Dann ergeben sich für die resultierenden Operatoren* T *und* S *alle Aussagen von Satz (0.3.1).*

Beweis. Sei $\phi \in \mathcal{C}^{\infty}_{p,q}(G)$, $0 \leq p, q \leq n$, und $z_0 \in G$. Da $\Omega(*, z_0)$ in einer Umgebung von ∂G glatt ist, reicht zur Konvergenz des Integrals $T\phi(z_0) = \int_{\zeta \in G} \Omega(\zeta, z_0) \wedge \phi(\zeta)$ die Integrabilität von ϕ (d.h. ϕ muss bzgl. der Koordinaten jeder Karte U eines $\overline{G}$ überdeckenden Atlas auf $U \cap G$ integrabel sein).

Nun gibt es wegen (H) eine von z_0 entfernte Umgebung $U(\partial G)$, so dass für jedes $\zeta \in U$ $\Omega(\zeta, *)$ eine holomorphe Differentialform in einer Umgebung von z_0 ist. Dies bedeutet, dass für $q > 1$ ($q = 0$: trivial) die für die Integration $\int \Omega(\zeta, z) \wedge \phi(\zeta)$ zuständige Komponente von $\Omega(\zeta, z_0)$ ihren Träger in $G \cap U$ hat, das Integral also auch ohne Wachstumsbeschränkung für ϕ konvergiert. Die Regularität der Fundamentallösung garantiert die Glattheit von $T\phi$. Damit ist (0.3.1.1) bewiesen.

Zum Nachweis von (0.3.1.2) beschränken wir uns auf den Fall $q = 1$. Seien im übrigen ϕ, z_0, U wie oben, zusätzlich sei $\kappa \in \mathfrak{D}(U)$, $\kappa \equiv 1$ in einer Umgebung $U'(\partial G) \subset\subset U$, $\phi' := \kappa\phi$, $\phi'' := \phi - \phi'$. Da ϕ'' kompakten Träger in G hat, folgt aus (1.1.2) $\phi'(z_0) = (-1)^n((\overline{\partial} T\phi'')(z_0) + (T\overline{\partial}\phi'')(z_0))$. Im Integral $\int \Omega(\zeta, z_0) \wedge \overline{\partial}\phi''(\zeta)$ ist die wirksame Komponente von $\Omega(\zeta, z_0)$ aber schon identisch Null, bevor $\overline{\partial}\phi''$ sich von $\overline{\partial}\phi$ zu unterscheiden beginnt, also ist $(T\overline{\partial}\phi'')(z_0) = (T\overline{\partial}\phi)(z_0)$, und natürlich gilt auch $\phi''(z_0) = \phi(z_0)$. Andererseits ist $(T\phi)(z_0) = (T\phi'')(z_0) + (T\phi')(z_0)$. Im Integral $(T\phi')(z) = \int \Omega(\zeta, z) \wedge \phi'(\zeta)$ ist aber dort, wo integriert wird, $\Omega(\zeta, *)$ holomorph in einer Umgebung von z_0; dasselbe gilt daher für $T\phi'$, daher ist $(\overline{\partial} T\phi')(z_0) = 0$ und somit $(\overline{\partial} T\phi)(z_0) = (\overline{\partial} T\phi'')(z_0)$. Insgesamt ergibt sich: $\phi(z_0) = (-1)^n((\overline{\partial} T\phi)(z_0) + (T\overline{\partial}\phi)(z_0))$. (Das Vorzeichen ist natürlich unerheblich.)

Für $q > 1$ ist schon bei der Bildung von $T\phi$ jede Wachstumsbeschränkung unnötig; im übrigen argumentiert man ähnlich wie oben. (0.3.1.3) folgt sofort aus (1.1.2) und (H). Darüberhinaus ist das Diagramm

$$\begin{array}{ccc} \Omega(\overline{G}) & \xrightarrow{\text{restr.}} & \mathcal{C}^{\infty}_{*}(\partial G) \\ & {\scriptstyle \text{restr.}} \searrow \qquad \swarrow {\scriptstyle S} & \\ & \Omega(G) & \end{array}$$

(Ω Garbe der holomorphen Differentialformen auf X)

kommutativ. q.e.d.

Ziel dieser Arbeit ist es, für ein streng pseudokonvexes Gebiet $G \subset\subset X$ eine reguläre Fundamentallösung für $\bar\partial$ auf $\bar G \times G$ mit der eben diskutierten Holomorphieeigenschaft (H) zu konstruieren.

(1.2) *Cauchy-Fantapié-Kerne und die Garbe* $\mathcal{Q}$. Man erinnere sich (vgl. [11]), dass die Konstruktion $\bar\partial$-lösender "holomorpher" Kerne für streng pseudokonvexe Gebiete im $\mathbb{C}^n$ durch Verschmelzung des Bochner-Martinelli-Kerns, der singulär auf der Diagonalen ist, mit dem holomorphen Ramirezkern, der eine relativ komplizierte nicht integrable Singularität besitzt, geschieht. Beide Kerne sind Cauchy-Fantapié-Kerne, d.h. $(n, n-1)$-Formen in ζ, z von der Form

$$\Omega = \frac{1}{F^n} \sum_{i=1}^{n} (-1)^{i-1} f_i \bigwedge_{j\neq i} \bar\partial f_j \wedge \omega$$

wobei $F(\zeta, z) = \sum_i f_i(\zeta, z)(\zeta_i - z_i)$ und $\omega = (d\zeta_1 - dz_1) \wedge \cdots \wedge (d\zeta_n - dz_n)$.

Für solche Formen gilt $\bar\partial\Omega = 0$ ausserhalb der Nullstellen von F [8]. Anders als im Falle des BM-Kerns, der ja lokal integrabel ist, gibt es im allgemeinen keine distributionelle Interpretation für Ω oder $\bar\partial\Omega$ in den Singularitäten. Um aber singuläre Kerne ohne Formalitäten problemlos addieren und differenzieren zu können, was wir im folgenden häufig tun müssen, wollen wir sie als Schnitte in einer Art "Lückengarbe" interpretieren.

Betrachten wir dazu die Garbe $\mathcal{C}^\infty$ der glatten Funktionskeime in $X \times X$. Für $x \in X \times X$ sei $\mathcal{J}_x$ das Primideal der einschliesslich aller Ableitungen verschwindenden Keime in $\mathcal{C}^\infty_x$. $\mathcal{C}^\infty_x \setminus \mathcal{J}_x$ ist ein multiplikatives System in $\mathcal{C}^\infty_x$; $\mathcal{Q}_x$ sei der zugehörige Ring von Quotienten, $\mathcal{Q}$ die Garbe mit den Halmen $\mathcal{Q}_x$. $\mathcal{Q}$ ist eine $\mathcal{C}^\infty$-Algebragarbe und enthält in natürlicher Weise $\mathcal{C}^\infty$ als Untergarbe. Über die Quotientenregel definieren wir den Dolbeault-Komplex bezüglich $\mathcal{Q}$

(1.2.1) $$0 \longrightarrow \mathcal{Q}_{p,0} \xrightarrow{\bar\partial} \mathcal{Q}_{p,1} \xrightarrow{\bar\partial} \cdots \xrightarrow{\bar\partial} \mathcal{Q}_{p,n} \longrightarrow 0 .$$

Wir werden nur solche Cauchy-Fantapié-Kerne Ω betrachten, die Schnitte in $\mathcal{Q}_{n,n-1}$ sind. Diese sind also auch in den Singularitäten definiert, und auch dort gilt $\bar\partial\Omega = 0$. Will man zur distributionellen Interpretation zurückkehren und z.b. $\bar\partial\Omega = \delta_\Delta$ beweisen, so ist die Singularität von Ω genauer zu analysieren.

(1.3) Dieser Abschnitt beschreibt, welche Schritte durchzuführen sind, um lokale Cauchy-Fantapié-Kerne auf $X\times X$ zu verschmelzen.

Gegeben seien bezüglich einer geeigneten (siehe §2) Steinschen Überdeckung $\mathfrak{U}$ von $X\times X$ für jedes $U\in\mathfrak{U}$ lokale Cauchy-Fantapié-Kerne $\omega_U\in\mathcal{Q}_{n,n-1}(U)$, die wir kollektiv durch eine Čech-Kokette $\omega^0_{n,n-1}\in C^0(\mathfrak{U},\mathcal{Q}_{n,n-1})$ beschreiben. Zunächst benötigen wir einen Prozess (s. §2), der die Gleichungen

(1.3.1)
$$\begin{aligned}\bar\partial\omega^1_{n,n-2} &= \delta\omega^0_{n,n-1}\\ \bar\partial\omega^i_{n,n-i-1} &= \delta\omega^{i-1}_{n,n-i} \qquad 2\le i\le n-1\,,\end{aligned}$$

wobei $\omega^i_{n,n-i-1}\in C^i(\mathfrak{U},\mathcal{Q}_{n,n-i-1})$, zu lösen erlaubt, einschliesslich einer genauen Kontrolle der beim Lösen auftretenden Singularitäten. Zum Schluss (für $n=1$ ist es der Anfang) ist zu zeigen, dass $\omega^{n-1}_{n,0}$ ein Čech-Kozykel von holomorphen n-Formen ist. Da auch $X\times X$ Steinsch ist, muss eine holomorphe Lösung

(1.3.2)
$$\delta\gamma^{n-1}_{n,0} = \delta\omega^{n-1}_{n,0}$$

existieren, und mit Hilfe einer Teilung der 1 (3.1) oder über Kohomologie (3.2) sind weitere Lösungen

(1.3.3)
$$\begin{aligned}\delta\Omega^{n-2}_{n,0} &= \omega^{n-1}_{n,0} - \gamma^{n-1}_{n,0}\\ \delta\Omega^{n-i}_{n,i-2} &= \omega^{n-i+1}_{n,i-2} - \bar\partial\Omega^{n-i+1}_{n,i-3} \qquad 2< i\le n\,,\end{aligned}$$

$\Omega^{n-i}_{n,i-2}\in C^{n-i}(\mathfrak{U},\mathcal{Q}_{n,i-2})$ zu konstruieren, wobei auch hier auf die Singularitäten zu achten ist. Es ergibt sich

$$\delta(\omega^0_{n,n-1} - \bar\partial\Omega^0_{n,n-2}) = 0$$

(im Falle $n = 1$ $\delta(\omega^0_{n,0} - \gamma^0_{n,0}) = 0$); somit erhalten wir einen globalen Kern $\Omega = \Omega_{n,n-1} \in \mathcal{Q}_{n,n-1}(X \times X)$ mit $\bar\partial\Omega = 0$. Die Eigenschaften dieses Kerns hängen natürlich wesentlich von den lokalen Ausgangskernen ab.

In (3.1) werden wir von lokalen Bochner-Martinelli- und in (3.2) von lokalen Ramirezkernen ausgehen; es zeigt sich, dass die resultierenden globalen Kerne deren charakteristische Eigenschaften beibehalten.

§2. *Der Verklebungsprozess von Toledo-Tong*

(2.1) Zunächst vereinbaren wir folgende Bezeichnungen: X sei eine n-dimensionale Steinsche Mannigfaltigkeit, $\mathcal{O}$ sei die Garbe der holomorphen Funktionskeime auf $X \times X$, $\mathcal{J}$ sei die Idealgarbe der auf der Diagonalen $\Delta \subset X \times X$ verschwindenden Keime, $e^1, \cdots, e^n$ seien die kanonischen Basisschnitte in $\Gamma(X \times X, \mathcal{O}^n)$, $\mathcal{K} = \Lambda \mathcal{O}^n = \bigoplus_{i \in \mathbf{Z}} \Lambda^i \mathcal{O}^n$ sei die von $\mathcal{O}^n$ über $\mathcal{O}$ erzeugte Äussere-Algebra-Garbe; wir setzen $\mathcal{K}^i := \Lambda^i \mathcal{O}^n$; $\tilde{\mathcal{K}}$ entstehe aus $\mathcal{K}$ durch Nullsetzen des O-ten Summanden, also $\tilde{\mathcal{K}}^i = \mathcal{K}^i$, falls $i \neq 0$, $\tilde{\mathcal{K}}^0 = 0$. Aus der exakten Sequenz $0 \to \mathcal{J} \to \mathcal{O}$ entsteht durch Tensorieren die exakte Sequenz $0 \to \mathcal{J}^i \to \mathcal{K}^i$, wobei $\mathcal{J}^i := \mathcal{K}^i \otimes_{\mathcal{O}} \mathcal{J}$; ausserdem setzen wir $\tilde{\mathcal{J}}^i := \mathcal{J}^i \cap \tilde{\mathcal{K}}^i$, $\mathcal{L}^j_i := \mathcal{H}om_{\mathcal{O}}(\mathcal{K}^j, \tilde{\mathcal{K}}^{j-i})$, $\mathcal{L}_i := \bigoplus_{j \in \mathbf{Z}} \mathcal{L}^j_i$ und $\mathcal{L} := \bigoplus_{i \in \mathbf{Z}} \mathcal{L}_i$.

(2.2) Die lokalen Koordinatenkomplexe. Wir wählen eine Steinsche Überdeckung $\mathfrak{U}$ von $X \times X$ so, dass alle Mengen $U \in \mathfrak{U}$, die die Diagonale schneiden, die Gestalt $U = U' \times U'$ haben, wobei U' eine Koordinatenumgebung in X ist. Sind $\zeta = (\zeta_1, \cdots, \zeta_n)$ die Koordinaten auf U', (ζ, z) die assoziierten Koordinaten auf U, so setzen wir

$$u_U = \sum_i (\zeta_i - z_i) e^i .$$

Für die übrigen $U \in \mathfrak{U}$ dürfen wir annehmen, dass sie von der Diagonale entfernt liegen; wir setzen hier $u_U = \sum_i e^i$. In beiden Fällen ist $u_U \in \mathcal{J}^1(U)$. Multiplikation mit u_U macht $\mathcal{K}|U$ zu einem Komplex

$$(2.2.1) \qquad 0 \longrightarrow \mathcal{K}^0|U \xrightarrow{d_U} \cdots \xrightarrow{d_U} \mathcal{K}^{n-1}|U \xrightarrow{d_U} \mathcal{J}^n|U \longrightarrow 0\,;$$

dieser Komplex ist für alle $U \in \mathfrak{U}$ exakt (über komplexe Koordinaten gegebener Koszulkomplex). Ist jetzt $\tilde{U} \subset U \in \mathfrak{U}$ eine Steinsche offene Menge, so bleibt, weil alle auftretenden Garben kohärent sind, die induzierte Schnittmodulsequenz

$$(2.2.1') \qquad 0 \longrightarrow \mathcal{K}^0(\tilde{U}) \xrightarrow{d_{\tilde{U}}} \cdots \xrightarrow{d_{\tilde{U}}} \mathcal{J}^n(\tilde{U}) \longrightarrow 0$$

exakt.

Sind zwei Mengen $U, V \in \mathfrak{U}$ gegeben, so schreiben wir kurz UV statt $U \cap V$. Die Homomorphismen d_U und d_V induzieren einen Homomorphismus D_{UV} auf $\mathcal{L}|UV$ durch

$$\mathcal{L}_i|UV \xrightarrow{D_{UV}} \mathcal{L}_{i-1}|UV$$

$$\psi_{UV} \longrightarrow d_U\psi_{UV} - (-1)^i \psi_{UV} d_V\,.$$

Es ergibt sich der Komplex

$$(2.2.2)\ 0 \longrightarrow \mathcal{L}_n|UV \xrightarrow{D_{UV}} \cdots \xrightarrow{D_{UV}} \mathcal{L}_0|UV \xrightarrow{D_{UV}} \cdots \xrightarrow{D_{UV}} \mathcal{L}_{-n}|UV \longrightarrow 0\,.$$

Aus der Exaktheit von (2.2.1) erhält man mittels einer kanonischen Diagrammjagd folgende Exaktheitsaussagen für diesen Komplex:

(2.2.3) Die Sequenz (2.2.2) ist exakt an den Stellen $i \geq 0$. Für $i < 0$ ergibt sich: ist ψ ein Keim in $\mathcal{L}_i|UV$, $\psi = \sum \psi^j$, $\psi^j \in \mathcal{L}_i^j$, mit $D_{UV}\psi = 0$, so existiert eine Lösung $D_{UV}\chi = \psi$ genau dann, wenn $\psi^0 \in \mathcal{H}om_{\mathcal{O}}(\mathcal{O}, \mathcal{J}^{-i})$.

Der Beweis ist in [15] durchgeführt.

Entsprechend (2.2.1′) erhalten wir die zu (2.2.3) analogen Exaktheitsaussagen auch für den Schnittmodulkomplex

(2.2.2′) $$0 \to \mathfrak{L}_n(W) \to \cdots \to \mathfrak{L}_{-n}(W) \to 0\,,$$

wenn $W \subset UV$ Steinsch ist.

Sei d_U^0 der Homomorphismus $\mathcal{K}^0(U) \to \mathcal{K}^1(U)$ aus (2.2.1′). d_U^0 lässt sich als Element von $\Gamma(U, \mathcal{H}om_{\mathcal{O}}(\mathcal{O}, \tilde{\mathcal{I}}^1)) \subset \mathfrak{L}_{-1}(U)$ auffassen. Über UV gilt offensichtlich $D_{UV}(d_U^0) = 0$. Damit folgt aus der Exaktheitsaussage fur (2.2.2′) die Existenz eines Schnittes

(2.2.4) $$\phi_{UV}^0 \in \mathfrak{L}_0(UV) \quad \text{mit} \quad D_{UV}\phi_{UV}^0 = d_U^0\,.$$

Die Elemente ϕ_{UV}^0 seien für den Rest der Arbeit fest gewählt.

(2.3) Der d- und der D-Komplex. Sei $\mathfrak{U}$ die Überdeckung aus (2.2). Wir bilden die üblichen Čech-Kokettengruppen $C^i(\mathfrak{U}, \mathfrak{L}_j)$, $C^i(\mathfrak{U}, \mathcal{K}^j)$. Es gibt natürliche bilineare Produkte (cup, cap)

(2.3.1) $$C^i(\mathfrak{U}, \mathfrak{L}_j) \times C^{i'}(\mathfrak{U}, \mathfrak{L}_{j'}) \ni (\psi, \chi) \to \psi\chi \in C^{i+i'}(\mathfrak{U}, \mathfrak{L}_{j+j'})$$

(2.3.2) $$C^i(\mathfrak{U}, \mathfrak{L}_j) \times C^{i'}(\mathfrak{U}, \mathcal{K}^{j'}) \ni (\psi, x) \to \psi \times C^{i+i'}(\mathfrak{U}, \tilde{\mathcal{K}}^{j'-j})\,,$$

definiert durch

$$(\psi\chi)_{k_0 \cdots k_{i+i'}} = \psi_{k_0 \cdots k_i} \circ \chi_{k_i \cdots k_{i+i'}}$$

$$(\psi x)_{k_0 \cdots k_{i+i'}} = \psi_{k_0 \cdots k_i}(x_{k_i \cdots k_{i+i'}})\,.$$

Man sieht sofort, dass diese Produkte assoziativ sind, d.h.

(2.3.3) $$(\psi\chi)\rho = \psi(\chi\rho)\,, \quad (\psi\chi)x = \psi(\chi x)\,.$$

Die Homomorphismen d_U aus (2.2.1′) lassen sich zu einer Kokette

(2.3.4) $$d = d_0^{-1} \in C^0(\mathfrak{U}, \mathfrak{L}_{-1})$$

und die in (2.2.4) gefundenen Homomorphismen ϕ^0_{UV} zu einer Kokette

$$\phi = \phi^0_1 \in C^1(\mathfrak{U}, \mathcal{L}_0) \tag{2.3.5}$$

zusammenfassen. Setzen wir für U, V, W $\in$ U $R^0_{UVW} = \phi^0_{UW} - \phi^0_{UV}\phi^0_{VW}$, so erhalten wir ein Element

$$R = R^0_2 \in C^2(\mathfrak{U}, \mathcal{L}_0)\,. \tag{2.3.6}$$

"Multiplikation" mit d induziert Homomorphismen $C^i(\mathfrak{U}, \mathcal{K}^j) \xrightarrow{d} C^i(\mathfrak{U}, \mathcal{K}^{j+1})$, und (2.2.1′) besagt gerade, dass die folgende Sequenz exakt ist:

$$0 \longrightarrow C^i(\mathfrak{U}, \mathcal{K}^0) \xrightarrow{d} \cdots \xrightarrow{d} C^i(\mathfrak{U}, \mathcal{J}^n) \longrightarrow 0\,. \tag{2.3.7}$$

Ebenso erhalten wir Homomorphismen $C^i(\mathfrak{U}, \mathcal{L}_j) \xrightarrow{D} C^i(\mathfrak{U}, \mathcal{L}_{j-1})$ durch $\psi \to d\psi - (-1)^j \psi d$, und aus der Exaktheitsaussage fur (2.2.2′) folgt, dass die Sequenz

$$0 \longrightarrow C^i(\mathfrak{U}, \mathcal{L}_n) \xrightarrow{D} \cdots \xrightarrow{D} C^i(\mathfrak{U}, \mathcal{L}_0) \xrightarrow{D} \cdots \xrightarrow{D} C^i(\mathfrak{U}, \mathcal{L}_{-n}) \longrightarrow 0 \tag{2.3.8}$$

exakt ist an den Stellen $j \geq 0$. Eine Aussage für die Stellen $j < 0$ wird im folgenden nicht mehr benötigt.

Für ψ, χ, x wie in (2.3.1), (2.3.2) rechnet man mit Hilfe der Assoziativgesetze (2.3.3) leicht folgende Produktregeln nach

$$\begin{aligned} D(\psi\chi) &= (D\psi)\chi + (-1)^j \psi(D\chi) \\ d(\psi x) &= (D\psi)x + (-1)^j \psi(dx)\,. \end{aligned} \tag{2.3.9}$$

(2.4) Der modifizierte Čech-Komplex. Zur Definition der Čech-Korandoperatoren unterscheiden wir drei Fälle. Während wir im Falle

a) $C^i(\mathfrak{U}, \mathcal{K}^0) \to C^{i+1}(\mathfrak{U}, \mathcal{K}^0)$

den üblichen Operator wählen, definieren wir mit Hilfe der Kokette $\phi \in C^1(\mathfrak{U}, \mathcal{L}_0)$ aus (2.3.5) modifizierte Operatoren

b) $C^i(\mathfrak{U}, \mathcal{L}_j) \xrightarrow{\delta} C^{i+1}(\mathfrak{U}, \mathcal{L}_j)$ c) $C^i(\mathfrak{U}, \tilde{\mathcal{K}}^j) \xrightarrow{\delta} C^{i+1}(\mathfrak{U}, \tilde{\mathcal{K}}^j)$

durch

$$\text{b)}\quad (\delta\psi)_{k_0\cdots k_{i+1}} = \phi_{k_0k_1}\circ\psi_{k_1\cdots k_{i+1}} + \sum_{p=1}^{i}(-1)^p\psi_{k_0\cdots\hat{k}_p\cdots k_{i+1}}$$

$$+ (-1)^{i+1}\psi_{k_0\cdots k_i}\phi_{k_ik_{i+1}}$$

$$\text{c)}\quad (\delta x)_{k_0\cdots k_{i+1}} = \phi_{k_0k_1}(x_{k_1\cdots k_{i+1}}) + \sum_{p=1}^{i+1}(-1)^p x_{k_0\cdots\hat{k}_p\cdots k_{i+1}}$$

Damit gilt nun aber nicht mehr $\delta\delta = 0$ ausser im Fall a), sondern

$$(2.4.1)\qquad \delta\delta\psi = R\psi - \psi R, \qquad \delta\delta x = Rx,$$

d.h. R ist die "Krümmung" zum "Zusammenhang" δ. Wir wollen aber diese Analogie hier nicht weiter präzisieren. Entsprechend (2.3.9) hat man die Produktregeln

$$(2.4.2)\qquad \begin{aligned} \delta(\psi\chi) &= (\delta\psi)\chi + (-1)^i\psi(\delta\chi) \\ \delta(\psi x) &= (\delta\psi)x + (-1)^i\psi(\delta x). \end{aligned}$$

Weiterhin ergibt sich sofort aus der Konstruktion von ϕ (2.3.5), dass δ, D bzw. δ, d kommutieren.

(2.5) *Konstruktion der Operatoren* ψ^j. Ziel dieses Abschnitts ist die induktive Konstruktion von Elementen $\psi^j \in C^{j+1}(\mathfrak{U}, \mathfrak{L}_j)$, die die folgenden Gleichungen erfüllen

$$\psi^j = 0 \quad \text{für} \quad j \le 0$$

$$D\psi^1 = R$$

$$D\psi^{j+1} = \delta\psi^j + \sum_{k+p=j}(-1)^k\psi^k\psi^p.$$

Setzt man $\psi = \sum_k \psi^k$ und $\bar{\psi}^k = (-1)^k\psi^k$, so lässt sich kurz schreiben $D\psi = \delta\psi + \bar{\psi}\psi + R$, eine Formel, die wieder differentialgeometrische Assoziationen weckt.

Der Induktionsanfang ist klar. Zunächst findet man $DR = 0$, wegen der Exaktheitsaussage (2.3.8) existiert also ψ. Weiter ist $D\delta\psi^1 = \delta D\psi^1 = \delta R = 0$ eine Art Bianchiidentität. Wie vorher existiert jetzt ψ^2 mit $D\psi^2 = \delta\psi^1$.

Nehmen wir nun an ψ^{i+1} sei bereits konstruiert. Dann gilt:

$$\begin{aligned}
D\delta\psi^{i+1} &= \delta D\psi^{i+1} \\
&= \delta\delta\psi^i + \sum_{j+k=i} (-1)^j[(\delta\psi^j)\psi^k + (-1)^{j+1}\psi^j\delta\psi^k] \\
&= R\psi^i - \psi^i R + \sum_{j+k=i} (-1)^j(D\psi^{j+1})\psi^k \\
&\quad - \sum_{\lambda+\mu+k=i} (-1)^{(\lambda+\mu)+\lambda}\psi^\lambda\psi^\mu\psi^k - \sum_{j+k=i} \psi^j D\psi^{k+1} \\
&\quad + \sum_{j+\lambda+\mu=i} (-1)^\lambda \psi^j\psi^\lambda\psi^\mu \\
&= (D\psi^1)\psi^i - \psi^i(D\psi^1) + \sum_{j+k=i} (-1)^j(D\psi^{j+1})\psi^k - \psi^j D\psi^{k+1} \\
&= \sum_{j+k=i+1} (-1)^{j+1}[(D\psi^j)\psi^k + (-1)^j\psi^j D\psi^k] \\
&= \sum_{j+k=i+1} (-1)^{j+1} D(\psi^j\psi^k) \\
&= D\Big(\sum_{j+k=i+1} (-1)^{j+1}\psi^j\psi^k\Big).
\end{aligned}$$

Also ist $D\Big(\delta\psi^{i+1} + \sum_{j+k=i+1} (-1)^j\psi^j\psi^k\Big) = 0$, und mit der Exaktheitsaussage (2.3.8) erhält man die Existenz von ψ^{i+2}. q.e.d.

(2.6) Der $\bar{d}$-Komplex. Um die im folgenden auftretenden Singularitäten zu erfassen, ziehen wir die Garbe $\mathfrak{Q}$ aus (1.2) hinzu. Es sei

(2.6.1) $$\overset{(\sim)}{\mathcal{Q}}{}^{i}_{p,q} = (\overset{(\sim)}{\mathcal{K}}{}^{i} \otimes_{\mathcal{O}} \mathcal{C}^{\infty}_{p,q}) \otimes_{\mathcal{C}^{\infty}} \mathcal{Q} .$$

Die Garbe $\mathcal{L}_j$ operiert natürlich auf $\mathcal{Q}_{p,q}$ genau wie auf $\mathcal{K}^i$; wir haben also ein natürliches Produkt

(2.6.2) $$C^i(\mathfrak{U}, \mathcal{L}_j) \times C^{i'}(\mathfrak{U}, \mathcal{Q}^{j'}_{p,q}) \to C^{i+i'}(\mathfrak{U}, \tilde{\mathcal{Q}}^{j'-j}_{p,q})$$

entsprechend (2.3.2). Auch haben wir, indem wir die Differentiation über die Quotientenregel erklären, einen $\bar{\partial}$-Komplex

(2.6.3) $$0 \longrightarrow C^k(\mathfrak{U}, \mathcal{Q}^i_{p,0}) \xrightarrow{\bar{\partial}} \cdots \xrightarrow{\bar{\partial}} C^k(\mathfrak{U}, \mathcal{Q}^i_{p,n}) \longrightarrow 0 .$$

Für das Produkt (2.6.2) erhalten wir nun wegen der Holomorphie im ersten Faktor die Produktregel

(2.6.4) $$\bar{\partial}(\chi x) = \chi \bar{\partial} x .$$

Den Čechoperator

(2.6.5) $$C^k(\mathfrak{U}, \mathcal{Q}^i_{p,q}) \xrightarrow{\delta} C^{k+1}(\mathfrak{U}, \mathcal{Q}^i_{p,q})$$

definieren wir analog (2.4) und erhalten die (2.4.1) und (2.4.2) entsprechenden Regeln. Analog (2.3.7) erhalten wir einen d-Komplex

(2.6.6) $$0 \longrightarrow C^k(\mathfrak{U}, \mathcal{Q}^0_{p,q}) \xrightarrow{d} \cdots \xrightarrow{d} C^k(\mathfrak{U}, \mathcal{Q}^n_{p,q}) \longrightarrow 0 ;$$

für den d-Operator gilt die zu (2.3.9) analoge Produktregel, und die Operatoren d, $\bar{\partial}$, δ kommutieren.

Entscheidend ist nun aber die Exaktheit des Komplexes (2.6.6). Um dies zu beweisen, konstruieren wir auf folgende Weise einen Lösungsoperator für die d-Gleichung: man nehme eine Kokette $v \in C^0(\mathfrak{U}, \mathcal{E}^1)$ ($\mathcal{E}^1 = \mathcal{K}^1 \otimes_{\mathcal{O}} \mathcal{C}^{\infty}$). Für jedes $U \in \mathfrak{U}$ betrachte man die lokalen Koeffizienten $v_U = \sum_i v^i_U e^i$, $v^i_U \in \mathcal{C}^{\infty}(U)$ und bilde unter Hinzuziehung der Kokette u aus (2.2) die lokalen Skalarprodukte $\langle v_U, u_U \rangle = \sum_i v_U^i u_U^i$ $\in \mathcal{C}^\infty(U)$. An v stellen wir die zusätzliche Forderung, dass für kein $U \in \mathfrak{U}$ dieses Produkt in einem Punkt aus U zusammen mit allen Ableitungen verschwinde, also im Sinne von (1.2) als Nenner eines Schnittes in $\Gamma(U, \mathcal{Q})$ zulässig sei.

Sei jetzt eine Kokette $f \in C^k(\mathfrak{U}, \mathcal{Q}^i_{p,q})$ gegeben, $\alpha = (\alpha_0 \cdots \alpha_k)$ und $f_\alpha = \sum_{j_1 < \cdots < j_i} f_\alpha^{j_1 \cdots j_i} e_{j_1} \wedge \cdots \wedge e_{j_i}$, $f_\alpha^{j_1 \cdots j_i} \in \Gamma(U_\alpha, \mathcal{Q}_{p,q})$. Sei

$$(2.6.7) \quad \theta_v^\alpha(f) = \sum_{j_1 < \cdots < j_{i-1}} \sum_{m=1}^n \frac{v_{\alpha_0}^m}{\langle v_{\alpha_0}, u_{\alpha_0} \rangle} f_\alpha^{m, j_1 \cdots j_{i-1}} e_{j_1} \wedge \cdots \wedge e_{j_{i-1}} .$$

Damit ist $\theta_v^\alpha(f) \in \Gamma(U_\alpha, \mathcal{Q}^{i-1}_{p,q})$. In dieser Definition treten übrigens die Brüche auf, die die Einführung der Garbe $\mathcal{Q}$ erforderlich machen. Die θ_v^α fassen wir nun zusammen zu einem Operator

$$(2.6.8) \qquad C^k(\mathfrak{U}, \mathcal{Q}^i_{p,q}) \xrightarrow{\theta_v} C^k(\mathfrak{U}, \mathcal{Q}^{i-1}_{p,q}) .$$

Man rechnet leicht nach, dass θ_v ein Homotopieoperator ist, d.h. dass

$$(2.6.9) \qquad d\theta_v + \theta_v d = \mathrm{id} .$$

θ_v kann also zum Lösen der d-Gleichung verwandt werden. Allerdings muss man erst ein geeignetes $v \in C^0(\mathfrak{U}, \mathcal{E}^1)$ finden. Das einfachste Beispiel ist $v = \bar{u}$, d.h. wir konjugieren die Koeffizienten von u. Für $U \in \mathfrak{U}$ gilt dann $\langle v_U, u_U \rangle = \|\zeta - z\|^2$, falls U über der Diagonalen liegt, anderenfalls ist $\langle v_U, u_U \rangle = n$.

Wir gehen jetzt über zum d, $\bar\partial$-Doppelkomplex, führen dort eine totale Graduierung ein, indem wir setzen

$$(2.6.10) \qquad \mathcal{R}^j := \bigoplus_{i+q=j} \mathcal{Q}^i_{n,q} \qquad j \in \mathbf{Z} ,$$

und betrachten den durch $x \to dx + (-1)^j \bar{\partial} x$ definierten totalen Operator

(2.6.11) $$C^k(\mathfrak{U}, \mathfrak{K}^j) \xrightarrow{\bar{d}} C^k(\mathfrak{U}, \mathfrak{K}^{j+1}) .$$

δ, $\bar{d}$ kommutieren, und aus (2.3.9) und (2.6.4) ergibt sich sofort die Produktregel

(2.6.12) $$\bar{d}(\chi x) = (D\chi) x + (-1)^j \chi \bar{d} x$$

fur $\chi \in C^i(\mathfrak{U}, \mathfrak{L}_j)$, $x \in C^{i'}(\mathfrak{U}, \mathfrak{K}^{j'})$. Da die d-Zeilen des $d, \bar{\partial}$-Doppelkomplexes exakt sind, muss auch der Totalkomplex

(2.6.13) $$0 \longrightarrow C^k(\mathfrak{U}, \mathfrak{K}^0) \xrightarrow{\bar{d}} \cdots \xrightarrow{\bar{d}} C^k(\mathfrak{U}, \mathfrak{K}^{2n}) \longrightarrow 0$$

exakt sein. Da wir (bei gegebenem v) in θ_v einen konkreten Operator zur Lösung der $\bar{d}$-Gleichung haben, ergibt eine Analyse der zur Lösung der $\bar{d}$-Gleichung nötigen Diagrammjagd, dass der Operator

(2.6.14) $$\bar{\theta}_v = \sum_{i=0}^{n} (-1)^i (\theta_v \bar{\partial})^i \theta_v$$

ein Lösungsoperator für die $\bar{d}$-Gleichung ist.

(2.7) *Die Hauptkonstruktion.* In diesem Abschnitt sei ein beliebiges aber festes Element $v \in C^0(\mathfrak{U}, \mathfrak{S}^1)$ vorgegeben, das den in (2.6) geforderten Bedingungen genügt und somit einen Lösungsoperator $\bar{\theta}_v$ für die $\bar{d}$-Gleichung definiert. Nun sei für ein $U \in \mathfrak{U}$, das über der Diagonalen liegt,

$$\alpha_0^U = \frac{1}{(2\pi i)^n} e_1 \wedge \cdots \wedge e_n \otimes (d\zeta_1 - dz_1) \wedge \cdots \wedge (d\zeta_n - dz_n) .$$

Fur die ubrigen $U \in \mathfrak{U}$ sei $\alpha_0^U = 0$. Somit ist ein Element

(2.7.1) $$\alpha_0 \in C^0(\mathfrak{U}, \mathfrak{K}^n \otimes_{\mathcal{O}} \Omega^n) \subset C^0(\mathfrak{U}, \mathfrak{K}^n)$$

definiert. Eine zu (2.5) analoge Konstruktion ergibt

(2.7.2) SATZ. *Es existieren Elemente* $a_i \in C^i(\mathfrak{U}, \mathcal{K}^{n-i} \otimes_{\mathcal{O}} \Omega^n)$, $b_i \in C^i(\mathfrak{U}, \mathfrak{R}^{n-i-1})$ $(i \in \mathbf{Z})$, *mit folgenden Eigenschaften*:

$$a_i,\ b_i = 0 \quad \textit{für} \quad i < 0\,; \ a_0 \quad \textit{wie in} \quad (2.7.1)$$

$$da_{i+1} = \delta a_i + \sum_{k+m=i} (-1)^k \psi^k a_m \quad \textit{für} \quad i \geq 0\,,$$

Setz man für $i \geq -1$ $\tilde{b}_i := \delta b_i + \sum_{k+m=i} (-1)^k \psi^k b_m + (-1)^{i+1} a_{i+1}$, *so ist* $\bar{d}\tilde{b}_i = 0$, $b_{i+1} = \bar{\theta}_v(\tilde{b}_i)$, *somit auch* $\bar{d} b_{i+1} = \tilde{b}_i$. *Schliesslich ist* $\delta b_{n-1} = (-1)^{n-1} a_n$.

Mit zu den im Anschluss an (2.5.1) eingeführten analogen Abkürzungen lässt sich kurz schreiben: $da = \delta a + \bar{\psi} a$; $\bar{d} b = \delta b + \bar{\psi} b + \bar{a}$. Man beachte, dass die Konstruktion der b_i von v, bzw. $\bar{\theta}_v$ abhängt, während die a_i bei welchselnden v dieselben bleiben.

Beweis von (2.7.2). Die induktive Konstruktion beginnt, indem wir $b_0 := \bar{\theta}_v a_0$ setzen. Im Induktionsschritt setzen wir die Existenz von $a_0 \cdots a_i$, $b_0 \cdots b_i$ voraus. Dann gilt:

$$\bar{d}(\delta b_i) = \delta \bar{d} b_i = \delta\delta b_{i-1} + \sum_{j+k=i-1} (-1)^j \delta(\psi^j b_k) + (-1)^i \delta a_i$$

$$= (D\psi^1)\, b_{i-1} + \sum_{j+k=i-1} (-1)^j (\delta\psi^j)\, b_k - \sum_{j+k=i-1} \psi^i \delta b_k + (-1)^i \delta a_i$$

$$= \bar{d}(\psi^1 b_{i-1}) + \psi^1 \bar{d} b_{i-1} + \sum_{j+k=i-1} (-1)^j (D\psi^{j+1})\, b_k -$$

$$- \sum_{j+k=i-1} (-1)^j \sum_{\mu+\nu=j} (-1)^\mu \psi^\mu \psi^\nu b_k - \sum_{j+k=i-1} \psi^i \delta b_k + (-1)^i \delta a_i$$

$$= \bar{d}(\psi^1 b_{i-1}) + \psi^1 \bar{d} b_{i-1} + \sum_{j+k=i-1} (-1)^j \bar{d}(\psi^{j+1} b_k) + \sum_{j+k=i-1} \psi^{j+1} b_k -$$

$$- \sum_{\mu+\nu+k=i-1} (-1)^{\nu}\psi^{\mu}\psi^{\nu} b_k - \sum_{j+k=i-1} \psi^i \delta b_k + (-1)^i \delta a_i$$

$$= \bar{d}\Big(\sum_{j+k=i} (-1)^{j-1}\psi^j b_k\Big) + \sum_{j+k=i} \psi^j\Big(\bar{d} b_k - \sum_{\lambda+\kappa=k-1} (-1)^{\lambda}\psi^{\lambda} b_{\kappa} - \delta b_{k-1}\Big)$$

$$+ (-1)^i \delta a_i =$$

$$= \bar{d}\Big(\sum_{j+k=i} (-1)^{j-1}\psi^j b_k\Big) + \sum_{j+k=i} \psi^j (-1)^k a_k + (-1)^i \delta a_i \,.$$

Nun ist $\delta a_i + \sum_{j+k=i} (-1)^j \psi^i a_k \in C^{i+1}(\mathfrak{U}, \mathcal{K}^{n-i} \otimes_{\mathcal{O}} \Omega^n)$ $\bar{d}$- also auch d-geschlossen. Im Falle $i = 0$ rechnet man leicht nach, dass die Koeffizienten von δa_0 auf der Diagonalen verschwinden, also $\delta a_0 \in C^1(\mathfrak{U}, \mathcal{J}^n \otimes_{\mathcal{O}} \Omega^n)$. Aus der Exaktheit von (2.7.3)

(2.7.3) $0 \longrightarrow C^i(\mathfrak{U}, \Omega^n) \xrightarrow{d} C^i(\mathfrak{U}, \mathcal{K}^1 \otimes_{\mathcal{O}} \Omega^n) \xrightarrow{d} \cdots \xrightarrow{d} C^i(\mathfrak{U}, \mathcal{J}^n \otimes_{\mathcal{O}} \Omega^n) \longrightarrow 0$

($i \geq 0$, $d = d \otimes id$), die man aus (2.3.7) unter Berücksichtigung der Tatsache erhalt, dass die Garbe Ω^n über allen Mengen $U \in \mathfrak{U}$ als frei angenommen werden darf, folgt nun für alle i, $0 \leq i \leq n-1$, die Existenz von a_{i+1} mit

$$d a_{i+1} = \delta a_i + \sum_{j+k=i} (-1)^j \psi^j a_k \,.$$

Im Falle $i = n$ folgt, dass bereits $d\delta a_n = 0$ und wegen der Injektivität von d an dieser Stelle auch $\delta a_n = 0$, a_n also ein Čech-Kozykel ist. Für $0 \leq i \leq n-1$ gilt daher

$$\bar{d}(\delta b_i + \sum_{j+k=i} (-1)^j \psi^j b_k + (-1)^{i+1} a_{i+1}) = 0 \,,$$

so dass wir für $i \leq n-2$

$$b_{i+1} := \bar{\theta}_v \Big(\delta b_i + \sum_{j+k=i} (-1)^j \psi^j b_k + (-1)^{i+1} a_{i+1}\Big)$$

setzen können. Im Falle $i = n-1$ ist schon $\bar{d}(\delta b_{n-1} + (-1)^n \alpha_n) = 0$, also in diesem Fall auch $d(\delta b_{n-1} + (-1)^n \alpha_n) = 0$. Nun ist $\alpha_n \in C^n(\mathfrak{U}, \Omega^n) \subset C^n(\mathfrak{U}, \mathcal{Q}_{n,0})$ und $\delta b_{n-1} \in C^n(\mathfrak{U}, \mathfrak{K}^0) = C_n(\mathfrak{U}, \mathcal{Q}_{n,0})$. Wegen der Exaktheit von (2.6.13) muss also gelten:

$$\delta b_{n-1} = (-1)^{n-1} \alpha_n . \qquad \text{q.e.d.}$$

Bezeichnen wir die Komponenten der b_i, die in $C^i(\mathfrak{U}, \mathcal{Q}_{n,n-i-1})$ liegen, mit $\omega^{i}_{n,n-i-1}$ und beachten, dass die $\psi^i b_j$ keine Komponenten in $C^i(\mathfrak{U}, \mathcal{Q}_{n,n-i-1})$ haben, so ergibt sich aus (2.7.2)

$$\begin{aligned} \bar{\partial}\omega^{0}_{n,n-1} &= 0 \\ (-1)^{n-i-2}\, \bar{\partial}\omega^{i+1}_{n,n-i-2} &= \delta\omega^{i}_{n,n-i-1} \qquad (0 \le i \le n-2) \\ \delta\omega^{n-1}_{n,0} &= (-1)^{n+1} \alpha_n \in C^n(\mathfrak{U}, \Omega^n) . \end{aligned} \tag{2.7.4}$$

Die lokalen Komponenten von $\omega^{0}_{n,n-1}$ ergeben sich schon bei der Konstruktion von b_0. Liegt $U \in \mathfrak{U}$ über der Diagonalen und setzt man $F_U = \langle v_U, u_U \rangle$, so lässt sich leicht nachrechnen, dass

$$\omega^{U}_{n,n-1} = \left(\frac{1}{2\pi i}\right)^n \frac{(n-1)!}{F^n_U} \sum_{i=1}^{n} (-1)^{i-1} v^i_U \bigwedge_{j \neq i} \bar{\partial} v^j_U \wedge (d\zeta_1 - dz_1) \wedge \cdots \wedge (d\zeta_n - dz_n)$$

d.h. die $\omega^{U}_{n,n-1}$ sind lokale Cauchy-Fantapié-Kerne! Mit der Herleitung der Gleichungen (2.7.4) ist daher das in (1.3.1) aufgestellte Programm durchgeführt.

Wie schon in (1.3.2) bemerkt, muss auch eine holomorphe Lösung $\gamma = \gamma^{n-1}_{n,0} \in C^{n-1}(\mathfrak{U}, \Omega^n)$ der Gleichung

$$\delta\gamma = (-1)^{n+1} \alpha_n \tag{2.7.5}$$

existieren.

§3.

(3.1) *Globale Kerne vom Bochner-Martinelli-Typ.* Wir legen jetzt das spezielle $v = \bar{u}$, das im Anschluss an (2.6.9) diskutiert wurde, der

Konstruktion (2.7) zugrunde. Die Kokette

$$\omega^{0}_{n,n-1} \in C^{0}(\mathfrak{U}, \mathcal{Q}_{n,n-1})$$

aus (2.7.4) ist dann die Kollektion der lokalen Bochner-Martinelli-Kerne. Bekanntlich sind die lokalen BM-Kerne jeweils reguläre Fundamentallösungen für $\bar{\partial}$ im Sinne von §1.

Zur Konstruktion einer globalen regulären Fundamentallösung müssen noch die (1.3.3) entsprechenden Gleichungen

$$\delta B^{n-2}_{n,0} = \omega^{n-1}_{n,0} - \gamma^{n-1}_{n,0}$$

$$\delta B^{n-i}_{n,i-2} = \omega^{n-i+1}_{n,i-2} + (-1)^{i-2}\,\bar{\partial} B^{n-i+1}_{n,i-3} \qquad (2 < i \leq n+1)$$

gelöst werden, wobei $B^{k}_{n,n-k-2} \in C^{k}(\mathfrak{U}, \mathcal{Q}_{n,n-k-2})$ und wir wie üblich $C^{-1}(\mathfrak{U}, \mathcal{Q}_{n,n-1}) := \Gamma(X \times X, \mathcal{Q}_{n,n-1})$ setzen, so dass $B_{n,n-1} = B^{-1}_{n,n-1}$ ein globaler Kern mit $\bar{\partial} B_{n,n-1} = 0$ wird. Zur Lösung der Gleichungen (3.1.1) benutzen wir eine Teilung der 1 für $\mathfrak{U}$, und man sieht sofort, dass mit den $\omega^{k}_{n,n-k-1}$ auch die $B^{k}_{n,n-k-2}$ und somit der globale Kern Singularitäten nur auf der Diagonalen haben kann. Darüberhinaus gilt:

(3.1.2) SATZ. $B_{n,n-1}$ *ist eine reguläre Fundamentallösung der* $\bar{\partial}$*-Gleichung auf* $X \times X$.

Zum Beweis muss man den Aufbau der Singularitäten der $\omega^{k}_{n,n-k-1}$ bzw. der b_k in der induktiven Konstruktion (2.7) verfolgen. Dies ist leicht möglich, da der Operator $\bar{\theta}_v$ konkret vorliegt. Es ergibt sich, dass $B_{n,n-1}$ eine "reguläre Singularität ([15], §2) vom Grade $2n-1$ auf der Diagonalen" besitzt. Dies impliziert die Regularität von $B_{n,n-1}$ ([15], Lemma 2.4). Dass $B_{n,n-1}$ auch Fundamentallösung ist, folgt aus der lokalen Darstellung

$$B^{U}_{n,n-1} = \omega^{U}_{n,n-1} + (-1)^{n-1}\,\bar{\partial} B^{U}_{n,n-2}$$

(bzw. $B^{U}_{1,0} = \omega^{U}_{1,0} - \gamma^{U}_{1,0}$ im Falle $n = 1$), denn für den lokalen BM-Kern gilt ja im distributionellen Sinne $\bar\partial(\omega^U_{n,n-1}) = \delta_\Delta|U$, und $B^U_{n,n-2}$ ist lokal integrabel, so dass wir $\bar\partial$ auch hier im distributionellen Sinn interpretieren können.

Im übrigen ist der Fall $n=1$ jetzt erledigt: da die eindimensionalen BM-Kerne gerade die lokalen Cauchykerne sind, ist auch der globale Kern eine holomorphe Fundamentallösung für $\bar\partial$. Natürlich war für diesen Fall der Toledo-Tong Prozess überflüssig.

(3.2) *Globale Ramirezkerne.* Als nächstes betrachten wir ein streng pseudokonvexes Gebiet $G \subset\subset X$ ($\dim X \geq 2$) mit glattem Rand. Für die folgenden Überlegungen nehmen wir o.B.d.A. an, dass X selbst relativ kompakt in einer grösseren Mannigfaltigkeit X' liegt. Nach einem Satz von Fornaess ([2], Th. 16) finden wir eine Randfunktion $\rho \in \mathcal{C}^\infty(X)$ für G mit $G = \{\rho < 0\}$ und $d\rho \neq 0$ auf ∂G, eine Funktion $\phi \in \mathcal{C}^\infty(X \times X)$, die holomorph in der zweiten Variablen ist, auf der Diagonalen verschwindet und der Abschätzung

$$\text{(3.2.1)} \qquad \operatorname{Re} \phi(\zeta, z) \geq \rho(\zeta) - \rho(z) \qquad (\zeta, z) \in X \times X$$

genügt.

Sei jetzt U ein Element unserer Überdeckung $\mathfrak{U}$ von X, das über der Diagonalen liegt. Nach dem Ramirezschen Divisionssatz lassen sich Funktionen $g_i^U \in \mathcal{C}^\infty(U)$ finden, die ebenfalls holomorph in der zweiten Variablen sind und für die gilt:

$$\text{(3.2.2)} \qquad \phi(\zeta, z) = \sum_i g_i^U(\zeta, z)(\zeta_i - z_i), \qquad (\zeta, z) \in U .$$

Für $U \cap \Delta = \phi$ setzen wir $g_i^U = \phi/n$ und definieren in beiden Fällen $g_U = \sum_i g_i^U e_i$. Somit erhalten wir eine Čech-Kokette

$$\text{(3.2.3)} \qquad g \in C^0(\mathfrak{U}, \mathcal{E}^1) .$$

Es ist $<g,u> = \phi$. Wegen der Holomorphieeigenschaft von ϕ und der Abschätzung (3.2.1) kann jetzt der Operator $\overline{\theta}_g$ für die Konstruktion (2.7.2) herangezogen werden. Dabei schreiben wir diesmal

$$h_i \text{ statt } b_i \text{ und } \eta^{i-1}_{n,n-i} \text{ statt } \omega^{i-1}_{n,n-i} .$$

Die Konstruktion mit $\overline{\theta}_g$ garantiert, dass die h_i und $\eta^{i-1}_{n,n-i}$ holomorph in der zweiten Variablen bleiben. Mann stellt fest, dass die lokalen Koeffizienten von $\eta^0_{n,n-1}$ die üblichen Ramirez-Henkinschen Kerne sind. Die in den h_i und $\eta^{i-1}_{n,n-i}$ auftretenden Singularitäten lassen sich durch Potenzen von ϕ "wegmultiplizieren." Genauer beweist man durch Induktion leicht, dass $\phi^{n-i+1}\eta^{i-1}_{n,n-i}$ singularitätenfrei ist, also in $C^{i-1}(U, \mathcal{C}^\infty_{n,n-i})$ liegt. Darüberhinaus ergibt sich sofort, dass die lokalen Koeffizienten dieser Koketten holomorphe Differentialformen in der zweiten Variablen sind, so dass sich schreiben lässt:

$$\phi^{n-i+1}\eta^{i-1}_{n,n-i} \in C^{i-1}(\mathfrak{U}, \bigoplus_{p+q=n} \mathcal{C}^\infty_{p,n-i} \hat{\otimes} \Omega^q)$$

$(1 \leq i \leq n)$ (bzgl. Tensorprodukten von Fréchetgarben siehe [8]). Aus der Gleichung $\delta(\eta^{n-1}_{n,0} - \gamma^{n-1}_{n,0}) = 0$ (2.3.2) folgt dann

$$\delta(\phi\eta^{n-1}_{n,0} - \phi\gamma^{n-1}_{n,0}) = 0 .$$

Wir haben also einen Čech-Kozykel mit Koeffizienten in $\bigoplus_{p+q=n} \mathcal{C}^\infty_{p,0} \hat{\otimes} \Omega^q$.

Da alle Mengen $U \in \mathfrak{U}$ ohne weiteres als Produkte Steinscher Mengen $U' \times U''$ angenommen werden dürfen, gilt nach der Kaupschen Künneth-formel [8] für alle $U \in \mathfrak{U}$ und $r \geq 1$:

$$H^r(U, \mathcal{C}^\infty_{p,0} \hat{\otimes} \Omega^q) = \bigoplus_{\mu+\nu=r} H^\mu(U', \mathcal{C}^\infty_{p,0}) \hat{\otimes} H^\nu(U'', \Omega^q) .$$

Da aber mindestens ein Faktor in jedem Summanden Null ist, verschwindet diese Kohomologiegruppe. Die Überdeckung U ist also azyklisch für die betrachteten Garben, und nach dem Satz von Leray sind die Kohomologiegruppen für $X \times X$ gleich den Čech-Gruppen für $\mathfrak{U}$.

Da aber mit derselben Argumentation wie oben auch

$$H^{n-1}(X \times X, \bigoplus_{p+q=n} \mathcal{C}^{\infty}_{p,0} \hat{\otimes} \Omega^q) = 0,$$

gibt es ein Element

$$\tilde{H}^{n-2}_{n,0} \in C^{n-2}(\mathfrak{U}, \bigoplus_{p+q=n} \mathcal{C}^{\infty}_{p,0} \hat{\otimes} \Omega^q), \text{ so dass } \delta\tilde{H}^{n-2}_{n,0} = \phi\eta^{n-1}_{n,0} - \phi\gamma^{n-1}_{n,0}.$$

Setzen wir jetzt $H^{n-2}_{n,0} := \frac{1}{\phi} H^{n-2}_{n,0} \in C^{n-2}(\mathfrak{U}, \mathfrak{Q}_{n,0})$, so haben wir eine Lösung

$$\text{(3.2.3)} \qquad \delta H^{n-2}_{n,0} = \eta^{n-1}_{n,0} - \gamma^{n-1}_{n,0}$$

mit den "richtigen" Holomorphieeigenschaften.

Mit denselben Argumenten konstruieren wir jetzt sukzessiv Lösungen

$$\text{(3.2.3')} \qquad \delta H^{n-i}_{n,i-2} = \eta^{n-i+1}_{n,i-2} + (-1)^{i-2}\,\bar{\partial} H^{n-i+1}_{n,i-3} \qquad (2 < i \leq n+1)$$

die ausserhalb der Singularitäten holomorph in der zweiten Variablen sind und für die $\phi^{i+1}H^{n-i}_{n,i-2}$ singularitätenfrei ist.

Für den resultierenden globalen Kern $H = H_{n,n-1} \in \Gamma(X \times X, \mathfrak{Q}_{n,n-1})$, der auf den Mengen $U \in \mathfrak{U}$ die Gestalt

$$H^U_{n,n-1} = \eta^U_{n,n-1} + (-1)^{n-1}\,\bar{\partial} H^U_{n,n-2}$$

hat, ist also $\phi^n H$ eine singularitätenfreie glatte $(n, n-1)$-Form, holomorph in der zweiten Variablen. Ausserhalb der Singularitäten gilt im üblichen Sinne $\bar{\partial}H = 0$.

(3.3) *Lösung der Gleichung* $\bar{\partial}L_{n,n-2} = B_{n,n-1} - H_{n,n-1}$. Bezeichnungen wie in (3.1), (3.2). Genau wie den Satz (2.7.2) beweisen wir

(3.3.1) SATZ. *Es gibt Elemente* $1_k \in C^k(\mathfrak{U}, \mathfrak{K}^{n-k-2})$, $k \in \mathbf{Z}$, *mit den Eigenschaften*:

a) $1_k = 0$ *für* $k < 0$.

b) *Setzen wir* $\tilde{1}_k := \delta 1_k + \sum_{i+j=k} (-1)^i \psi^i 1_j + (-1)^{k+1}(b_{k+1} - h_{k+1})$, *so ist* $\bar{d}\tilde{1}_k = 0$, $1_{k+1} = \bar{\theta}_{\bar{u}}(\tilde{1}_k)$ *und somit* $\bar{d}1_{k+1} = \tilde{1}_k$ *für* $-1 \leq k \leq n-3$

c) $1_{n-2} = (-1)^n(b_{n-1} - h_{n-1})$.

In unserer Kurzschrift dürfen wir schreiben: $\bar{d}1 = \delta 1 + \bar{\psi} 1 + \overline{(b-h)}$. Diejenigen Komponenten der 1_k, die in $C^k(\mathfrak{U}, \mathfrak{Q}_{n,n-k-2})$ liegen, bezeichnen wir mit $\lambda^k_{n,n-k-2}$. Damit ergibt sich aus (3.3.1):

(3.3.2) KOROLLAR.

a) $(-1)^{n-2}\bar{\partial}\lambda^0_{n,n-2} = \omega^0_{n,n-1} - \eta^0_{n,n-1}$

b) $(-1)^{n-k-2}\bar{\partial}\lambda^k_{n,n-k-2} = \delta\lambda^{k-1}_{n,n-k-1} + (-1)^k(\omega^k_{n,n-k-1} - \eta^k_{n,n-k-1})$ *für* $1 \leq k \leq n-2$.

c) $\delta\lambda^{n-2}_{n,0} = (-1)^n(\omega^{n-1}_{n,0} - \eta^{n-1}_{n,0})$.

Indem wir für die Konstruktion (3.3.1) den Operator $\bar{\theta}_{\bar{u}}$ verwendet haben, ist garantiert, dass auch die 1, λ nur Singularitäten auf der Diagonalen bzw. den Nullstellen von ϕ haben.

Betrachten wir nun den δ, $\bar{\partial}$-Doppelkomplex $C^i(\mathfrak{U}, \mathfrak{Q}_{n,j})$, wobei $C^{-1}(\mathfrak{U}, \mathfrak{Q}_{n,j}) := \Gamma(X \times X, \mathfrak{Q}_{n,j})$, und setzen

$$M^j := \bigoplus_{i+q=j} C^i(\mathfrak{U}, \mathfrak{Q}_{n,q}), \qquad M := \bigoplus_{j \geq -1} M^j$$

sowie auf der Komponente $C^i(\mathfrak{U}, \mathfrak{Q}_{n,q})$ $\bar{\delta} := \delta + (-1)^q \bar{\partial}$. Damit ist der totale Komplex

$$(3.3.3) \qquad 0 \longrightarrow M^{-1} \xrightarrow{\bar{\delta}} M^0 \xrightarrow{\bar{\delta}} \cdots \xrightarrow{\bar{\delta}} M^k \xrightarrow{\bar{\delta}} \cdots$$

exakt, denn dazu genügt bekanntlich bereits die Exaktheit der δ-Zeilen des Doppelkomplexes, und die ist durch eine Teilung der Eins gegeben. Fassen wir jetzt die BM-Koketten aus (3.1) zusammen zu

$$\omega := \sum_i \omega^{i-1}_{n,n-i} \in M^{n-1}, \qquad B := \sum_i B^{i-1}_{n,n-i-1} \in M^{n-2},$$

definieren analog die totalen Ramirezketten $\eta \in M^{n-1}$, $H \in M^{n-2}$ und fassen die Elemente aus (3.3.2) zu $\lambda \in M^{n-2}$ zusammen, so folgt aus (2.7.4) bzw. (2.7.5)

$$\overline{\delta}\omega = \overline{\delta}\eta = \overline{\delta}\gamma = (-1)^{n+1} a_n .$$

Die Gleichungen (3.1.1), (3.2.3) und (3.3.2) bedeuten gerade, dass

$$\overline{\delta}B = \omega - \gamma, \quad \overline{\delta}H = \eta - \gamma, \quad \overline{\delta}\lambda = \omega - \eta,$$

wobei aus (3.3.2 a) speziell folgt, dass sich $\Gamma(X \times X, \mathcal{Q}_{n,n-2}) \ni \lambda_{n,n-2} = \lambda^{-1}_{n,n-2} = 0$ setzen lässt. Zunächst ist dann $\overline{\delta}(\lambda - (B-H)) = 0$, und wegen der Exaktheit von (3.3.3) finden wir ein Element $L \in M^{n-3}$ mit

(3.3.4) $$\overline{\partial}L = (B-H) - \lambda .$$

In der Komponente $\Gamma(X \times X, \mathcal{Q}_{n,n-2}) = C^{-1}(\mathfrak{U}, \mathcal{Q}_{n,n-2})$ von M^{n-3} gilt daher die Gleichung

$$(-1)^n \overline{\partial}L_{n,n-2} = B_{n,n-1} - H_{n,n-1} .$$

Da diese globale Lösung $L_{n,n-2}$ mit Hilfe einer Teilung der 1 aus B, H, λ konstruiert wird, kann sie wie diese nur Singularitäten in den Nullstellen von ϕ haben.

(3.4) *Verklebung von B, H und L; Ende der Konstruktion.* Die Ergebnisse von (3.1)-(3.3) lassen sich wie folgt zusammenfassen:

1. Es gibt eine reguläre Fundamentallösung B für $\overline{\partial}$ auf $X \times X$.

2. Zu der in (3.2) betrachteten "Stützflächenfunktion" ϕ für G gibt es eine singuläre $(n, n-1)$-Form H auf $X \times X$, die glatt ausserhalb

der Nullstellen von ϕ, dort holomorph in der zweiten Variablen und $\bar\partial$-geschlossen ist.

3. Es gibt eine ausserhalb der Nullstellen von ϕ glatte (n, n–2)-Form L, die dort die Gleichung $\bar\partial L = B - H$ erfüllt.

Die Abschätzung (3.2.1) für Re ϕ erlaubt nun die Konstruktion einer Funktion $\phi \in \mathcal{C}^\infty(\bar G \times G)$ mit folgenden Eigenschaften:

a) $0 \le \phi \le 1$.

b) $\phi \equiv 0$ in einer Umgebung der Nullstellen von ϕ.

c) Zu kompaktem $K \subset G$ gibt es eine Umgebung $U(\partial G)$ mit $\phi \equiv 1$ auf $U \times K$.

Setzen wir jetzt

$$\Omega = \phi H + (1-\phi) B + \bar\partial\phi \wedge L ,$$

so ist Ω offenbar eine reguläre Fundamentallösung für $\bar\partial$ auf $\bar G \times G$. Wegen c) hat Ω die in (1.1) diskutierte Holomorphieeigenschaft (H). q.e.d.

Nachtrag

Zum selben Thema gibt es jetzt eine Arbeit von Henkin und Leiterer "Global Integral Formulas for Solving the $\bar\partial$-Equation on Stein Manifolds," die in den Annales Polonici Mathematicae erscheinen soll.

Darin werden auch Integralformeln für allgemeine analytische Polyeder hergeleitet. Anders als in unserer Arbeit enthält jedoch die Lösungsformel für $\bar\partial u = f$ auch im streng pseudokonvexen Fall Randintegrale, so dass f auch auf dem Rand definiert und stetig sein muss.

Während wir zur Herleitung der globalen Kerne die Koppelman-Formel lokal benutzten und vor der Aufgabe standen, die entstehenden lokalen Kerne zu verkleben, gehen Henkin und Leiterer von einer Globalisierung der Koppelman-Formel aus, wobei sie die Funktionen $\zeta - z$, denen auf der Mannigfaltigkeit ja keine globale Bedeutung gegeben werden kann, durch einen globalen Schnitt in einem geeigneten Bündel ersetzen.

UNIVERSITÄT KAISERSLAUTERN,
FACHBEREICH MATHEMATIK
PFAFFENBERGSTRASSE, D-6750 KAISERSLAUTERN

LITERATURVERZEICHNIS

[1] Behnke, H., Stein, K., Entwicklung analytischer Funktionen auf Riemannschen Flächen, Math. Ann. 120, 430-461.

[2] Fornaess, J. E., Embedding strictly pseudokonvex domains in convex domains, Amer. J. Math. 98 (1976), no. 2, 529-569.

[3] Grauert, H., Lieb, I., Das Ramirezsche Integral und die Lösung der Gleichung $\bar{\partial}f = \alpha$ im Bereich der beschränkten Formen, Rice University Studies, Vol. 56, no. 2 (1970), 29-50.

[4] Henkin, G. M., Integral representations of functions holomorphic in strictly pseudoconvex domains, and some applications, Math. USSR-Sb 7 (1969), 597-616.

[5] ———, Integral representations of functions in strictly pseudoconvex domains and applications to the $\bar{\partial}$-problem, Math. USSR-Sb 11 (1970), 273-282.

[6] Hörmander, L., An introduction to complex analysis in several variables, Princeton 1966.

[7] Hortmann, M., Uber die Lösbarkeit der $\bar{\partial}$-Gleichung auf Ringgebieten mit Hilfe von L^p-, $\mathcal{C}^k$- und $\mathcal{D}'$-stetigen Integral-operatoren, Math. Ann. 223, Nr. 2, 1976, 139-156.

[8] Kaup, L., Eine Künnethformel für Fréchetgarben, Math. Z. 97 (1967), 157-168.

[9] Kerzman, N., Hölder and L^p estimates for solutions of $\bar{\partial}u = f$ in strongly pseudoconvex domains, Comm. Pure Appl. Math. 24 (1971), 301-380.

[9a] Kohn, J. J., Folland G. B., The Neumann Problem for the Cauchy-Riemann Complex, Annals of Mathematics Studies 75.

[10] Lieb, I., Die Cauchy-Riemannschen Differentialgleichungen auf streng pseudokonvexen Gebieten: Beschränkte Lösungen, Math. Ann. 190 (1970), 6-44.

[11] Øvrelid, N., Integral representation formulas and L^p-estimates for the $\bar{\partial}$-equation. Math. Scand. 29 (1971), 137-160.

[12] ———, Integral representation formulae for differential forms, and solutions of the $\bar{\partial}$ equation, Fonctions analytiques de plusieurs variables et analyse complexe, Paris 1972.

[13] Palm, A., An integral formula for holomorphic functions on strictly pseudoconvex hypersurfaces, Duke Mathe. J. 42 (1975), 347-356.

[14] Ramirez de Arellano, E., Ein Divisionsproblem und Randintegraldarstellungen in der komplexen Analysis, Math. Ann. 184 (1970), 172-187.

[15] Toledo, D., Tong, Y. L., A parametrix for $\bar{\partial}$ and Riemann-Roch in Čech-Theory, Topology 15, (1976), 273-301.

MAPPINGS BETWEEN CR MANIFOLDS

Howard Jacobowitz*

The object of this paper is to prove the following result (see the next section for the relevant definitions).

THEOREM. *Let* M_1 *and* M_2 *be strictly pseudo-convex, integrable* CR *manifolds of the same dimension and let* $\phi: M_1 \to M_2$ *be a* CR *map.*

(a) *If* ϕ *and the* CR *structures are real analytic and* M_1 *is connected, then either* ϕ *is a local diffeomorphism or* $\phi(M_1)$ *is a single point.*

(b) *If* ϕ *and the* CR *structures are* C^∞, *or even* C^k *for* k *large enough, and if* q *is a non-flat point, then* ϕ *is either a local diffeomorphism at* q *or maps an open set containing* q *to a single point.*

This result is known when M_1 and M_2 are real hypersurfaces in $\mathbb{C}^N$ [6], [2, page 549]. In this case the alternative of (a) holds without an assumption of real analyticity. The result is established using the relation between strictly pseudo-convex domains and pluri-sub-harmonic functions. In particular a distinction between flat and non-flat points is not encountered. Note that (a) of our theorem thus also follows from the fact

*This work was started while the author was supported by the NSF grant MCS 77-18723 at the Institute for Advanced Study.

Recent Developments in Several Complex Variables
0-691-08285-5/81/000227-15$00.75/0 (cloth)
0-691-08281-2/81/000227-15$00.75/0 (paperback)

that any real analytic CR structure may be realized as a hypersurface. It is not known if for abstract CR manifolds it is actually necessary to restrict q to be non-flat. This is related to a well-known conjecture which we shall now describe. In light of the above remarks, our theorem would be of interest only if there exist strictly pseudo-convex CR manifolds which cannot be realized as real hypersurfaces. Fortunately there do exist such manifolds, at least in dimension 3 [4], [5]. The examples of non-realizable structures occur in neighborhoods of flat points. It is reasonable to conjecture that every CR structure is realizable in a neighborhood of each non-flat point.* Since every real analytic structure is realizable, this conjecture fits in nicely with our result. Clearly it would be very interesting to have an example of a CR map which is neither a local diffeomorphism nor a constant map.

It is known that the theorem is false for M_1 only weakly pseudo-convex [3], [6]. However one may speculate that a CR map between weakly pseudo-convex manifolds is either open or reduces to a constant map. The theorem is also false if M_1 is strongly pseudo-convex but M_2 is only non-degenerate. For consider the map $\phi: \{z \in \mathbf{C}^3 | \operatorname{Im} z_3 = |z_1|^2 + |z_2|^2\} \to \{z \in \mathbf{C}^3 | \operatorname{Im} z_3 = |z_1|^2 - |z_2|^2\}$ given by $\phi(z_1, z_2, z_3) = (z_1, z_1, 0)$. This example also shows that our theorem is not merely a consequence of the existence of Cartan connections.

Finally each CR structure has an underlying contact structure and so it is natural to ask if every nontrivial contact map must be a diffeomorphism. By a contact map we mean one under which the contact form on the range is pulled back to a multiple, zero not excluded, of the contact form on the domain. A simple example shows this is not so. Indeed the following general result is valid.

LEMMA. *Let* ω_1 *be a one-form in a neighborhood of* $0 \in \mathbf{R}^p$ *and* ω_2 *a one-form in a neighborhood of* $0 \in \mathbf{R}^q$ *with* $p > 1$. *There exists a map* $f: U \to V$ *where* U *and* V *are neighborhoods of* 0 *in* $\mathbf{R}^p$ *and* $\mathbf{R}^q$

**Added in proof:* This conjecture is now known to be false. Any CR structure may be perturbed in the neighborhood of any point to obtain a non-realizable CR structure. See a forthcoming paper with F. Treves.

such that $f^*\omega_2 = \lambda\omega_1$ *but* f *is neither a local diffeomorphism nor a constant map.*

Proof. We construct such a map with $f^*\omega_2 = 0$. Let Y be a vector field near $0 \in \mathbf{R}^q$ which is not zero at the origin and which is annihilated by ω_2. Introduce coordinates $(y_1, \cdots, y_q)$ near the origin with $\frac{\partial}{\partial y_1} = Y$. Let $(x_1, \cdots, x_p)$ be coordinates near $0 \in \mathbf{R}^p$. Consider the map $f(x_1, \cdots, x_p) = (x_1, 0, \cdots, 0) \in \mathbf{R}^q$. Then $f^*\omega_2 = 0$.

§2. *Basic definitions*

Let M be a manifold of dimension $2n+1$. The following three structures are locally equivalent.

I) Let θ be a real valued one form on M and $\theta^1, \cdots, \theta^n$ complex valued one forms with

$$\theta \wedge \theta^1 \wedge \cdots \wedge \theta^n \wedge \overline{\theta}^1 \wedge \cdots \wedge \overline{\theta}^n \neq 0 .$$

II) Let $H \subset TM$ be a smooth distribution of planes of dimension $2n$ and let $J: H \to H$ be a smoothly varying linear map preserving the fibres of H with $-J^2$ equal to the identity.

III) Let $V \subset \mathbf{C} \otimes TM$ be a sub-bundle of complex fibre dimension n with $V \cap \overline{V}$ equal to the zero sub-bundle.

The equivalence $I \Rightarrow II \Rightarrow III \Rightarrow I$ is easily established. Given the structure in I let $H = \theta^{\perp}$. Extend the action of each θ^j and $\overline{\theta}^j$ to $\mathbf{C} \otimes TM$. For each $X \in H$ there is a unique $Y \in H$ for which $X + iY$ is annihilated by each θ^j. Set $JX = Y$. Given H and J extend J to a map of $\mathbf{C} \otimes H$ to itself and let V be its $+i$ eigenspace. Finally given V let θ be a real form annihilating $\{\mathrm{Re}\, V\} \cup \{\mathrm{Im}\, V\}$ and let $\{\theta^1, \cdots, \theta^n\}$ be some basis for $\overline{V}^{\perp}$ chosen such that $\theta \wedge \theta^1 \wedge \cdots \wedge \theta^n \neq 0$. Note that II and III actually correspond to global structures and that the equivalence with I involves various choices.

DEFINITION. A CR manifold is an odd dimensional manifold together with one of the above structures.

CR is an abbreviation for Cauchy-Riemann and emphasizes the role played by the induced Cauchy-Riemann equations on such manifolds. These manifolds are also called pseudo-conformal manifolds.

For our work I is the most useful description of the CR structure, and the one we primarily use.

We say that a CR manifold is real analytic if for some choice of local coordinates the forms $\theta, \theta^1, \cdots, \theta^n$ are real analytic.

The most important class of CR manifolds is provided by the real hypersurfaces in $\mathbf{C}^{n+1}$. Let M^{2n+1} be such a hypersurface and let r be a smooth function on $\mathbf{C}^{n+1}$ with $M = \{x \in \mathbf{C}^{n+1} | r(x) = 0\}$ and $dr \neq 0$ on M. Assume that in terms of the usual coordinates and near some point $p \in M$ we have $\frac{\partial r}{\partial x_{n+1}}(p) \neq 0$. Set $\theta = i\left(\frac{\partial r}{\partial z_k} dz_k - \frac{\partial r}{\partial \bar{z}_k} d\bar{z}_k\right)$ and $\theta^j = dz^j$, $j = 1, \cdots, n$ (all forms are restricted to M). Then $\{\theta, \theta^1, \cdots, \theta^n\}$ define a CR structure on a neighborhood of p in M. If instead of M we were to consider $\Phi(M)$ where Φ is some local biholomorphism defined on a neighborhood of p in $\mathbf{C}^{n+1}$, then the CR structure on $\Phi(M)$ would be isomorphic in a natural way to that on M. The importance of the CR geometry is derived from this—invariants of the CR structure provide local biholomorphic invariants for domains in $\mathbf{C}^{n+1}$.

The structure on the sphere, $S^{2n+1} \subset \mathbf{C}^{n+1}$, serves as a model for all CR geometries. To describe this structure it is easiest to work with the image of S^{2n+1} under a certain Cayley transform. Let $z_{n+1} = u + iv$ and $r(z_1, \cdots, z_{n+1}) = v - |z_1|^2 - \cdots - |z_n|^2$. This image, called the hyperquadric, is given by

$$Q = \{z \in \mathbf{C}^{n+1} | r(z) = 0\} .$$

Thus our forms on Q are

$$\begin{aligned} \theta &= du + iz_\alpha d\bar{z}^\alpha - i\bar{z}_\alpha dz^\alpha \\ \theta^\alpha &= dz^\alpha . \end{aligned}$$

In this paper we use the convention that the indices $a, b, \cdots, \alpha, \beta, \cdots$ run from 1 to n and that repeated indices are summed.

We note for later that the subspace $\overline{V} \subset \mathbf{C} \otimes TM$ which is annihilated by each of $\theta, \theta^1, \cdots, \theta^n$ is spanned by the vectors $L^\alpha = \frac{\partial}{\partial \bar{z}^\alpha} - iz_\alpha \frac{\partial}{\partial u}$, $\alpha = 1, \cdots, n$. These partial differential operators are the Lewy operators. They provided the first example of non-solvable linear differential equations.

We also note for later that

$$d\theta = 2i\theta^k \wedge \overline{\theta}^k .$$

The CR structure of a real hypersurface has an integrability property which we shall also require of abstract CR manifolds.

DEFINITION. A CR manifold is said to be integrable if $d\theta = d\theta^j = 0$ mod $\{\theta, \theta^1, \cdots, \theta^n\}$.

All the CR manifolds in this paper are assumed to be integrable and this adjective is often omitted. The manifold $\mathbf{C}^n \times \mathbf{R}$ has an obvious CR structure. But, as a consequence of its co-dimension one foliation by complex manifolds, its properties are quite different from those of generic CR manifolds. Note that since θ is real one has, for an integrable structure, that $d\theta = ig_{\alpha\beta}\theta^\alpha \wedge \overline{\theta}^\beta$ mod θ where $g_{\alpha\beta}$ is an hermitian matrix.

DEFINITION. A CR manifold is *non-degenerate* if the matrix $g_{\alpha\beta}$ is invertible. A CR manifold is *strictly pseudo-convex* if the matrix $g_{\alpha\beta}$ is definite (and then replacing θ by $-\theta$ if necessary we take it to be positive definite).

Again all CR manifolds in this paper are assumed to be strictly pseudo-convex.

The CR structure is said to be weakly pseudo-convex if $g_{\alpha\beta}$ is semi-definite.

Note that the CR structure of the hyperquadric Q is strictly pseudo-convex.

Let $\{\theta_k, \theta_k^1, \cdots, \theta_k^n\}$ be a CR structure on M_k, $k = 1, 2$.

DEFINITION. A smooth map $\phi: M_1 \to M_2$ is a *CR map* if $\phi^*(\theta_2) \in$ linear span $\{\theta_1\}$ and $\phi^*(\theta_2^\alpha) \in$ linear span $\{\theta_1, \theta_1^1, \cdots, \theta_1^n\}$.

As we shall see a CR map ϕ fails to be a diffeomorphism only at those points where it is a constant up to infinite order. Our starting point will be the observation that on any strictly pseudo-convex CR manifold coordinates may be chosen in such a way that the CR structure agrees with the structure of the hyperquadric Q up to some order. The following result provides a measure of how different the given structure actually is from that of Q. As we show in the last section the map ϕ can fail to be a diffeomorphism and yet still be non-trivial only at those points where the structure agrees with that of Q to high order.

The point of the next result is that if M has a CR structure in the sense of II or III, then we may choose forms satisfying I. By going to a bundle over M which represents all choices of these forms we may define a set of differential forms (more precisely a connection) which is inherent to the structure II, or III, and does not depend on any choices.

THEOREM ([1]). *Let* M *be a strictly pseudo-convex CR manifold. There exists a principal fibre bundle* Y *over* M *and a Cartan connection* $\omega: Y \to \mathfrak{su}(1,n)$ *with the property that if a diffeomorphism* $\phi: M_1 \to M_2$ *is a CR map, then* ϕ *induces a connection preserving map of the corresponding bundles.*

The curvature of this connection provides a measure of the difference between M and Q. If the curvature vanishes along all the fibres of Y over some open set U (i.e. the connection is flat over U), then U is locally diffeomorphic to Q via CR maps. If the curvature vanishes at each point of the fibre of Y over $p \in M$, then p is called a flat point. (In some other contexts such a point is called an umbilic point.)

§3. *The equations*

We introduce special coordinates on a CR manifold. Fix some point $q \in M$. Let $S: \mathbf{C}^n \times \mathbf{R}^1 \to M$ with $S(0) = q$ and $S_* T\mathbf{C}^n|_0$ the 2n plane of H at q. The map S gives coordinates $(z_1, \cdots, z_n, u)$ on M for which V at q is $\left\{\frac{\partial}{\partial z_1}, \cdots, \frac{\partial}{\partial z_n}\right\}$. A function $k(z_1, \cdots, z_n, \overline{z}_1, \cdots, \overline{z}_n, u)$ on a neighborhood of $q \in M$ is said to be of weight r (with respect to the given coordinate system) if $k(tz, t\overline{z}, t^2u) = t^r k(z, \overline{z}, u)$, $t \in \mathbf{R}$. For any function $k(z, \overline{z}, u)$ we may regroup the terms in its Taylor series, with remainder, to obtain a decomposition $k = \sum_1^N k_j + R$ where k_j is of weight j. The function k is said to have weight greater than r if $k_j = 0$ for $j = 0, 1, \cdots, r$.

Note that a given CR structure $\{\tilde{\theta}, \tilde{\theta}^1, \cdots, \tilde{\theta}^n\}$ may also be described by any other forms $\{\theta, \theta^1, \cdots, \theta^n\}$ where θ is a non-zero real multiple of $\tilde{\theta}$ and $\{\theta, \theta^1, \cdots, \theta^n\}$ has the same linear span as $\{\tilde{\theta}, \tilde{\theta}^1, \cdots, \tilde{\theta}^n\}$. The conditions for integrability and pseudo-convexity are unaffected by such changes in the defining forms. The next lemma shows that it is possible to choose the forms and the local coordinates so as to simplify the defining equations. We omit the proof. It is a simple consequence of the integrability condition.

LEMMA. *Let* M *be a strictly pseudo-convex CR manifold and* q *some point of* M. *There exists a local coordinate system* $(z^1, \cdots, z^n, u)$ *with origin at* q *and forms defining the CR structure such that*

$$\theta = du + (iz^\alpha + k^\alpha)\, d\overline{z}^\alpha + (-i\overline{z}^\alpha + \overline{k}^\alpha)\, dz^\alpha$$

$$\theta^\alpha = dz^\alpha + h^{\alpha\beta}\, d\overline{z}^\beta$$

where $k^\alpha = k^\alpha(z, \overline{z}, u)$ *and* $h^{\alpha\beta} = h^{\alpha\beta}(z, \overline{z}, u)$ *are of weight greater than* 1 *with respect to this coordinate system.*

Note that, as we have already seen, we can take $k = h = 0$ for the hyperquadric.

We derive the equations that must be satisfied by a CR map between strictly pseudo-convex manifolds in terms of these coordinates. Assume the CR structure on M_1 is given by

$$\theta = du + (iz^\alpha + k^\alpha)\,d\bar{z}^\alpha + (-i\bar{z}^\alpha + \bar{k}^\alpha)\,dz^\alpha$$

$$\theta^\alpha = dz^\alpha + h^{\alpha\beta}\,d\bar{z}^\beta$$

with $h(z, \bar{z}, u)$ and $k(z, \bar{z}, u)$ of weight greater than 1. Assume the CR structure on M_2 is given by

$$\Theta = dU + (iZ^\alpha + K^\alpha)\,d\bar{Z}^\alpha + (-i\bar{Z}^\alpha + \bar{K}^\alpha)\,dZ^\alpha$$

$$\Theta^\alpha = dZ^\alpha + H^{\alpha\beta}\,d\bar{Z}^\beta$$

with $H(Z, \bar{Z}, U)$ and $K(Z, \bar{Z}, U)$ of weight greater than 1. Let $\Phi^*(\Theta^\alpha) = 0 \mod\{\theta, \theta^\beta\}$ and $\Phi^*(\Theta) = \lambda\theta$. In terms of our local coordinates we have $\Phi = (F_1, \cdots, F_n, f)$ where f is real. These functions depend on the variables $(z_1, \cdots, z_n, u)$. It is easy to see that the equations Φ satisfies take the form

$$L^c F^\alpha = k^c F^\alpha_u + H^{\alpha\beta} \circ \Phi(-L^c \bar{F}^\beta + k^c \bar{F}^\beta_u)$$
$$+ h^{\beta c}(\bar{L}^\beta F^\alpha - F^\alpha_u \bar{k}^\beta + (H^{\alpha\delta} \circ \Phi)(\bar{L}^\beta \bar{F}^\delta - k^\beta \bar{F}^\delta_u))$$

and

$$L^c f = k^c f_u + (iF^\alpha + K^\alpha \circ \Phi)(-L^c \bar{F}^\alpha + k^c \bar{F}^\alpha_u)$$
$$+ (i\bar{F}^\alpha - \bar{K}^\alpha \circ \Phi)(L^c F^\alpha - k^c F^\alpha_u)$$

where again L^c is the Lewy operator.

In the second equations f is required to be real. This means that the above system is overdetermined even when $n = 1$. We now examine the compatibility conditions.

LEMMA. *The system*

$$L^\gamma f = A^\gamma$$
$$f \ \text{real}$$

has a solution only if A *satisfies*

$$L^bA^a - L^aA^b = 0 \qquad \textit{all } a \textit{ and } b$$

$$\bar{L}^bA^a - L^a\bar{A}^b = 0 \qquad \textit{all } a \textit{ and } b \textit{ with } a \neq b$$

$$L^c(L^b\bar{A}^b - \bar{L}^bA^b) = 2iA^c_u \qquad \textit{all } c \textit{ and } b, \textit{ no summation over } b.$$

Proof. Because f is real we also have the equations

$$\bar{L}^\gamma f = \bar{A}^\gamma .$$

The lemma is a simple consequence of the following commutativity relations for the Lewy operator

$$\left[L^c, \frac{\partial}{\partial u}\right] = 0 \qquad ($$

$$[L^a, L^b] = 0$$

$$[L^a, \bar{L}^b] = 0 \qquad \text{if } a \neq b$$

$$[L^b, \bar{L}^b] = 2i\frac{\partial}{\partial u} \qquad \text{(no summation)}.$$

We shall be particularly interested in this system of compatibility equations when A^c has the special form $A^c = iL^c\|F\|^2$. Here $\|F\|^2 = F^\beta\bar{F}^\beta$. In this case the compatibility conditions become

$$L^a\bar{L}^b\|F\|^2 = 0 \qquad \text{for } a \neq b$$

$$L^cL^b\bar{L}^b\|F\|^2 = 0 \qquad \text{for all } c \text{ and } b \text{ (no summation)}.$$

We will also need the following simple uniqueness result.

LEMMA. *If* f *is a real valued function of* $(z, \bar{z}, u)$ *and satisfies*

$$L^\gamma f = 0, \quad \gamma = 1, \cdots, n,$$

then f *is a constant.*

§4. *Determination of the coefficients*

We first show that if a CR map fails to be a diffeomorphism at some point, then all its derivatives must vanish at that point.

THEOREM. *Let* $\Phi: M_1 \to M_2$ *be a CR map between strictly pseudoconvex manifolds of the same dimension. If* $d\Phi$ *is not an isomorphism at* $q \in M_1$, *then the derivatives of all orders of* Φ *at* q *are zero.*

Proof. We introduce coordinate systems as in §3 with origins at q and $\Phi(q)$. *Write* $\Phi = (F^1, \cdots, F^n, f)$. *Then*

$$L^\gamma F^\alpha = k^\gamma F_u^\alpha + (H^{\alpha\beta} \circ \Phi)(-L^\gamma \bar{F}^\beta + k^\gamma \bar{F}_u^\beta)$$
$$+ h^{\beta\gamma}(\bar{L}^\beta F^\alpha - F_u^\alpha \bar{k}^\beta + H^{\alpha\delta}(\bar{L}^\beta \bar{F}^\delta - \bar{k}^\beta \bar{F}_u^\delta))$$

and

$$L^\gamma f = k^\gamma f_u + (iF^\alpha + K^\alpha \circ \Phi)(-L^\gamma \bar{F}^\alpha + k^\gamma \bar{F}_u^\alpha)$$
$$+ (i\bar{F}^\alpha - \bar{K}^\alpha \circ \Phi)(L^\gamma F^\alpha - k^\gamma F_u^\alpha).$$

Let us use a weight decomposition

$$F^\alpha = \sum F_j^\alpha, \quad f = \sum f_j .$$

Since h, k, H and K are of weight at least two, we see that

$$L^\gamma f_1 = \bar{L}^\gamma f_1 = 0$$
$$L^\gamma F_1^\alpha = 0$$
$$L^\gamma f_2 = -iF_1^\alpha L^\gamma \bar{F}_1^\alpha .$$

From the first two equations we see that without assumptions on $d\Phi$, at the origin $\frac{\partial f}{\partial z^\gamma} = \frac{\partial f}{\partial \bar{z}^\gamma} = \frac{\partial F^\alpha}{\partial \bar{z}^\gamma} = 0$. Further since f is real and $[L^\gamma, \bar{L}^\beta] = 2i\delta_{\gamma\beta} \frac{\partial}{\partial u}$ the third equation implies

$$f_u(0)\,\delta_{\beta\gamma} \;=\; \frac{\partial F^\alpha}{\partial z^\beta}(0)\,\frac{\partial \overline{F}^\alpha}{\partial \overline{z}^\gamma}(0)\,.$$

(In particular note that if $\Phi^*(\theta_2) = \lambda\theta_1$, then λ is non-negative.)

Thus for $\mathfrak{M}_{\beta\alpha} = \dfrac{\partial F^\alpha}{\partial z^\beta}$ we have

$$\det \mathfrak{M}_{\beta\alpha}(0) \;=\; (f_u(0))^{n/2}\,.$$

Let $z^\alpha = x^\alpha + iy^\alpha$ and $F^\alpha(x,y,u) = X^\alpha(x,y,u) + iY^\alpha(x,y,u)$. If $\mathfrak{J}$ is the Jacobian matrix of the map Φ thought of as a transformation $(x,y,u) \to (X,Y,U)$, then $\det \mathfrak{J} = (\det \mathfrak{M})^2 f_u(0) = f_u(0)^{n+1}$.

So if $d\Phi$ is not a diffeomorphism at q, then f_{z^γ}, $f_{\overline{z}^\gamma}$, f_u, $F^\alpha_{z^\beta}$ and $F^\alpha_{\overline{z}^\beta}$ are all zero at the origin. We now iteratively prove that the derivatives of F^α and f of all orders are zero at the origin.

LEMMA. *Let the CR map* $\Phi = (F,f)$ *have a weight decomposition*

$F = \sum F_k$, $f = \sum f_j$ *with respect to the above coordinates. If* $F_k = 0$,

$k = 1, \cdots, R$, *then*

(1) $f_j = 0$, $j = 1, \cdots, 2R+1$

(2) $\overset{\bullet}{F}_{R+1} = 0$.

Proof. Under the hypotheses of the lemma, we see that f satisfies the system of equations

$$L^\gamma f \;=\; k^\gamma f_u + K^\alpha(F,f)(-L^\gamma \overline{F}^\alpha + k^\gamma \overline{F}^\alpha_u)$$

$$+\; \overline{K^\alpha(F,f)}(L^\gamma F^\alpha - k^\gamma F^\alpha_u) + Q$$

where Q consists of terms of weight greater than $2R$. Let f_p be the first non-zero term in the weight decomposition of f. Since $K(Z,U)$ is of weight greater than 1, weight $k(F,f) \geq \min(2(R+1), p)$. Assume $p < 2R+2$ and look at the terms of weight $p-1$ in the above equation to obtain $L^\gamma f_p = 0$. We use a previous lemma to conclude that $f_p = 0$, and this holds for all p, $p = 1, \cdots, 2R+1$.

We now turn to part (2) of our lemma.

For F_{R+1} and f_{2R+2} we have the equations

$$L^\gamma F^\alpha_{R+1} = 0$$

$$L^\gamma f_{2R+2} = -iF^\alpha_{R+1} L^\gamma \overline{F}^\alpha_{R+1} .$$

The second equation may be rewritten, using the first equation, as

$$L^\gamma(-f_{2R+2}) = iL^\gamma \|F_{R+1}\|^2 .$$

As we have seen, the compatibility conditions for such an equation take the form

$$L^a \overline{L}^b \|F_{R+1}\|^2 = 0 \qquad \text{for } a \neq b$$

$$L^c L^b \overline{L}^b \|F_{R+1}\|^2 = 0 \qquad \text{(no summation)}.$$

This last equation may be rewritten, again using the first equation above, as

$$L^c \|\overline{L}^b F_{R+1}\|^2 = 0 .$$

By considering the conjugate of this equation and the relation $[L^c, \overline{L}^c] = 2i \frac{\partial}{\partial u}$ we derive that $\|\overline{L}^b F_{R+1}\|^2$ is a constant. It follows that $F^\beta_{R+1} = 0$ unless $R = 0$.

This concludes the proof of the lemma and by induction the proof of the theorem at the start of this section. Now (a) of our main theorem in §1 follows by analytic continuation. In the next section we establish (b).

§5. *Non-flat points*

We want to show that if ϕ is neither locally a constant nor a local diffeomorphism at a point $q \in M_1$, then q is a flat point of the CR structure of M_1. Our proof is based on Webster's work on pseudo-hermitian structures [7] which we first briefly summarize. We restrict ourselves to the case $\dim M > 3$. A slightly different argument is used for $\dim M = 3$.

Let M have an integrable strongly pseudo-convex CR structure and let θ be some choice of the real one form which annihilates H. (Any other choice is then given by $\tilde{\theta} = \lambda\theta$.) Then (M, θ) is called a pseudo-hermitian structure. A CR diffeomorphism $\Phi: M_1 \to M_2$ which satisfies $\Phi^*\theta_2 = \theta_1$ is called a pseudo-hermitian map. We write $\Phi: (M_1, \theta_1) \to (M_2, \theta_2)$.

Let P be the bundle of coframes $\{\theta, \theta^\alpha, \overline{\theta^\alpha}\}$ with $\theta^\alpha \in T^{1,0^*}(M)$ and $d\theta = ig_{\alpha\beta}\theta^\alpha \wedge \overline{\theta^\beta} \bmod \theta$ for some positive definite matrix g. The forms $\theta, \theta^\alpha, \overline{\theta^\alpha}$ on M lift to intrinsic global forms on P. These can be extended to an intrinsic basis $\{\theta, \theta^\alpha, \overline{\theta^\alpha}, \omega^\alpha_\beta, \overline{\omega^\alpha_\beta}\}$. We also obtain extra globally defined forms τ^α and the functions $g_{\alpha\beta}$. Consider the curvature $\Omega^\alpha_\beta = d\omega^\alpha_\beta - \omega^\gamma_\beta \wedge \omega^\alpha_\gamma - i\theta_\beta \wedge \tau^\alpha + i\tau_\beta \wedge \theta^\alpha$. It satisfies $\Omega_{\beta\bar{\alpha}} = R_{\beta\bar{\alpha}\rho\bar{\sigma}}\theta^\rho \wedge \overline{\theta^\sigma}$ (mod θ). R is a pseudo-hermitian invariant but not a pseudo-conformal one since the above construction depended on a choice of θ. However the pseudo-conformal curvature tensor $S_{\beta\rho\bar{\sigma}}$ introduced in [1] can be expressed in terms of R using a formula similar to Weyl's formula for the conformal curvature tensor of a Riemannian metric.

Choosing a section $M \to P$ allows one to think of S as being a set of functions on M. These functions transform tensorially. So the norm of S is well defined. It is denoted by $\|S\|_\theta$ to emphasize that it depends only on the original choice of θ. This norm has the following three properties.

1. If $\tilde{\theta} = \lambda\theta$, then $\|S\|_\theta = |\lambda|\, \|S\|_{\tilde{\theta}}$.

2. If at a point $q \in M$, $\|S\|^2_\theta$ and sufficiently many of its derivatives vanish, then the CR structure of M is flat at that point.

3. If $\Phi: M_1 \to M_2$ is a CR diffeomorphism and $\Phi^*(\theta_2) = \theta_1$, then $\|S_1\|_{\theta_1} = \|S_2\|_{\theta_2}$ where S_i is our tensor defined on M_i.

Now consider a CR map $\Phi : M_1 \to M_2$. Fix some choice for θ_1 and θ_2 and consider (M_1, θ_1) and (M_2, θ_2) as pseudo-hermitian structures. We have, by the definition of a CR map,

$$\Phi^*(\theta_2) = \lambda\theta_1 .$$

Let $Q \subset M_1$ be the set on which Φ is not a local diffeomorphism. Thus all the derivatives of Φ vanish on Q and so λ vanishes to infinite order on Q. Let $q \in Q$. If q is an interior point of Q, then Φ is a constant on the connected component of Q containing q. We must only show that if q is a boundary point of Q, then M_1 is flat at q. We do this by showing that at q, $\|S_1\|^2_{\theta_1}$ vanishes along with all its derivatives.

So let $q \in \partial Q$ and take some sequence $q_i \to q$ with Φ a local diffeomorphism in an open neighborhood of each q_i. Let P be the union of these open neighborhoods.

The map $\Phi : (M_1, \lambda\theta_1) \to (M_2, \theta_2)$ is pseudo-hermitian on P. Therefore $\|S_1\|_{\lambda\theta_1} = \|S_2\|_{\theta_2}$ on P. But $\|S_1\|_{\theta_1} = |\lambda|\,\|S_1\|_{\lambda\theta_1} = |\lambda|\,\|S_2\|_{\theta_2}$. As we approach the point q along any path in P, we have that λ and all its derivatives approach 0. Therefore $\|S_1\|_{\theta_1}$ and its derivatives vanish at q and so q is a flat point.

For $\dim M = 3$, i.e. $n = 1$, we must work with a different component of the curvature. The component R introduced by Cartan satisfies $\|R\|_{\lambda\theta} = |\lambda|^{-2}\,\|R\|_\theta$. Again, if R and enough of its derivatives are 0 at a point q then all the components of the curvature are 0 at q and q is a flat point.

RUTGERS UNIVERSITY
CAMDEN, NEW JERSEY 08003

REFERENCES

[1] Chern, S. S. and Moser, J., Real hypersurfaces in complex manifolds, Acta Math. 133 (1974), 219-271.

[2] Fornaess, J., Embedding strictly pseudoconvex domains in convex domains, Amer. J. Math. 98 (1976), 529-569.

[3] ———, Biholomorphic mappings between weakly pseudo-convex domains, Pacific J. Math. 74 (1978), 63-65.

[4] Nirenberg, L., Lectures on linear partial differential equations, Regional Conference Series in Mathematics, No. 17, Amer. Math. Soc., 1973.

[5] ———, On a question of Hans Lewy, Russian Math. Surveys 29 (1974), 251-262.

[6] Pinchuk, S., On proper holomorphic mappings of strictly pseudoconvex domains, Siberian Math. J., 15 (1974), 644-649.

[7] Webster, S., Pseudo-hermitian structures on a real hypersurface, Journal of Differential Geometry 13 (1978), 25-41.

BOUNDARY REGULARITY OF $\bar{\partial}$

J. J. Kohn

This article presents a brief summary of some recent results on boundary regularity of $\bar{\partial}$ as well as a discussion of some of the open problems in this area. For simplicity, I will restrict myself to domains in $\mathbf{C}^n$, although most of this material has generalizations to domains in complex manifolds. Let $\Omega \subset \mathbf{C}^n$ be a bounded domain with a C^∞ boundary $b\Omega$. $L_2^{p,q}(\Omega)$ will denote the space of (p,q)-forms with square-integrable components with Lebesgue measure and standard Euclidian inner product on forms. (p doesn't play any role in the regularity results discussed here.) Consider the $\bar{\partial}$-equation

$$\bar{\partial}\phi = a \tag{1}$$

where $\phi \in L_2^{p,q-1}(\Omega)$, $a \in L_2^{p,q}(\Omega)$ and

$$\bar{\partial}a = 0 . \tag{2}$$

Given $a \in L_2^{p,q}(\Omega)$ satisfying (2), we are concerned with the boundary regularity of the solution ϕ of (1).

DEFINITION. $\bar{\partial}$ is *hypoelliptic for* (p,q)-*forms* on Ω if for each $a \in L_2^{p,q}(\Omega)$ which satisfies (2) there exists $\phi \in L_2^{p,q-1}(\Omega)$ such that (1) holds and

$$\text{sing. supp. } (\phi) = \text{sing. supp. } (a) , \tag{3}$$

Recent Developments in Several Complex Variables
0-691-08285-5/81/000243-18 \$00.90/0 (cloth)
0-691-08281-2/81/000243-18 \$00.90/0 (paperback)
For copying information, see copyright page.

where sing. supp. denotes the C^∞ singular support with respect to the closed domain $\overline{\Omega}$. More precisely, if $u \in L_2(\Omega)$ then the complement of sing. supp. (u) consists of all the points $x \in \overline{\Omega}$ such that x has a neighborhood U with the property that the restriction of u to $U \; \overline{\Omega}$ is in $C^\infty(U \cap \overline{\Omega})$. The singular support of a form is the union of the singular supports of its components.

PROBLEM I. When is $\bar{\partial}$ hypoelliptic for (p,q)-forms on Ω?

For domains Ω in which $\bar{\partial}$ is hypoelliptic consider the following problem.

PROBLEM II. If $\phi \in L_2^{p,q-1}(\Omega)$ is the unique solution of (1) which is orthogonal to the null space of $\bar{\partial}$, does ϕ satisfy (3)?

In the case $q = 1$ the solution of these problems can be used to study boundary behavior of holomorphic functions. In particular, the solution of problem II is connected with the Bergman projection operator as follows. Let $P : L_2(\Omega) \to L_2(\Omega)$ be the orthogonal projection onto the subspace $\mathcal{H}$ of square-integrable holomorphic functions. Suppose that $f \in L_2(\Omega)$ and that $\bar{\partial} f \in L_2^{0,1}(\Omega)$, let u be the unique solution of $\bar{\partial} u = \bar{\partial} f$ which is orthogonal to $\mathcal{H}$, then $Pf = f - u$. Thus, if u has property (3) then sing. supp. (Pf) $\subset$ sing. supp. (f). Furthermore, the regularity of the Bergman kernel function is intimately related to regularity properties of P (see [22b]).

Observe that if $z^0 \in \Omega$ and if α is C^∞ in a neighborhood of z^0, then every solution ϕ of (1) is C^∞ in a neighborhood of z^0. This follows from the fact that (1) is an elliptic system. If $z^0 \in b\Omega$ then, in general, there are many solutions ψ of $\bar{\partial}\psi = 0$ which are not smooth in a neighborhood of z^0, thus if ϕ is a solution of (1) which is smooth around z^0 then $\phi + \psi$ is a solution which violates (3).

The fact that $b\Omega$ is smooth means that there exists a C^∞ real-valued function r defined in a neighborhood of $b\Omega$ and that $dr \neq 0$, $r > 0$ outside of $\overline{\Omega}$, $r = 0$ on $b\Omega$ and $r < 0$ in Ω. We recall that Ω is *pseudo-*

convex if for every $P \in b\Omega$ we have

$$\sum r_{z_i\bar{z}_j}(P)\zeta^i\zeta^j \geq 0\,, \tag{4}$$

whenever

$$\sum r_{z_i}(P)\zeta^i = 0\,. \tag{5}$$

The existence of square-integrable solutions of (1) on weakly pseudo-convex domains was proved by L. Hörmander in [20a]. The existence of globally regular solutions (see [23a] and [23b]) is given by the following result.

THEOREM. *If* $\Omega \subset \mathbf{C}^n$ *is pseudo-convex with a* C^∞ *boundary and if* a *is a* (p,q)*-form on* Ω, *with* $q \geq 1$, $\bar{\partial}a = 0$ *and sing. supp.* $(a) = \emptyset$ (*i.e. the components of* a *are in* $C^\infty(\bar{\Omega})$), *then there exists a solution* $\phi \in C^\infty(\bar{\Omega})$, *i.e. such that* sing. supp. $(\phi) = \emptyset \cdot (\phi) = \emptyset$.

This gives rise to the following open problem.

PROBLEM III. Given $\Omega \subset \mathbf{C}^n$ pseudo-convex with C^∞ boundary. If a is a (p,q)-form in $C^\infty(\bar{\Omega})$ satisfying (2), is the solution ϕ of (1), which is orthogonal to the null space of $\bar{\partial}$, in $C^\infty(\bar{\Omega})$?

This, in my opinion, is a very important problem because in conjunction with the following recent result of Bell and Ligocka (see [2]) it can be used to generalize the result of C. Fefferman [12a] on boundary regularity of holomorphic mappings. Recently Nirenberg, Webster and Yang found a proof of Fefferman's theorem which does not use "$\bar{\partial}$ methods" (see [28]).

THEOREM (Bell and Ligocka). *Suppose that* Ω *and* Ω' *are domains in* $\mathbf{C}^n$ *such that* $b\Omega$ *and* $b\Omega'$ *are* C^∞ *and such that the Bergman projection operators* P *and* P' *preserve smoothness up to the boundary (that is,* $PC^\infty(\bar{\Omega}) \subset C^\infty(\bar{\Omega})$ *and* $P'C^\infty(\bar{\Omega}) \subset C^\infty(\bar{\Omega}')$). *Furthermore, suppose that there is a one-to-one holomorphic map* $h: \Omega \to \Omega'$. *Then* h *has a unique* C^∞ *extension to a mapping from* $\bar{\Omega}$ *to* $\bar{\Omega}'$.

If the answer to the question posed in problem III is affirmative for $p = 0$, $q = 1$ then the above theorem would show that a one-one holomorphic mapping between pseudo-convex domains has a C^∞ extension to the boundary.

Hypoellipticity does not hold, in general, on pseudo-convex domains. For example, it has been recently proved, by D. Catlin and by K. Diederich and P. Pflug (see [6b] and [10]) that if Ω is pseudo-convex and if $b\Omega$ contains a complex-analytic variety of dimension q then there exists an $\alpha \in L_2^{0,q}(\Omega)$ such that (3) does not hold for any solution of (1). Furthermore, Catlin (see [6a]) has given an example of a pseudo-convex domain in $\mathbb{C}^3$, where the boundary does not contain any complex curves but on which $\bar{\partial}$ is not hypoelliptic for (0, 1)-forms. In his thesis [6a] Catlin shows that a necessary condition for hypoellipticity of (0, 1)-forms on a pseudo-convex domain Ω is that for any compact set $K \subset \bar{\Omega}$ we have $\hat{K} \cap b\Omega = K \cap b\Omega$, where $\hat{K}$ denotes the hull with respect to holomorphic functions in $C^\infty(\bar{\Omega})$. In dimension two this condition is equivalent to not having any complex curves in $b\Omega$. These considerations suggest the following problem.

PROBLEM IV. Describe the propagation of singularities of $\bar{\partial}$. That is, given sing. supp. (α), what is the "smallest" set in which sing. supp. (ϕ) lies, for some solution ϕ of (1). Also, what is the sing. supp. (ϕ) for ϕ orthogonal to the null space of $\bar{\partial}$?

Problems II and III may also be addressed via the solution of the "$\bar{\partial}$-Neumann problem" which we recall briefly here.

Let $\mathrm{Dom}^{p,q}(\bar{\partial})$ denote the domain of the closure of $\bar{\partial}$ on $L_2^{p,q}(\Omega)$, let $\bar{\partial}^*$ denote the L_2-adjoint of $\bar{\partial}$ and $\mathrm{Dom}^{p,q}(\bar{\partial}^*)$ denotes domain of $\bar{\partial}^*$ in $L_2^{p,q}(\Omega)$. We set $\mathcal{D}^{p,q} = \mathcal{D}^{p,q}(\Omega) = \mathrm{Dom}^{p,q}(\bar{\partial}) \cap \mathrm{Dom}^{p,q}(\bar{\partial}^*)$. The $\bar{\partial}$-Neumann problem may be then formulated as follows. Given $\alpha \in L_2^{p,q}(\Omega)$

find $\psi \in \mathcal{D}^{p,q}$ such that $\bar\partial\psi \in \mathcal{D}^{p,q+1}$, $\bar\partial^*\psi \in \mathcal{D}^{p,q-1}$ and

$$\bar\partial\bar\partial^*\psi + \bar\partial^*\bar\partial\psi = a .$$

If $\bar\partial a = 0$ then $\bar\partial\bar\partial^*\bar\partial\psi = 0$ so that

$$(\bar\partial\bar\partial^*\bar\partial\psi, \bar\partial\psi) = \|\bar\partial^*\bar\partial\psi\|^2 = 0 ,$$

when (,) and $\|\quad\|$ denote the L_2 inner product and norm respectively. Hence $\bar\partial^*\bar\partial\psi = 0$ and $\bar\partial\bar\partial^*\psi = a$. Since $\bar\partial^*\psi$ is orthogonal to the null space of $\bar\partial$ we see that $\phi = \bar\partial^*\psi$ is the solution of (1) which solves problem II.

The $\bar\partial$-Neumann problem can also be formulated as follows. Given $a \in L_2^{p,q}(\Omega)$ find $\psi \in \mathcal{D}^{p,q}$ such that

$$(\bar\partial\psi, \bar\partial\gamma) + (\bar\partial^*\psi, \bar\partial^*\gamma) = (a, \gamma)$$

for all $\gamma \in \mathcal{D}^{p,q}$.

DEFINITION. If $x_0 \in b\Omega$ then the $\bar\partial$-Neumann problem is *subelliptic* on (p,q)-forms at x if there exists a neighborhood U of x_0 and constants $C > 0$ and $\varepsilon > 0$ such that:

$$\|\phi\|_\varepsilon^2 \le C(\|\bar\partial\phi\|^2 + \|\bar\partial^*\phi\|^2) , \tag{6}$$

for all $\phi \in D_U^{p,q}(\Omega)$, where $D_U^{p,q}(\Omega)$ is the space of forms with components in $C^\infty(\bar\Omega)$ which are contained in $\mathcal{D}^{p,q}(\Omega)$ and whose support lies in $U \cap \bar\Omega$. We will often set $D_U^{p,q} = D_U^{p,q}(\Omega)$. If the components of ϕ are in $C^\infty(\bar\Omega)$ then $\phi \in \mathrm{Dom}(\bar\partial^*)$ if and only if $\phi \lrcorner \bar\partial r = 0$ on $b\Omega$. In terms of local coordinates for (0, 1)-forms the condition $\phi \lrcorner \bar\partial r = 0$ is equivalent to

$$\sum r_{z_j}\phi_j = 0 \quad \text{on} \quad b\Omega . \tag{7}$$

The norm $\|\quad\|_2^2$ denotes the sum of the squares of the Sobolev ε-norms of the components. The following consequence of subellipticity if important in connection with the above discussion (see [25] and [13]).

THEOREM. *If the $\bar{\partial}$-Neumann problem is subelliptic on* (p,q)*-forms at* $x_0 \in b\Omega$ *and if* $\alpha \in L_2^{p,q}(\Omega)$, $\bar{\partial}\alpha = 0$ *and* $x_0 \notin$ sing. supp. (α) *then* $x_0 \notin$ sing. supp. (ϕ), *where* ϕ *is the unique solution of (1) orthogonal to the null space of* $\bar{\partial}$. *Hence if subellipticity holds at all points of* $b\Omega$ *then (3) is satisfied.*

If (6) is satisfied with $\varepsilon = 1$ then the estimate is called elliptic. This happens if and only if $q = n$ in which case $\phi \in D_U^{p,n}$ means that $\phi = 0$ on $b\Omega$. The estimate (6) holds with $\varepsilon = 1/2$ if and only if the Levi-form (4) at x_0 has at least $n-q$ positive eigenvalues. If $n > q$ then the estimate does not hold with $\varepsilon > \frac{1}{2}$ (see [13]). In the case of (p,n–1)-forms we have an exact description of subellipticity (see [23c], [15] and [23d]). To formulate the result we need the notion of order of contact.

DEFINITION. If $x_0 \in b\Omega$ and V is a q-dimensional complex-analytic variety containing x_0 then we define $O_V^q(x_0)$ the *order of contact* of V with $b\Omega$ at x_0 as follows.

$m = O_V^q(x_0)$ if for every sequence $\{x_\nu\}$, such that $x_\nu \in V$ and $\lim x_\nu = x_0$ we have $r(x_\nu) = O(|x_0 - x_\nu|^m)$ and if there is a sequence $\{x_\nu\}$ such that $|x_0 - x_\nu|^{-p} r(x_\nu) \to \infty$ whenever $p > m$. Let $v^q(x_0)$ be the set of all q-dimensional complex analytic varieties containing x_0. The q-*order* of Ω at x_0, denoted by $O^q(x_0)$, is the maximum of the set $\{O_{x_0}^q(V) | V \in v^q(x_0)\}$ when this set is bounded, it is infinite otherwise. The *regular* q-*order* of Ω at x_0, denoted by reg $O^q(x_0)$ is the maximum of the set $\{O_{x_0}^q(V) | V \in W^q(x_0)\}$, when $W^q(x_0)$ is the set q-dimensional complex manifolds containing x_0.

THEOREM. *Suppose that* $\Omega \subset \mathbf{C}^n$ *is pseudo-convex and that* $x_0 \in b\Omega$. *Then the $\bar{\partial}$-Neumann problem is subelliptic on* (p, n–1)*-forms at* x_0 *if and only if* $O^{n-1}(x_0) < \infty$. *If* $O^{n-1}(x_0) < \infty$ *then (6) holds for all* $\varepsilon \leq (O^{n-1}(x_0))^{-1}$ *and (6) does not hold for* $\varepsilon > (O^{n-1}(x_0))^{-1}$.

In [23d] this theorem is reduced to the study of Hörmander's equation (see [20c]), to obtain the best value of ε we use the results of Stein and Rothschild, see [30]. The number $O^{n-1}(x_0)$ can be described in several ways as shown below.

PROPOSITION (see [23d]). *If* $x_0 \in b\Omega$ *and* $\Omega \subset \mathbb{C}^n$ *is pseudo-convex then the following are equivalent.*

(a) $O^{n-1}(x_0) < \infty$.

(b) $\operatorname{reg} O^{n-1}(x_0) < \infty$. *We also have* $\operatorname{reg} O^{n-1}(x_0) = O^{n-1}(x_0)$.

(c) *If* f, g *and* h *are germs of* C^∞ *complex-valued functions at* x_0, *we define the integer* $P = P(f, g, h)$ *to be the largest integer such that*

$$r(x) = f(x)h(x) + g(x)\overline{h(x)} + O(|x-x_0|^P).$$

Let $S = \{P = P(f, g, h) | h$ *holomorphic with* $dh \neq 0\}$. *Then* (a) *is equivalent to*: S *bounded. Further* $O^{n-1}(x_0) = \max S$.

(d) *Let* $\mathfrak{L}^1(x_0)$ *be the set of germs of complex vector fields at* x_0 *defined by*: $L \in \mathfrak{L}^1(x_0)$ *if and only if* $\langle L, \partial r\rangle = \langle L, \bar{\partial} r\rangle = 0$, *where* $\langle\ \rangle$ *denotes contraction. We define inductively* $\mathfrak{L}^k(x_0)$ *by*

$$\mathfrak{L}^k(x_0) = \mathfrak{L}^{k-1}(x_0) + [\mathfrak{L}^1(x_0), \mathfrak{L}^{k-1}(x_0)]. \tag{8}$$

Denote $\dot{\mathfrak{L}}^k(x_0)$ *the set of tangent vectors at* x_0 *obtained by evaluating all elements of* $\mathfrak{L}^k(x_0)$ *at* x_0. *Then (a) is equivalent to the existence of a* k *such that* $\dim_{\mathbb{C}} \dot{\mathfrak{L}}^k(x_0) = 2n-1$ *and* $O^q(x_0)$ *is the smallest such* k.

Recently, D. Catlin has established the following necessary condition for subellipticity.

THEOREM (Catlin). *Suppose that the* $\bar{\partial}$*-Neumann subelliptic estimate holds for* (p,q)*-forms on* $\Omega \subset \mathbb{C}^n$ *at* $x_0 \in b\Omega$ *and that* Ω *is pseudo-convex. Then* $O^q(x_0) \le \frac{n+2}{\varepsilon}$, *where* ε *is given in (6).*

When $q = n-1$ we know (see (b) above) that $O^{n-1}(x_0) = \operatorname{reg} O^{n-1}(x_0)$. (We always have $O^q(x_0) \ge \operatorname{reg} O^q(x_0)$). Consider the following example (see [5]), let $\Omega \subset \mathbb{C}^3$ be given by:

$$r(z_1, z_2, z_3) = \mathrm{Re}(z_3) + |z_1^2 - z_2^3|^2 .$$

In this case $\mathrm{reg}\, O^1(x_0) = 6$ and $O^1(x_0) = \infty$, where $x_0 = (0,0,0)$. The following result was obtained by J. D'Angelo (see [7a]).

THEOREM (D'Angelo). *If Ω is pseudo-convex and $x_0 \in b\Omega$ and if $\mathrm{reg}\, O^q(x_0) \le 5$ then $\mathrm{reg}\, O^q(x_0) = O^q(x_0)$.*

These considerations imply that, if $q \le n-2$ the condition $O^q(x_0) < \infty$ cannot be described in terms of vector fields as in (d) above.

In [23d] we introduce certain ideals to study the contact between complex-analytic varieties and real pseudo-convex hypersurfaces. For $x_0 \in b\Omega$ we will define a set $I^q(x_0)$ of germs of C^∞ functions at x_0 as follows. If f is a germ of a C^∞ function at x_0 then $f \in I^q(x_0)$ if and only if there exists a neighborhood U of x_0 and constants $\varepsilon > 0$, $C > 0$ such that

$$\|f\phi\|_\varepsilon^2 \le C(\|\bar\partial\phi\|^2 + \|\bar\partial^*\phi\|^2) \tag{9}$$

for all $\phi \in D_U^{0,q}$. If (9) holds for all $\phi \in D^{0,q}$ then it also holds for all $\phi \in D_U^{p,q}$ for all p. Evidently then the inequality (6) holds if and only if $1 \in I^q(x_0)$. The following result is proved in [23d].

THEOREM. *If Ω is pseudo-convex and if $x_0 \in b\Omega$ then $I^q(x_0)$ has the following properties.*

(i) *$I^q(x_0)$ is an ideal.*

(ii) *$I^q(x_0) = \sqrt[R]{I^q(x_0)}$, where $\sqrt[R]{I^q(x_0)}$ is the real radical of $I^q(x_0)$ which is defined as follows. If J is an ideal of germs of functions then a germ $f \in \sqrt[R]{J}$ if and only if there exists $g \in J$ and an integer m such that $|f|^m \le |g|$.*

(iii) *$r \in I^q(x_0)$ and $\mathrm{coeff}\,\{\partial r \wedge \bar\partial r \wedge (\partial\bar\partial r)^{n-q}\} \subset I^q(x_0)$, where $\mathrm{coeff}\,\{\ \}$ is the set of coefficients of the forms in $\{\ \}$.*

(iv) *If* $f_1, \cdots, f_{n-q} \in I^q(x_0)$ *then*

$$\text{coeff}\{\partial f_1 \wedge \cdots \wedge \partial f_j \wedge \partial r \wedge \bar{\partial} r \wedge (\partial\bar{\partial} r)^{n-q-j}\} \subset I^q(x_0)$$

for $j = 1, \cdots, n-q$.

For each q *we define a sequence of ideals*

$$I_1^q(x_0) \subset I_2^q(x_0) \subset \cdots \subset I_k^q(x_0) \subset I^q(x_0)$$

inductively as follows.

$$I_1^q(x_0) = \sqrt[R]{(r, \text{coeff}\{\partial r \wedge \bar{\partial} r \wedge (\partial\bar{\partial} r)^{n-q}\})}\,, \tag{10}$$

where (A) *denotes the ideal generated by* A.

$$I_{k+1}^q(x_0) = \sqrt[R]{(I_k^q(x_0), \text{coeff}\{\partial f_1 \wedge \cdots \wedge \partial f_j \wedge \partial r \wedge \bar{\partial} r \wedge (\partial\bar{\partial} r)^{n-q-j}\})}\,, \tag{11}$$

for all $f_1, \cdots, f_j \in I_k^q(x_0)$ *with* $j \leq n-q$.

In view of the above theorem it is clear that $I \in I_k^q(x_0)$ is a sufficient condition for subellipticity. However, the following problem remains.

PROBLEM IV. What is the best value for ε in (6)?

To study that question we would have to study a "tree" of ideals as follows. Given an ideal J of germs of C^∞ function at x_0, we define the kth real radical $R^k(J)$ as follows:

$$R^k(J) = \{f \in \underline{C^\infty(x_0)} \mid \text{there exists } g \in J \text{ such that } |f|^k \leq |g|\}\,, \tag{12}$$

where $\underline{C^\infty(x_0)}$ denotes the set of C^∞ functions at x_0. For each q with $1 \leq q \leq n$ we define $B_q(J)$ by

$$B_q(J) = (J, \text{coeff}\{\partial f_1 \wedge \cdots \wedge \partial f_j \wedge \partial r \wedge \bar{\partial} r \wedge (\partial\bar{\partial} r)^{n-q-j}, \text{ with } f_j \in J\})\,. \tag{13}$$

Then we set $J^q(x_0) = (r, \text{coeff}\{\partial r \wedge \bar{\partial} r \wedge (\partial\bar{\partial} r)^{n-q-j}\})$ and

$$J^q_{a_1 b_1 \cdots a_j b_j}(x_0) = B_q^{b_j} R^{a_j} \cdots B_q^{b_1} R^{a_1}(J^q(x_0))\,. \tag{14}$$

We then have

$$J^q_{a_1 b_1 \cdots a_j b_j}(x_0) \supset J^q_{a_1 b_1 \cdots a_j b'_j} \quad \text{if} \quad b_j \geq b'_j$$

and

$$J^q_{a_1 b_1 \cdots a_j 0}(x_0) \supset J^q_{a_1 b_1 \cdots a'_j 0} \quad \text{if} \quad a_j \geq a'_j \,.$$

The size of the optimal ε in (6) ought to be determined by the set $S = \{a_1 b_1 \cdots a_j b_j\}0$ such that $1 \in J^q_{a_1 b_1 \cdots a_j b_j}(x_0)$. The proof of the above theorem shows that (6) holds for $\varepsilon \leq 2^{-\Sigma b_k}(\Pi a_k)^{-1}$ when $a_1 b_1 \cdots a_j b_j \in S$. There is some evidence that suggests that the optimal ε is given by $\varepsilon = \max\left(\sum_1^j a_k + \sum_1^j b_k\right)^{-1}$. In any case, the determination of the best ε would surely involve a detailed study of the contact properties of complex analytic varieties with pseudo-convex hypersurfaces. Thus we are led to pose the following problem.

PROBLEM V. If $\Omega \subset \mathbb{C}^n$ is pseudo-convex and if $x_0 \in b\Omega$ what are the necessary and sufficient conditions for subellipticity of the $\bar{\partial}$-Neumann problem at x_0. In particular what is the relationship between the conditions $O^q(x_0) < \infty$ and $1 \in I^q_k(x_0)$ for some k.

In [23d] we prove the following lemma in the real-analytic case. The proof makes heavy use of a theorem of Diederich and Fornaess (see [9]).

LEMMA. *If Ω is pseudo-convex and if r is real-analytic in a neighborhood of $x_0 \in b\Omega$, then the condition: $1 \in I^q_k(x_0)$ for some k, is equivalent to the condition: there does not exist any q-dimensional complex-analytic variety V with $x_0 \in V\, b\Omega$.*

Combining this with the previous results we obtain the following result.

THEOREM. *Suppose that Ω is pseudo-convex and that r is real-analytic in a neighborhood of $x_0 \in b\Omega$. Then the $\bar{\partial}$-Neumann problem is subelliptic* *for* (p,q)-*forms at* x_0 *if and only if there exists no* q-*dimensional complex-analytic variety* V *with* $x_0 \in V \subset b\Omega$.

In [9], Diederich and Fornaess proved the following theorem.

THEOREM (Diederich and Fornaess). *If* $\Omega \subset \mathbb{C}^n$ *is a bounded pseudo-convex domain with a real-analytic boundary (i.e. the defining function* r *is real-analytic), then* $b\Omega$ *does not contain any complex-analytic varieties of dimension greater than zero.*

These two theorems then imply the following.

THEOREM. *If* $\Omega \subset\subset \mathbb{C}^m$ *has smooth, real-analytic, pseudo-convex boundary, then the* $\bar{\partial}$-*Neumann problem is subelliptic on* (p,q)-*forms, if* $q > 0$, *at every point of* $b\Omega$.

Now applying the previously mentioned result of Bell and Ligocka (see [2]) we have

THEOREM. *If* Ω *and* Ω' *are two bounded pseudo-convex domains in* $\mathbb{C}^n$ *with real-analytic boundaries and if* $h : \Omega \to \Omega'$ *is a one-one holomorphic map then* h *has a unique extension to a* C^∞ *map from* $\bar{\Omega}$ *to* $\bar{\Omega}'$.

This leads to the following problem.

PROBLEM VI. If Ω and Ω' are two bounded pseudo-convex domains in $\mathbb{C}^n$ with real-analytic boundaries and if $h : \Omega \to \Omega'$ is a one-one holomorphic map, is the C^∞ extension of h to a mapping from Ω to $\bar{\Omega}$ real-analytic?

In case Ω is strongly pseudo-convex the answer to the problem is affirmative. This has been proved by H. Lewy in [26] and Pincuk in [30]. This result can also be proved by the arguments of Bell and Ligocka since one knows that the Bergman projection operator preserves real-analyticity on the boundary. This property of the Bergman projection operator on strongly pseudo-convex domains has been proved via the global real-analyticity of the $\bar{\partial}$-Neumann problem by Derridj and Tartakoff in [11] and by Komatsu in [A2], the local result was proved by Treves in [34] and Tartakoff (see [33] for an account of these developments). It has also been proved independently by use of hyperfunctions by Kashiwara in [21]. Thus we are led to the following open problem.

PROBLEM VII. Problems I and II in the real-analytic category, where sing. supp. in (3) is taken with respect to real-analytic functions on $\overline{\Omega}$. In particular, if the $\bar{\partial}$-Neumann problem is subelliptic at x_0 and r is real-analytic does (3) hold with respect to real-analytic functions?

For pseudo-convex domains on which the $\bar{\partial}$-Neumann problem is subelliptic with $\varepsilon = \frac{1}{2}$, solutions of (1) have been obtained satisfying Hölder and L_p estimates (see Henkin [19] and Kerzman [22], such estimates (see Henkin [19] and Kerzman [22], such estimates have also been obtained for the solution orthogonal to the null space of $\bar{\partial}$ (see Folland and Stein [14] and Greiner and Stein [17]). Such results have been obtained by Krantz [27] in the case of $(p, n-1)$-forms whenever the $\bar{\partial}$-Neumann problem is subelliptic.

PROBLEM VIII. Suppose that the $\bar{\partial}$-Neumann problem is subelliptic on a pseudo-convex domain. Prove L_p and Hölder estimates for the solution of (1) which is orthogonal to the null space of $\bar{\partial}$.

Next we will consider the case when Ω is not pseudo-convex. The necessary and sufficient conditions for subellipticity with $\varepsilon = \frac{1}{2}$ is that the Levi form has either at least $n-q$ positive or at least $q+1$ negative eigenvalues. When this condition is not satisfied and Ω is neither pseudo-convex nor pseudo-concave little is known, although there are some suggestive examples found by Derridj (see [8]). The pseudo-concave case (that is when the Levi form is negative) can be reduced to the pseudo-convex case as follows.

Let $\Omega \subset \mathbf{C}^n$ be a domain with a smooth boundary (no requirements are made on the Levi form), let $x_0 \in b\Omega$, U a neighborhood of x_0 and denote by ${}^b\mathcal{D}_U^{p,q}$ the restriction of $\mathcal{D}_U^{p,q}$ to $b\Omega$. In [20b], Hörmander shows that subellipticity for a boundary value problem is an equivalent to subellipticity for a system of pseudo-differential operators in the boundary. This theory has been worked out explicitly for differential complexes by Sweeney, see [31].

In particular, if we let $B_{U'}$ be the space of restrictions of all forms with support in U', to $b\Omega$, then, if $U' \supset U$, there exists a first order pseudo-differential operator $P^{p,q}: {}^bD_U^{p,q} \to B_{U'}$ such that (6) holds if and only if

(14) $$^b\|\phi\|_\varepsilon \leq \text{const. } {}^b\|P^{p,q}\phi\| ,$$

where $^b\| \quad \|_\varepsilon$ and $^b\| \quad \|$ denote the Sobolev and L_2-norms on $b\Omega$, ${}^bD_U^{p,q}$ denotes the restriction of $D_U^{p,q}$ to $b\Omega$. It also follows that $f \in I^q(x_0)$ if and only if $^b\|f\phi\|_\varepsilon \leq \text{const. } {}^b\|P^{p,q}\phi\|$. Let $\tilde{\Omega} \subset \mathbf{C}^n$ be a domain with a smooth boundary $b\tilde{\Omega}$, let $\tilde{r}$ be the corresponding defining function (i.e. $\tilde{r} < 0$ in $\tilde{\Omega}$, $\tilde{r} = 0$ on $b\tilde{\Omega}$ and $\tilde{r} > 0$ inside of $\bar{\tilde{\Omega}}$) and suppose that in a neighborhood V of x_0 we have $\tilde{r} = -r$, where r is the defining function for Ω. If $U \subset V$ then ${}^bD_U^{p,q}(\Omega) = {}^bD_U^{p,q}(\tilde{\Omega})$. We define $\#: {}^bD_U^{p,q} \to D_U^{p,n-q-1}$ by

(15) $$\#\phi = {}^*(\phi \wedge \bar{\partial} r) .$$

Then, if $U' \subset V$, a straightforward calculation gives

(16) $$|P^{p,q}\phi| = |\tilde{P}^{p,n-q-1}\#\phi| ,$$

for all $\phi \in {}^bD_U^{p,q}$, $\bar{U} \subset U' \subset V$. We also have

(17) $$^b\|\phi\|_\varepsilon \leq \text{const. } {}^b\|\#\phi\|_\varepsilon \leq \text{const. } {}^b\|\phi\|_\varepsilon$$

for all $\phi \in D_U^{p,q}$. We therefore have the following result.

PROPOSITION. *If Ω and $\tilde{\Omega}$ are domains as above then $I^q(x_0) = \tilde{I}^{n-q-1}(x_0)$. Thus the $\bar{\partial}$-Neumann is subelliptic for (p,q)-forms at x_0 on Ω if and only if it is subelliptic for $(p, n-q-1)$-forms at x_0 in $\tilde{\Omega}$. Furthermore, the optimal ε for which (6) holds is the same in both these cases.*

COROLLARY. *If* Ω *is pseudo-concave in a neighborhood of* x_0 *then* $\tilde{\Omega}$ *is pseudo-convex in that neighborhood and hence if a subelliptic estimate holds for* (p,q)*-forms on* Ω *at* x_0 *we have* $O^{n-q-1}(x_0) < \infty$. *Also, if* $1 \in \tilde{I}^{n-q-1}(x_0)$ *then a subelliptic estimate holds for* (p,q)*-forms on* Ω *at* x_0. *In particular, if* $b\Omega$ *is real analytic in a neighborhood of* x_0 *then subellipticity for* (p,q)*-forms holds if and only if there are no* $n-q-1$ *dimensional complex varieties contained in* $b\Omega$ *and containing* x_0.

We will mention briefly some problems connected with the $\bar{\partial}_b$ on boundaries of domains and on CR manifolds. We recall (see [26]), that if $\mathcal{A}^{p,q}$ is the space of C^∞ (p,q)-forms on $\bar{\Omega}$ we define $\mathcal{C}^{p,q} = \{\phi \mid \phi \wedge \bar{\partial} r = 0$ on $b\Omega\}$ and $\mathcal{B}^{p,q} = \mathcal{A}^{p,q}/\mathcal{C}^{p,q}$. The operator $\bar{\partial}_b : \mathcal{B}^{p,q} \to \mathcal{B}^{p,q+1}$ is then defined by means of the commutative diagram

$$\begin{array}{ccccccccc}
0 & \longrightarrow & \mathcal{C}^{p,q+1} & \longrightarrow & \mathcal{A}^{p,q+1} & \longrightarrow & \mathcal{B}^{p,q+1} & \longrightarrow & 0 \\
 & & \uparrow{\scriptstyle \bar{\partial}} & & \uparrow{\scriptstyle \bar{\partial}} & & \uparrow{\scriptstyle \bar{\partial}_b} & & \\
0 & \longrightarrow & \mathcal{C}^{p,q} & \longrightarrow & \mathcal{A}^{p,q} & \longrightarrow & \mathcal{B}^{p,q} & \longrightarrow & 0 .
\end{array}$$

We can now pose the same questions for $\bar{\partial}_b$ as we did for $\bar{\partial}$. In particular, subellipticity at x_0 is given by

$$ {}^b\|\phi\|_\varepsilon^2 \le C({}^b\|\bar{\partial}_b\phi\|^2 + {}^b\|\bar{\partial}_b^*\phi\|^2) , \tag{18}$$

for all $\phi \in \mathcal{B}_U^{p,q}$, where U is a neighborhood of x_0 and $\mathcal{B}_U^{p,q}$ consists of those elements of $\mathcal{B}^{p,q}$ with support in U. The same reasoning as in [25] then gives the following result.

PROPOSITION. *Subellipticity (18) holds whenever the* $\bar{\partial}$*-Neumann problems for* (p,q)*-forms on* Ω *and* (p, n−q−1)*-forms on* $\tilde{\Omega}$ *at* x_0 *are subelliptic. In particular if* Ω *is pseudo-convex, then: (18) holds whenever* $1 \in I_k^{\min(q,n-q-1)}(x_0)$ *for some* k, *and (18) does not hold if* $O^{\min(q,n-q-1)}(x_0) = \infty$.

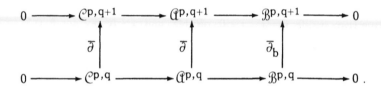

The orthogonal projection on the null space of $\bar{\partial}_b$ in $L_2(b\Omega)$ is given by the Szego kernel. The analyticity property of this kernel (which in the strongly pseudo-convex case is proved in [34a], [35], [11] and [21]) can be used to study Lewy type equations, see [16].

To conclude this article we wish to call attention to the fact that on strongly pseudo-convex domains very precise understanding of boundary regularity is obtained by giving explicit formulae for the operators involved in solving the $\bar{\partial}$ and $\bar{\partial}_b$ equations (see [19], [14], [17]) as well as for the Bergman and Szego kernels (see [12a], [12b], [20], [3]). The starting point in all these formulas is the existence of the Levi-polynomial whose zeros locally intersect the boundary at only one point. In [25] Nirenberg and I have given an example of a pseudo-convex domain Ω for which no such function exists (nevertheless the $\bar{\partial}$-Neumann problem is subelliptic on Ω). However, the existence of peak functions, which might play a role analogous to the reciprocal of polynomials has been investigated by Bloom [4], by Bedford and Fornaess [1], and by Hakim and Sibony in [A1]. It is an open problem how to represent these various operators in the weakly pseudo-convex case; for some special classes of domains this has been done by Range [31], Harvey and Polking [18] and D'Angelo [7b].

REFERENCES

[1] Bedford, E. and Fornaess, J. E., "A construction of peak functions on weakly pseudo-convex domains." Annals of Math 107 (1978), 555-568.

[2] Bell, S. and Ligocka, E., "A simplification and extension of Fefferman's theorem on biholomorphic mappings." Invent. Math. (to appear).

[3] Boutet de Monvel, L. and Sjöstrand, J., "Sur la singularité des noyanx de Bergman et de Szegö." Soc. Math. de France Astérisque, 34-35 (1976), 123-164.

[4] Bloom, T., "Remarks on type conditions for real hypersurfaces in $\mathbb{C}^n$." Proc. of Int. Confs. on Several Complex Variables, Cortona, Italy 1976-77, Sc. Norm. Sup. Pisa (1978), 14-24.

[5] Bloom, T. and Graham, I., "A geometric characterization of points of type m on real hypersurfaces." J. of Diff. Geom., Vol. 12 (1977), 171-182.

[6] Catlin, D.: (a) "Boundary behavior of holomorphic functions on weakly pseudo-convex domains." Thesis, Princeton Univ. 1978. (b) "Necessary Conditions for Subellipticity and Hypoellipticity for the $\bar{\partial}$-Neumann problem on pseudoconvex domains." Appears in these proceedings.

[7] D'Angelo, J. P.: (a) "Finite type conditions for real hypersurfaces." J. of Diff. Geom. (to appear). (b) "A note on the Bergman kernel." Duke Math. J., Vol. 45 No. 2 (1978), 259-265. (c) "Order of contact of real and complex subvarieties." Illinois J. of Math (to appear).

[8] Derridj, M., "Estimations pour $\bar{\partial}$ dans des domains vous pseudo-convexes." Ann. de L'Inst. Fourier, Vol. 38 No. 4 (1978), 239-254.

[9] Diederich, K. and Fornaess, J. E., "Pseudoconvex domains with real-analytic boundary." Ann. of Math. 107 (1978), 371-384.

[10] Diederich, K. and Pflug, P., "Necessary Conditions for Hypoellipticity of the $\bar{\partial}$ Problem." Appears in these proceedings.

[11] Derridj, M. and Tartakoff, D., "On the global real-analyticity of the solutions of the $\bar{\partial}$-Neumann problem." Comm. Part. Diff. Eqs. 1 (1976), 401-435.

[12] Fefferman, C.: (a) "The Bergman kernel and biholomorphic mappings of pseudoconvex domains." Invent. Math. 26 (1974), 1-65. (b) "Parabolic invariant theory in complex analysis." Adv. in Math. Vol. 31 No. 2 (1979), 131-262.

[13] Folland, G. B. and Kohn, J. J., "The Neumann problem for the Cauchy-Riemann complex." Ann. of Math. Studies, No. 75, Princeton Univ. Press, 1972.

[14] Folland, G. B. and Stein, E. M., "Estimates for the $\bar{\partial}_b$ complex and analysis on the Heisenberg group." Comm. Pure and App. Math., 27 (1974), 429-522.

[15] Greiner, P. C., "On subelliptic estimates of the $\bar{\partial}$-Neumann problem in $\mathbb{C}^2$." J. Diff. Geom. 9 (1974), 239-250.

[16] Greiner, P. C., Kohn, J. J. and Stein, E. M., "Necessary and sufficient conditions for solvability of the Lewy equation." Proc. Nat. Acad. of Sci. 72 (1975), 3287-3289.

[17] Greiner, P. C. and Stein, E. M., "Estimates for the $\bar{\partial}$-Neumann problem." Math. Notes No. 19, Princeton Univ. Press (1977).

[18] Harvey, R. and Polking, J., "Fundamental solutions in complex analysis." Duke J. of Math. (to appear).

[19] Henkin, G. M. and Cirka, E. M., "Boundary properties of holomorphic functions of several complex variables." Problems of Math. Vol. 4, Moscow (1975), 13-142.

[20] Hörmander, L.: (a) "L_2 estimates and existence theorems for the $\bar{\partial}$ operator." Acta Math. 113 (1965), 89-152.
(b) "Pseudo-differential operators and non-elliptic boundary value problems." Ann. of Math. 83 (1966), 129-209.
(c) "Hypoelliptic second order differential equations." Acta Math. 119 (1967), 147-171.

[21] Kashiwara, M., "Analyse micro-locale du noyau de Bergman." Sem. Goulaouic-Schwartz 1976-77, Expose No. VIII.

[22] Kerzman, N.: (a) "Hölder and L_p estimates for solutions of $\bar{\partial}u = f$ in strongly pseudo-convex domains." Comm. Pure Appls. Math. 24 (1971), 301-379.
(b) "The Bergman kernel function: differentiability at the boundary," Math. Ann. 195 (1972), 149-158.

[23] Kohn, J. J.: (a) "Global regularity for $\bar{\partial}$ on weakly pseudo-convex manifolds." Trans. A.M.S. 181 (1973), 273-292.
(b) "Methods of partial differential equations in complex analysis." Proc. of Symp. in Pure Math. vol. 30, part 1 (1977), 215-237.
(c) "Boundary behaviour of $\bar{\partial}$ on weakly pseudo-convex manifolds of dimension two." J. Diff. Geom., 6 (1972), 523-542.
(d) "Subellipticity of the $\bar{\partial}$-Neumann problem on pseudo-convex domains: sufficient conditions." Acta Math. 142 (1979), 79-122.

[24] Kohn, J. J. and Nirenberg, L.: (a) "Non-coercive boundary value problems." Comm. Pure Appl. Math. 18 (1965), 443-492.
(b) "A pseudo-convex domain not admitting a holomorphic support function." Math. Ann. 201 (1973), 265-268.

[25] Kohn, J. J. and Rossi, H., "On the extension of holomorphic functions from the boundary of a complex manifold." Ann. of Math., 81 (1965), 451-472.

[26] Krantz, S. G.: (a) Characterization of various domains of holomorphy via $\bar{\partial}$ estimates and applications to a problem of Kohn." (preprint)
(b) "Optimal Lipschitz and L_p regularity for the equation $\bar{\partial}u = f$ on strongly pseudo-convex domains." Math. Ann. 219 (1976), 223-260.

[27] Lewy, H., "On the boundary behaviour of holomorphic mappings." Contrib. Centro Linceo Inter. Sc. Mat. e loro App. No. 35, Acad. Naz. dei Lincei (1977), 1-8.

[28] Nirenberg, L., Webster, S. and Yang, P., "Local boundary regularity of holomorphic mappings." Comm. Pure and App. Math. (to appear).

[29] Pincuk, S. I., "On the analytic continuation of biholomorphic mappings." Math. Sbornik, Vol. 27, No. 3 (1975), 375-392.

[30] Range, R. M., "On Hölder estimates for $\bar{\partial}u = f$ on weakly pseudo-convex domains." (preprint).

[31] Rothschild, L. P. and Stein, E. M., "Hypoelliptic differential operators and nilpotent groups." Acta Math. 137 (1976), 247-320.

[32] Sweeney, W. J., "The D-Neumann problem." Acta Math. 120 (1968), 223-277.

[33] Tartakoff, D., "A Survey of Some Recent Results in C and Real Analytic Hypoellipticity for Partial Differential Operators with Applications to Several Complex Variables." Appears in these proceedings.

[34] Treves, F., "Analytic hypo-ellipticity of a class of pseudodifferential operators with double characteristics and applications to the $\bar{\partial}$-Neumann problem." Comm. in P.D.E. Vol. 3, No. 6-7 (1978), 475-642.

Supplementary Bibliography

[A1] Hakim, M. and Sibony, N., "Quelques conditions pour l'existence de fonctions pics dans les domaines pseudoconvexes." Duke Math. J. 44 (1977), 399-406.

[A2] Komatsu, G., "Global analytic-hypoellipticity of the $\bar{\partial}$-Neumann problem," Tohoku Math. J. Ser. 2, vol 28 (1976), 145-156.

ON CP^1 AS AN EXCEPTIONAL SET

Henry B. Laufer*

I. *Introduction*

Artin [2] introduced the definition of when a two-dimensional singularity $p \in V$ is rational. This definition was generalized by Burns [6] to higher dimensions. Let M be a resolution of V in the two-dimensional case. Brieskorn [5] and Artin and Schlessinger [4], [3] related deformations of M to deformations of V. Some of Riemenschneider's results [19] about related deformations hold also in the higher dimensional case.

This paper will provide more examples for Burns' theory and higher dimensional examples for Riemenschneider's work.

Let M be a strictly pseudoconvex manifold of dimension n with exceptional set A isomorphic to P^1. Replacing M by a smaller neighborhood of A, if necessary, we may assume that M has a versal deformation [17]. Let N be the normal bundle for A in M. Then [11, Théorème 21 and Corollaire, p. 126], since $A = P^1$, N is the direct sum of line bundles. Let $(c_1, \cdots, c_{n-1})$ be the chern classes of these line bundles. Let $c = c_1 + \cdots + c_{n-1}$ be the chern class of the determinant of N. Given n, c is the sole topological invariant of N.

---

*This research was partially supported by NSF Grant MCS760496A01. The author also is a Sloan Fellow.

Recent Developments in Several Complex Variables
0-691-08285-5/81/000261-15 $00.75/0 (cloth)
0-691-08281-2/81/000261-15 $00.75/0 (paperback)
For copying information, see copyright page.

While for $n = 2$, necessarily $c_1 < 0$, Example 2.3 shows that $(c_1, c_2) = (-3, 1)$ is possible, even for a hypersurface. Grauert [9, p. 354] gave an example of an exceptional set which was not exceptional in its normal bundle. With $(c_1, c_2) = (-3, 1)$, the normal bundle is not even holomorphically convex. It is possible, Example 2.4, for (c_1, c_2) to be $(-30, 2)$, yielding a non-rational singularity.

For $n = 2$, it is well known that $c = c_1 \leq -1$. For general, n, $c \leq -n+1$ (Theorem 3.1). If $c_i \leq 0$ for all i, then V is rational, has embedding dimension $-c + n-1$, and multiplicity $-c$. The Hilbert function is also computed (Theorem 3.3).

With A irreducible and of dimension 1, the possibilities for V to be a hypersurface with $n \geq 3$ are severely limited. Necessarily $A = \mathbf{P}^1$, $n = 3$ and (c_1, c_2) equals $(-1,-1)$, $(-2, 0)$ or $(-3, 1)$; see Theorem 4.1.

II. *Examples*

Recall the following definition from Grauert [9, p. 342]. Let N be a vector bundle over the complex manifold A. Then N is weakly negative if A is exceptional as the 0-section in N. Let L be a line bundle over a compact Riemann surface A with chern class c. Then L is weakly negative (i.e., A is exceptional in L) if and only if $c < 0$ [9, p. 367].

PROPOSITION 2.1. *Let* N *be a vector bundle over the compact Riemann surface* A. *Suppose that* $N \approx \oplus L_i$, $1 \leq i \leq n-1$, *where* L_i *is a line bundle of chern class* c_i *on* A. *Then* A *is exceptional in* N *if and only if* $c_i < 0$, $1 \leq i \leq n-1$.

Proof. Suppose that A is exceptional in N. There exists a strictly pseudoconvex neighborhood U of A in N. There is a C^∞ exhaustion function $f \geq 0$ on U such that $f = 0$ precisely on A and f is strictly plurisubharmonic off A. Restrict f to L_i. Then L_i is weakly negative. So $c_i < 0$.

Conversely, suppose that $c_i < 0$, $1 \leq i \leq n-1$. Then there is an exhaustion function f_i, as in the previous paragraph, for each L_i. $f = \Sigma f_i$,

$1 \leq i \leq n-1$, is well defined on N using the direct sum decomposition $N = \oplus L_i$. f shows that N is weakly negative. This completes the proof of the proposition.

For the rest of this paper we specialize to the case $A = P^1$. A will be the exceptional set in the n-dimensional manifold M. Let N be the normal bundle to A in M. Then [11, Théorème 2.1 and Corollaire, p. 126] necessarily $N = \oplus L_i$, $1 \leq i \leq n-1$. Let c_i denote the chern class of L_i. Order the L_i so that $c_{i+1} \geq c_i$. Then $(c_1, c_2, \cdots, c_{n-1})$ is uniquely determined by N.

Proposition 2.1 tells when A is exceptional in N. Moreover, if A is exceptional in N, then A is exceptional in M; [9, Satz 8, p. 353]. As Grauert [9, p. 354] shows, A exceptional in M does not in general imply that A is exceptional in N. There are similar examples in our context. We shall let w and x be local coordinates for P^1 with $w = 1/x$. M will be given by two coordinate patches (w, z_1, z_2) and (x, y_1, y_2) and by a transition map between the coordinate patches. $A = P^1$ will be given by $\{z_1 = z_2 = 0\} = \{y_1 = y_2 = 0\}$.

EXAMPLE 2.2. Let $k \geq 1$ be an integer. Let M be the 3-manifold given near A by transition maps

$$\begin{cases} z_1 = x^2 y_1 + x y_2^k \\ z_2 = y_2 \\ w = 1/x . \end{cases}$$

Let $\pi : M \to V$ be the blowing down map. Then V may be given near $0 = (0,0,0,0) = \pi(A)$ as

$$V = \{(u_1, u_2, u_3, u_4) | u_1^2 + u_2^2 + u_3^2 + u_4^{2k} = 0\} .$$

Let N be the normal bundle for A in M. Then (c_1, c_2) for N equals $(-1,-1)$ for $k = 1$ and equals $(-2, 0)$ for $k \geq 2$.

Proof. Below we give four holomorphic functions on M: v_1, v_2, v_3 and v_4. $\tau = (v_1, v_2, v_3, v_4): M \to \mathbb{C}^4$ is a proper map which is biholomorphic off of A. Each v_i, $1 \leq i \leq 4$, is given in both the (w, z_1, z_2) and (x, y_1, y_2) coordinate systems.

$$v_1 = \quad = z_2 \quad = y_2$$

$$v_2 = \quad = z_1 \quad = x^2 y_1 + x y_2^k$$

$$v_3 = \quad = w z_1 \quad = x y_1 + y_2^k$$

$$v_4 = \quad = w^2 z_1 - w z_2^k = y_1 .$$

$\tau(M) = \{(v_1, v_2, v_3, v_4) | v_2 v_4 - v_3^2 + v_3 v_1^k = 0\}$. A change of coordinates shows that $\tau(M) \approx V$. Since V is normal, π preserves all holomorphic functions and so is the desired blowing down map.

The statements about (c_1, c_2) follow immediately from the linear terms in (y_1, y_2) in the transition functions.

EXAMPLE 2.3. Let $k \geq 3$ be an odd integer. Let M_k be the 3-manifold given near A by transition maps

$$\begin{cases} z_1 = x^3 y_1 + y_2^2 + x^2 y_2^k \\ z_2 = y_2/x \\ w = 1/x . \end{cases}$$

Let $\pi: M \to V$ be the blowing down map. Then near $(0,0,0,0) = \pi(A)$, V may be given as

$$V = \{(u_1, u_2, u_3, u_4) | u_1^2 + u_2^3 + u_3 u_4^2 + u_3^k u_2 = 0\} .$$

Also, for all $k \geq 3$, (c_1, c_2) for N equals $(-3, 1)$.

Proof. Let $k = 2n+1$, $n \geq 1$. The following functions, initially defined for $x \neq 0$, have obvious holomorphic extensions to all of M.

$$v_1 = z_1 \qquad = x^3y_1 + y_2^2 + x^2y_2^{2n+1}$$

$$v_2 = w^2z_1 - z_2^2 \qquad = xy_1 + y_2^{2n+1}$$

$$v_3 = w^3z_1 - wz_2^2 - z_1^n z_2 \qquad = y_1 + \frac{1}{x}[y_2^{2n+1} - y_2(x^3y_1 + y_2^2 + x^2y_2^{2n+1})^n]$$

$$v_4 = w^2z_1z_2 - z_2^3 - wz_1^{n+1} = y_1y_2 + \frac{1}{x}[y_2^{2n+2} - (x^3y_1 + y_2^2 + x^2y_2^{2n+1})^{n+1}] .$$

$\tau = (v_1, v_2, v_3, v_4)$ maps M to $V = \{f(v_1, v_2, v_3, v_4) = 0\}$ where $f(v_1, v_2, v_3, v_4) = v_4^2 + v_2^3 - v_1v_3 - v_1^{2n+1}v_2$. V has an isolated singularity at $0 = (0,0,0,0)$. $\tau^{-1}(0) = A$. τ is proper. Let

$$\omega = \frac{dv_1 \wedge dv_2 \wedge dv_3}{\partial f/\partial v_4} = -\frac{dv_2 \wedge dv_3 \wedge dv_4}{\partial f/\partial v_1}$$

$$= \frac{dv_3 \wedge dv_4 \wedge dv_1}{\partial f/\partial v_2} = -\frac{dv_4 \wedge dv_1 \wedge dv_2}{\partial f/\partial v_3} .$$

Then ω is a (non-zero) holomorphic 3-form on $V-0$. $\tau^*(\omega) = -dw \wedge dz_1 \wedge dz_2 = dx \wedge dy_1 \wedge dy_2$ is non-zero on M. So τ is a locally biholomorphic map off A. Since $V-0$ is simply-connected [18, Theorem 2.10, p. 18 and Theorem 5.2, p. 45]. τ is biholomorphic off A. Since V is normal, τ preserves all holomorphic functions and so is the blowing-down map. An easy change of coordinates puts the defining equation for V into the indicated form.

It follows immediately from the change of coordinates for M that $(c_1, c_2) = (3, -1)$.

The singularities of Example 2.3 are in series Q of Arnold's classification [1]. In particular, for $k = 1$, we get Q_{11}, and for $k = 3$, we get Q_{17}.

Wahl [22] has pointed out an alternative construction for the singularities of Example 2.2. Recall that $S = \{(x,y,z) | x^2+y^2+z^2 = 0\}$ is resolved with exceptional set $A = \mathbf{P}^1$. The versal deformation for S, $W = \{x^2 + y^2 + z^2 = t\}$, may be simultaneously resolved after base-change [5] and the base change is $t = u^2$. Example 2.2 comes from the base changes $t = u^{2k}$. Observe that the singularities of Example 2.3 do not come from such a construction.

EXAMPLE 2.4. Let M be given near A by the transition maps

$$\begin{aligned} z_1 &= x^{30}y_1 + x^{20}y_2^2 + x^{10}y_2^3 \\ z_2 &= y_2/x^2 \\ w &= 1/x\,. \end{aligned}$$

Then $(c_1, c_2) = (-30, 2)$. The following ten holomorphic functions on M show that A is exceptional in M.

$$\begin{aligned} z_1 &= x^{30}y_1 + x^{20}y_2^2 + x^{10}y_2^3 \\ wz_1 &= x^{29}y_1 + x^{19}y_2^2 + x^9y_2^3 \\ z_1z_2 &= x^{28}y_1y_2 + x^{18}y_2^3 + x^8y_2^4 \\ w^{16}z_1 - z_2^3 &= x^{14}y_1 + x^4y_2^2 \\ w^{17}z_1 - wz_2^3 &= x^{13}y_1 + x^3y_2^2 \\ w^{19}z_1 - w^3z_2^3 &= x^{11}y_1 + xy_2^2 \\ w^{20}z_1 - w^4z_2^3 &= x^{10}y_1 + y_2^2 \\ w^{28}z_1z_2 - w^4z_2^3 - w^{12}z_2^4 &= y_1y_2 \\ w^{29}z_1 - w^5z_2^2 - w^{13}z_2^3 &= xy_1 \\ w^{30}z_1 - w^6z_2^2 - w^{14}z_2^3 &= y_1\,. \end{aligned}$$

A calculation by the given Leray cover, or the proof of Theorem 4.1 below, shows that $H^1(M,\mathcal{O}) \neq 0$. Hence V is not rational.

One motivation for studying one-dimensional exceptional sets is that M is known to have a versal deformation near A [17]. Let Θ be the tangent sheaf to M. Then the versal deformation of M near A is parameterized by a manifold of dimension $\dim H^1(M,\Theta)$. Riemenschneider's work [19] may be applied to blow down deformations of M. Suppose that M is in fact the normal bundle N to A. Then, as in [9, p. 344], $H^1(M,\Theta)$ may be expanded in a power series along the fibers of N. Let $\mathcal{I}$ be the ideal sheaf of A on M. Then

$$H^1(M,\Theta) \approx \oplus H^1(M, \mathcal{I}^{\nu}\Theta/\mathcal{I}^{\nu+1}\Theta), \quad \nu \geq 0 .$$

Let $\mathcal{T}$ and $\mathcal{N}$ be the tangent sheaf and normal sheaf on A respectively. Then

$$0 \to \mathcal{T} \to \Theta/\mathcal{I}\Theta \to \mathcal{N} \to 0 \tag{2.1}$$

is exact. From the long exact sequence for (2.1) and Proposition 2.1, it follows that $H^1(M,\Theta) = 0$ if and only if $(c_1,\cdots,c_{n-1}) = (-1,\cdots,-1)$. For $n = 2$, this just gives A as an exceptional curve of the first kind. For $n \geq 3$, the blown-down singularities are given by

$$\frac{u_1}{u_2} = \frac{u_3}{u_4} = \cdots = \frac{u_{2n-3}}{u_{2n-2}} .$$

For $n \geq 4$, these singularities are known to be rigid [10, §5, pp. 255-259]. For $n = 3$, we get the hypersurface for $k = 1$ in Example 2.2. Using t as a parameter, the versal deformation [20] for this hypersurface is $\{u_1^2+u_2^2+u_3^2+u_4^2 = t\} = W$. The $H^1(M,\Theta) = 0$ result shows that W cannot be simultaneously resolved even after base change, *using the given resolution at* $t = 0$. The lack of unique minimal resolutions above dimension 2 [12, pp. 38-39] prevents the conclusion from this work that W cannot be simultaneously resolved using any initial resolution. For $t \neq 0$, the fiber in W has the homotopy type of S^3 [18]. Since any resolution of the $t = 0$ fiber must have non-zero homology in some even dimension, W in fact cannot be simultaneously resolved in the strictest sense. However, there remains the possibility that u may be simultaneously "resolved" allowing also modifications of the non-singular fibers.

One can also compute the versal deformations for the other resolutions in Example 2.2.

$$\begin{cases} z_1 = x^2 y_1 + x(y_2^k + t_{k-2} y_2^{k-2} + \cdots + t_0) \\ z_2 = y_2 \\ u = 1/x\,. \end{cases}$$

This deformation blows down to give the following versal deformation of V which can be simultaneously resolved with initial resolution M:

$$u_1^2 + u_2^2 + u_3^2 + (u_4^k + t_{k-2} u_4^{k-2} + \cdots + t_0)^2 = 0\,.$$

III. *Some general results*

Let $N = \oplus L_i$, $1 \le i \le n-1$, be a vector bundle over $A = \mathbf{P}^1$. Let $\mathfrak{N}$ be the sheaf of germs of sections of N. Let N^{-1} be the dual bundle to N. The usual deformation theory for complex structures shows that infinitesimal (= first order) deformations of N into nearby vector bundles are parameterized by $H^1(A, \mathfrak{N} \otimes \mathfrak{N}^{-1})$. Since A is one-dimensional, a versal deformation with a smooth parameter space of dimension $\dim H^1(A, \mathfrak{N} \otimes \mathfrak{N}^{-1})$ can be constructed as in [15] or [17]. $c = \sum c_i$, $1 \le i \le n-1$, as the chern class of the determinant bundle, is constant under deformation. Observe that $H^1(A, \mathfrak{N} \otimes \mathfrak{N}^{-1}) = 0$ if and only if $|c_i - c_j| \le 1$ for all i, j. This condition means simply that the set of values of the c_i is either one integer or else two integers which differ by 1. Moreover, one can write down one-dimensional deformations of N such that each fiber except for the initial fiber satisfies $|c_i - c_j| \le 1$, all i, j. In short, given $n-1$ and c, the condition $|c_i - c_j| \le 1$, all i, j, describes the generic vector bundle over P^1. This result has also been discussed in a more general category in [7].

THEOREM 3.1. *Let* $A = P^1$ *be an exceptional set in the* n*-dimensional manifold* M. *Let* c *be the chern class of the determinant of the normal bundle* N *to* A *in* M. *Then* $c \leq -n+1$.

Proof. Let Δ be the unit disc in one variable. There is a deformation $\omega' : \beta \to \Delta$ of the vector bundle $N = B_0 = (\omega')^{-1}(0)$ such that for $t \neq 0$, the vector bundle $B_t = (\omega')^{-1}(t) \approx \oplus L'_i$, $1 \leq 1 \leq n-1$, with $c' = c$ and $|c'_i - c'_j| \leq 1$ for all $1 \leq i$, $j \leq n-1$. Moreover, using a cover, as in Example 2.2 or [17], of a neighborhood M^* of A in M with only two open sets, we may induce ω' by a deformation $\omega : \mathfrak{M} \to \Delta$ of $M^* = \omega^{-1}(0)$. There is a submanifold $\mathfrak{A}$ of $\mathfrak{M}$ such that $\omega|_{\mathfrak{A}} : \mathfrak{A} \to \Delta$ is the trivial deformation of A. Since A is exceptional in M^*, we may choose ω to be 1-convex [19, Satz 1, p. 547]. Also, [19, p. 553], $\mathfrak{W}$, the union of all of the exceptional sets A_t in the $M_t = \omega^{-1}(t)$, $t \in \Delta$, is a subvariety of $\mathfrak{M}$. Since M^* has codimension one in $\mathfrak{M}$ and $A = \mathfrak{W} \cap M^*$ has codimension n–1 in M^*, $\mathfrak{W}$ has codimension n–1 in M_t, i.e., A_t is of dimension one.

Now suppose that $c \geq -n+2$. We shall reach a contradiction. $c'_1 \geq -1$, $1 \leq i \leq n-1$, and $c'_j \geq 0$ for at least one j. By [13] there is a non-trivial family of compact Riemann surfaces in M_t, $t \neq 0$, containing $\mathfrak{A} \cap M_t$. A_t is the maximal compact set in M_t. Thus A_t has dimension at least two. This contradicts the previous paragraph and completes the proof of Theorem 3.1.

THEOREM 3.2. *Let* M *be a strictly pseudoconvex manifold of dimension* n, $n \geq 2$, *with exceptional set* $A \approx P^1$. *Let* $N = \oplus L_i$, $1 \leq i \leq n-1$, *be the normal bundle of* A *in* M. *If* $|c_i - c_j| \leq 1$, $1 \leq i$, $j \leq n-1$, *then there is a neighborhood* U *of* A *in* M *which is isomorphic to a neighborhood of* A *as the* 0*-section in* N.

Proof. We are going to apply Grauert's obstruction theory [9]. While [9] is primarily written for a codimension one exceptional set, there are no difficulties in extending the results to non-singular A of arbitrary codimension. Let $\mathcal{O}$ be the structure sheaf on M. Let $\mathcal{I}$ be the ideal sheaf of A in M. Let $A(r)$ be the non-reduced space $(A, \mathcal{O}/\mathcal{I}^r)$. $A'(r)$ is defined similarly for N. Start with an isomorphism $\phi_1 : A = A(1) \to A' = A'(1)$. In general, suppose an isomorphism $\phi_r : A(r) \to A'(r)$ has been found. Then ϕ_r may be extended to an isomorphism $\phi_{r+1} : A(r+1) \to A'(r+1)$ if (and only if) the obstruction $[\phi_r]$ in $H^1(A, \mathcal{A}ut(\mathcal{I}^{r+1}; \mathcal{I}^r)$ vanishes, i.e., is the distinguished cohomology class.

We may compute $\mathcal{A}ut(\mathcal{I}^{r+1}; \mathcal{I}^r)$ as in [9]. For $r = 1$, there is an exact sequence (3.1) of sheaves of non-abelian groups on A.

$$1 \to \mathcal{A}ut(\mathcal{I}^2, \mathcal{I}) \to \mathcal{A}ut(\mathcal{I}^2; \mathcal{I}) \to \mathcal{A}n(\mathcal{I}^2, \mathcal{I}) \to 1 . \tag{3.1}$$

$\mathcal{I}/\mathcal{I}^2$ is isomorphic to $\mathcal{N}^{-1}$, the sheaf of germs of sections of N^{-1}. Let $\mathcal{T}$ be the tangent sheaf to A. Then $\mathcal{A}ut(\mathcal{I}^2, \mathcal{I}) \approx \mathcal{T} \otimes (\mathcal{I}/\mathcal{I}^2)$. Although we shall not need it, let us observe that with $\mathcal{O}_A = \mathcal{O}/\mathcal{I}$, $\mathcal{A}n(\mathcal{I}^2, \mathcal{I}) \approx GL(n-1, \mathcal{O}_A)$.

(3.1) yields (3.2), an exact sequence of pointed sets.

$$H^1(A, \mathcal{A}ut(\mathcal{I}^2, \mathcal{I})) \longrightarrow H^1(A, \mathcal{A}ut(\mathcal{I}^2; \mathcal{I})) \xrightarrow{\beta} H^1(A, \mathcal{A}n(\mathcal{I}^2, \mathcal{I}) . \tag{3.2}$$

By Theorem 3.1, $c_i < 0$, $1 \leq i \leq n-1$. Then $H^1(A, \mathcal{A}ut(\mathcal{I}^2, \mathcal{I}))$ has just one element. $\beta([\phi_1])$ depends only on N and N'. Since $N = N'$, $\beta([\phi_1])$ is the distinguished element in $H^1(A, \mathcal{A}n(\mathcal{I}^2, \mathcal{I}))$. Then the exactness of (3.2) implies that $[\phi_1]$ vanishes. Hence ϕ_2 exists.

Now let $r \geq 2$. Let N^{-r} denote the r-fold symmetric tensor product $N^{-1} \otimes \cdots \otimes N^{-1}$. $\mathcal{I}^r/\mathcal{I}^{r+1}$ is isomorphic to the sheaf of germs of sections of N^{-r}. N^{-r} is also isomorphic to a direct sum of line bundles of positive chern class. There is an exact sequence of sheaves of abelian groups on A:

$$0 \to \mathcal{T} \otimes (\mathcal{I}^r/\mathcal{I}^{r+1}) \to \mathcal{A}ut(\mathcal{I}^{r+1}; \mathcal{I}^r) \to \mathcal{N} \otimes (\mathcal{I}^r/\mathcal{I}^{r+1}) \to 0 . \tag{3.3}$$

$H^1(A, \mathcal{T} \otimes (\mathcal{I}^r/\mathcal{I}^{r+1})) = 0$. Since $|c_i - c_j| \leq 1$, $1 \leq i, j \leq n-1$, $H^1(A, \mathcal{N} \otimes (\mathcal{I}^r/\mathcal{I}^{r+1})) = 0$. Hence $H^1(A, \mathcal{A}ut(\mathcal{I}^{r+1}; \mathcal{I}^r)) = 0$. Hence ϕ_r exists for all r. The theorem now follows from [9, pp. 361-363] after blowing up M along A to achieve a codimension one exceptional set.

THEOREM 3.3. *Let* $A \approx P^1$ *be an exceptional set in the* n*-dimensional manifold* M. *Let* $N \approx \oplus L_i$ *be the normal bundle to* A *in* M. *Suppose that* $c_i \leq 0$ *for* $1 \leq i \leq n-1$. *Let* $\pi : M \to V$ *be the blow-down of* M, $0 = \pi(A)$. *Then* 0 *is a rational singularity of* V. *Let* $\mathfrak{m}$ *be the maximal ideal of* V *at* 0. *Let* $h(r) = \dim \mathfrak{m}^r/\mathfrak{m}^{r+1}$ *be the Hilbert function for* V *at* 0. *Then*

$$h(r) = \left(-c + \frac{n-1}{r}\right)\binom{n+r-2}{r-1}.$$

In particular, at 0 *the embedding dimension of* V *is* $-c + n-1$ *and the multiplicity of* V *is* $-c$.

Proof. The last two statements will follow immediately from the fact that the embedding dimension of V is $h(1)$ and from [20]. So we must calculate the Hilbert function and show rationality [6].

We may proceed as in [14, Chapter VII, pp. 134-142] or [16, Theorem 3.13, pp. 1270-1274].

We retain the notation of the proof of Theorem 3.2. $\mathcal{I}^0 = \mathcal{O}$. Consider the exact sheaf sequence

$$0 \to \mathcal{I}^{r+1} \to \mathcal{I}^r \to \mathcal{I}^r/\mathcal{I}^{r+1} \to 0.$$

Since $c_i \leq 0$ for all i, $H^1(M, \mathcal{I}^r/\mathcal{I}^{r+1}) = 0$ for all $r \geq 0$. Hence $\sigma_i : H^1(M, \mathcal{I}^{r+1}) \to H^1(M, \mathcal{I}^r)$ is onto for all r. By [9, p. 356] and [8], $H^1(M, \mathcal{I}^r) = 0$ for all r. Then, as in [16] $\mathfrak{m}^r \approx \Gamma(A, \mathcal{I}^r)$ and $\mathfrak{m}^r/\mathfrak{m}^{r+1} \approx \Gamma(A, \mathcal{I}^r/\mathcal{I}^{r+1})$.

N^{-r} is the direct sum of $\binom{n+r-2}{r}$ line bundles S_j, each of which is the tensor product of r of the L_i. The chern class of S_j is the sum of the chern classes of its factors. The total number of factors for all of the S_j is $r\binom{n+r-2}{r}$ and each L_i appears as a factor $\frac{r}{n-1}\binom{n+r-2}{r}$ times. So the sum of the chern classes of the S_j is $\frac{rc}{n-1}\binom{n+r-2}{r} = c\binom{n+r-2}{r-1}$. $\dim \Gamma(M, \mathcal{I}^r/\mathcal{I}^{r+1}) = \left(-c + \frac{n-1}{r}\right)\binom{n+r-2}{r-1}$, as claimed.

We showed above that $H^1(M, \mathcal{I}) = 0$. By Burns' definition [6], we also need that $H^i(M, \mathcal{O}) = 0$ for $i \geq 2$. But by [17, Theorem 2], for suitably small M, M has a Stein cover where all triple intersections are empty. Hence $H^i(M, \mathcal{F}) = 0$ for $i \geq 2$ and any coherent sheaf $\mathcal{F}$.

IV. *Gorenstein singularities*

THEOREM 4.1. *Let* V *be an analytic space of dimension* $n \geq 3$ *with an isolated singularity at* p. *Suppose that there exists a non-zero holomorphic* n*-form* ω *on* $V - p$. *Let* $\pi: M \to V$ *be a resolution of* V. *Suppose that* $A = \pi^{-1}(p)$ *is 1-dimensional and irreducible. Then* A *is isomorphic to* $\mathbf{P}^1$ *and* $n = 3$. *Also,* (c_1, c_2) *equals* $(-1, -1)$, $(-2, 0)$ *or* $(-3, 1)$.

Proof. $\pi^*(\omega)$ is holomorphic and non-zero on $M - A$. Since A has codimension at least two in M, $\pi^*(\omega)$ has a holomorphic, necessarily non-zero, extension to M.

Take V to be a small Stein neighborhood of p. By [6, Proposition 3.2, p. 239], $H^1(M, \mathcal{O}) = 0$. Let $\mathcal{I}$ be the ideal sheaf of A on M. Let $\mathcal{O}_A$ be the structure sheaf of A.

$$0 \to \mathcal{I} \to \mathcal{O} \to \mathcal{O}_A \to 0 \tag{4.1}$$

is exact. Since $H^2(M, \mathcal{F}) = 0$ for any coherent sheaf on M [17, Theorem 2], $H^1(A, \mathcal{O}_A) = 0$. Then A is indeed $\mathbf{P}^1$.

The only holomorphic functions on A are the constants, which necessarily extend to M. Hence, from the long exact cohomology sequence for (4.1), $H^1(M, \mathcal{I}) = 0$.

$$0 \to \mathcal{I}^2 \to \mathcal{I} \to \mathcal{I}/\mathcal{I}^2 \to 0 \tag{4.2}$$

is exact. Let $\mathfrak{N}$ be the normal sheaf to A in M. From (4.2), $H^1(M, \mathfrak{N}^{-1}) \approx H^1(M, \mathfrak{I}/\mathfrak{I}^2) = 0$. Then $c_i \leq 1$ for all i.

Since $\pi^*(\omega)$ is non-zero near A, K, the canonical bundle on M, is trivial near A. K restricted to A has chern class $-2-c$. Hence $c = -2$. By Theorem 3.1, $n = 3$. Also, the only possibilities for (c_1, c_2) are $(-1,-1)$, $(-2,0)$ and $(-3,1)$. This completes the proof of Theorem 4.1.

Example 2.3 falls under the $(-3,1)$ case.

Example 2.2 gives, in fact, all normal singularities in the $(-1,-1)$ and $(-2,0)$ cases. Namely, the $(-1,-1)$ case follows from Theorem 3.2. For $k \geq 3$, let $\mathfrak{m}$ be the maximal ideal of p. As in the proof of Theorem 3.3, $\mathfrak{m}/\mathfrak{m}^2 \approx \Gamma(A, \mathfrak{I}/\mathfrak{I}^2) \approx \Gamma(A, \mathfrak{I})/\Gamma(A, \mathfrak{I}^2)$. Let $v_1, v_2, v_3, v_4 \in \Gamma(A, \mathfrak{I})$ project onto a basis of $\mathfrak{m}/\mathfrak{m}^2$. We may assume that v_1, v_2, v_3 and v_4 have the same image in $\Gamma(A, \mathfrak{I}/\mathfrak{I}^2)$ as have v_1, v_2, v_3, v_4 from Example 2.2. Then $v_1 v_3 - v_2^2 \equiv 0$ in $\Gamma(A, \mathfrak{I}^2/\mathfrak{I}^3)$. That is

$$v_1 v_3 - v_2^2 = f(v_1, v_2, v_3, v_4) \tag{4.3}$$

with $f(v_1, v_2, v_3, v_4) \in \mathfrak{m}^3$. After a change of coordinates (4.3) becomes

$$u_1^2 + u_2^2 + u_3^2 + u_4^r = 0 \tag{4.4}$$

with $r \geq 3$. (4.4) is necessarily the defining equation for V. Let K be the boundary of a spherical neighborhood of $(0,0,0,0)$ in V. Then [18, Theorem 8.5, p. 68 and Theorem 9.1, p. 71], K has non-zero third Betti number for r even and zero third Betti number for r odd. Hence r must be even.

STATE UNIVERSITY OF NEW YORK AT STONY BROOK
LONG ISLAND, NEW YORK 11794

REFERENCES

[1] Arnold, V., Local normal forms of functions, Inv. Math. *35*(1976), 87-109.

[2] Artin, M., On isolated rational singularities of surfaces, Amer. J. Math. *88*(1966), 129-136.

[3] ———, Algebraic construction of Brieskorn's resolutions, J. Alg. *29*(1974), 330-348.

[4] Artin, M. and Schlessinger, M., Algebraic construction of Brieskorn's resolutions, Preprint.

[5] Brieskorn, E., Singular elements of semi-simple algebraic groups, Proc. Inter. Congr. Math., Nice 1970, Gauthier-Villars, Paris 1971, vol. 2, 274-284.

[6] Burns, D., On rational singularities in dimensions > 2, Math. Ann. *211*(1974), 237-244.

[7] Gohberg, I. and Krein, M., Systems of integral equations on a half line with a kernel depending on the difference of arguments, Uspehi Mat. Nauk 13(1958), no. 2(80) 3-72; A.M.S. Trans. (2) *14*(1960), 217-287.

[8] Grauert, H., Ein Theorem der analytischen Garbentheorie und die Modulräume komplexer Strukturen, Publ. Inst. Hautes Etudes Sci., no. 5, 1960.

[9] ———, Über Modifikationen und exzeptionelle analytische Mengen, Math. Ann. *146*(1962), 331-368.

[10] Grauert, H. and Kerner, H., Deformationen von Singularitäten komplexer Räume, Math. Ann. *153*(1964), 236-260.

[11] Grothendieck, A., Sur la classification des fibres holomorphes sur la sphere de Riemann, Amer. J. Math. *79*(1957), 121-138.

[12] Kempf, G., et al., *Toroidal Embeddings I*, Lecture Notes in Math., No. 339, Springer-Verlag, New York, 1973.

[13] Kodaira, K., On stability of compact submanifolds of complex manifolds, Amer. J. Math. *85*(1963), 79-94.

[14] Laufer, A., *Normal Two-Dimensional Singularities*, Ann. Math. Studies No. 71, Princeton University Press, Princeton 1971.

[15] ———, Deformations of resolutions of two-dimensional singularities, Rice Univ. Studies *59*(1973), vol. 1, 53-90.

[16] ———, On minimally elliptic singularities, Amer. J. Math. *99* (1977), 1257-1295.

[17] ———, Versal deformations for two-dimensional pseudoconvex manifolds, Annali della Scuola Norm. Sup. d. Pisa (4) *7*(1980), 511-521.

[18] Milnor, J., *Singular points of complex hypersurfaces*, Ann. Math. Studies, no. 61, Princeton Univ. Press, Princeton 1968.

[19] Riemenschneider, O., Familien komplexer Räume mit streng pseudokonvexer spezieller Faser, Comm. Math. Helv. *51*(1976), 547-565.

[20] Serre, J.-P., *Algèbra locale-multiplicités*, Springer-Verlag, Berlin 1965.

[21] Tjurina, G. N., Locally semi-universal flat deformations of isolated singularities of complex spaces (Russian). Izv. Akad. Nauk SSR Ser. Mat. *33* (1969), 1026-1058.

[22] Wahl, J., Personal communication, 1979.

ORTHOGONAL MEASURES FOR SUBSETS OF THE BOUNDARY OF THE BALL IN $\mathbb{C}^2$

Hsuan-Pei Lee and John Wermer

*Introduction**

Let X be a compact set in $\mathbb{C}^n$. We denote by $C(X)$ the algebra of all complex-valued continuous functions on X and by $R(X)$ the uniform closure on X of all rational functions P/Q on $\mathbb{C}^n$ with $Q \neq 0$ on X. Suppose $R(X) \neq C(X)$. Then there exists a non-zero finite Borel measure μ on X which is orthogonal to $R(X)$ in the sense that $\int_X f\, d\mu = 0$ for every $f \in R(X)$.

Suppose now that X is a subset of the complex plane, and let μ be a measure on X orthogonal to $R(X)$. It is a well-known consequence of the Cauchy-Green formula that the following relation holds:

$$\text{(A)} \qquad \int_X \phi\, d\mu = \int_{\mathbb{C}} \frac{\partial \phi}{\partial \bar{z}} K\, dz \wedge d\bar{z},$$

for every $\phi \in C_0^\infty(\mathbb{C})$, where

$$K(z) = \frac{-1}{2\pi i} \int \frac{d\mu(\zeta)}{\zeta - z},$$

and also that K satisfies

* An abstract of this paper was given in Notices Amer. Math. Soc. Vol. 26, No. 3(1979), A-271.

Recent Developments in Several Complex Variables
0-691-08285-5/81/000277-13 $00.65/0 (cloth)
0-691-08281-2/81/000277-13 $00.65/0 (paperback)
For copying information, see copyright page.

(B) $K \in L^1(C; dz \wedge d\bar{z})$ and $K = 0$ on $C \setminus X$.

(See, e.g., [7], pp. 8-10.)

(A) and (B) together "explain" why μ is orthogonal to $R(X)$, because if ϕ is holomorphic in some neighborhood of X, then by (A)

$$\int \phi \, d\mu = \int_X \frac{\partial \phi}{\partial \bar{z}} K \, dz \wedge d\bar{z} + \int_{C \setminus X} \frac{\partial \phi}{\partial \bar{z}} K \, dz \wedge d\bar{z}.$$

The first term on the right vanishes since $\frac{\partial \phi}{\partial \bar{z}} = 0$ on X while the second term vanishes because of (B).

To what extent is an analogue of (A) and (B) valid for a subset X of C^n with $n > 1$? In this paper, we study this question for the case when X is a subset of the boundary of the unit ball in C^2. We put $B = \{(z_1, z_2) \in C^2 \mid |z_1|^2 + |z_2|^2 < 1\}$, and we write ∂B for the boundary of B.

Let X be a compact subset of ∂B. We denote by $h_r(X)$ the rationally convex hull of X in C^2 which is the set $\{(z_1, z_2) \in C^2 \mid$ if P is a polynomial such that $P(z_1, z_2) = 0$ then P has a zero on $X\}$. We say that X is rationally convex if $h_r(X) = X$.

R. Basener in [1] exhibited a rationally convex compact subset X_0 of ∂B satisfying $R(X_0) \neq C(X_0)$. The following theorem concerns sets of this type.

THEOREM. *Let X be a compact subset of ∂B which is rationally convex. Let μ be a measure on X orthogonal to $R(X)$. Then there is a function K summable on ∂B which extends analytically to B from $\partial B \setminus X$, such that the following holds:*

Let X^+ be a smoothly bounded neighborhood of X on ∂B. Then for all $\phi \in C_0^\infty(C^2)$,

$$(1) \quad \int \phi \, d\mu = \frac{1}{4\pi^2} \int_{X^+} \bar{\partial}\phi \wedge K \, d\zeta_1 \wedge d\zeta_2 - \frac{1}{4\pi^2} \int_{\partial X^+} \phi K \, d\zeta_1 \wedge d\zeta_2 .$$

Formula (1) gives an explanation why μ is orthogonal to $R(X)$. To see this, fix a function ϕ which is analytic in a neighborhood N of X in $\mathbb{C}^2$. Choose a neighborhood X^+ of X on ∂B so that $N \supset X^+$. Then the first term on the right-hand side in (1) equals zero, since $\bar{\partial}\phi \equiv 0$ on X^+; the second term also vanishes, for the following reason. Let $X_r^+ = \{(rz_1, rz_2) : (z_1, z_2) \text{ in } X^+\}$, $r < 1$. By the above theorem K is holomorphic in a neighborhood of X_r^+, so Stokes' formula gives

$$\int_{\partial X_r^+} \phi K \, d\zeta_1 \wedge d\zeta_2 = \int_{X_r^+} d(\phi K \, d\zeta_1 \wedge d\zeta_2) = \int_{X_r^+} \bar{\partial}\phi \wedge K \, d\zeta_1 \wedge d\zeta_2 .$$

Hence

$$\int_{\partial X^+} \phi K \, d\zeta_1 \wedge d\zeta_2 = \lim_{r \to 1} \int_{\partial X_r^+} \phi K \, d\zeta_1 \wedge d\zeta_2 = \lim_{r \to 1} \int_{X_r^+} \bar{\partial}\phi \wedge K \, d\zeta_1 \wedge d\zeta_2 .$$

For r near 1, $\bar{\partial}\phi \equiv 0$ on X_r^+. Thus the integral $= 0$.

Our proof is based on a formula obtained by Henkin [2] ([5], [6]) which states that if μ is a measure on ∂B which is orthogonal to the polynomials, then for any $\phi \in C_0^\infty(\mathbb{C}^2)$ we have

$$\int_X \phi d\mu = \frac{1}{4\pi^2} \int_{\partial B} \bar{\partial}\phi \wedge K_\mu \, d\zeta_1 \wedge d\zeta_2 , \tag{2}$$

where

$$K_\mu(\zeta_1, \zeta_2) = \int \frac{\bar{\zeta}_1 \bar{z}_2 - \bar{\zeta}_2 \bar{z}_1}{|1 - \zeta_1 \bar{z}_1 - \zeta_2 \bar{z}_2|^2} \, d\mu(z_1, z_2) . \tag{3}$$

(See [3] for a direct, elementary proof of (2).) Our task is to show that K_μ is a desired K in the theorem.

The proof consists of four parts. The first one gives some properties of K_μ which are needed later. For any (a, b) in B, by the rational convexity of X, there exists a polynomial which vanishes at (a, b) but does not have any zero on X. In the second part of the proof we show that for a properly chosen such polynomial P, K_μ can be analytically extended on the variety $\{P=0\}$ from $\{P=0\} \cap \partial B$ to (a, b). Part three shows that such an extension of K_μ is unique. Finally, part four proves the resulting function is indeed analytic on B and extends K_μ from $\partial B \setminus X$.

I. *Properties of* K_μ

Recall (3) $\quad K_\mu(\zeta_1, \zeta_2) = \displaystyle\int \frac{\bar{\zeta}_1 \bar{z}_2 - \bar{\zeta}_2 \bar{z}_1}{|1 - \zeta_1 \bar{z}_1 - \zeta_2 \bar{z}_2|^2}\, d\mu(z_1, z_2)$, where (ζ_1, ζ_2) is a point in a small neighborhood of $\bar{B}$.

(i) K_μ is summable with respect to three-dimensional measure on ∂B. (This is shown in [2] and [3].)

(ii) The integrand on the right-hand side of (3) has a singularity only when $(\zeta_1, \zeta_2) = (z_1, z_2)$. If $(\zeta_1, \zeta_2) \not\in X$, then K_μ is smooth in a neighborhood of (ζ_1, ζ_2). Furthermore, since in this case the integrand is smooth, we can carry out the differentiation of K_μ with respect to ζ_1, ζ_2 inside the integral sign on the right-hand side of (3). Moreover, as we show below, we have that

(iii) K_μ satisfies the tangential Cauchy-Riemann equation $\left(\zeta_2 \dfrac{\partial K_\mu}{\partial \bar{\zeta}_1} - \zeta_1 \dfrac{\partial K_\mu}{\partial \bar{\zeta}_2} = 0\right)$ for ∂B at every $(\zeta_1, \zeta_2) \in \partial B \setminus X$. Thus, for $(\zeta_1, \zeta_2) \in \partial B \setminus X$, the form, on ∂B,

$$\bar{\partial} K_\mu \wedge d\zeta_1 \wedge d\zeta_2 |_{\partial B} = \frac{\partial K_\mu}{\partial \bar{\zeta}_1} d\bar{\zeta}_1 \wedge d\zeta_1 \wedge d\zeta_2 + \frac{\partial K_\mu}{\partial \bar{\zeta}_2} d\bar{\zeta}_2 \wedge d\zeta_1 \wedge d\zeta_2$$

$$= \begin{cases} \left(\zeta_2 \dfrac{\partial K_\mu}{\partial \bar{\zeta}_1} - \zeta_1 \dfrac{\partial K_\mu}{\partial \bar{\zeta}_2}\right) \dfrac{1}{\zeta_2}\, d\bar{\zeta}_1 \wedge d\zeta_1 \wedge d\zeta_2 & \text{if } \zeta_2 \neq 0, \text{ or} \\ -\left(\zeta_2 \dfrac{\partial K_\mu}{\partial \bar{\zeta}_1} - \zeta_1 \dfrac{\partial K_\mu}{\partial \bar{\zeta}_2}\right) \dfrac{1}{\zeta_1}\, d\bar{\zeta}_2 \wedge d\zeta_1 \wedge d\zeta_2 & \text{if } \zeta_1 \neq 0 \end{cases}$$

is zero. And, if X^+ is any smoothly bounded neighborhood of X on ∂B with ∂X^+ suitably oriented, Stokes' formula gives that

$$\int_{\partial B\setminus X^+} \bar\partial\phi \wedge K_\mu\, d\zeta_1 \wedge d\zeta_2$$

$$= \int_{\partial B\setminus X^+} \bar\partial\phi \wedge K_\mu\, d\zeta_1 \wedge d\zeta_2 + \int_{\partial B\setminus X^+} \phi\,\bar\partial K_\mu \wedge d\zeta_1 \wedge d\zeta_2$$

$$= \int_{\partial B\setminus X^+} d(\phi K_\mu\, d\zeta_1 \wedge d\zeta_2)$$

$$= -\int_{\partial X^+} \phi K_\mu\, d\zeta_1 \wedge d\zeta_2$$

which is the desired formula (1). So to prove our theorem it only remains to prove that K_μ extends analytically to B from $\partial B\setminus X$.

(iv) Let T be any compact subset of $\partial B\setminus X$. The function

$$\text{(4)} \qquad \int \frac{\bar\zeta_1\bar z_2 - \bar\zeta_2\bar z_1}{(|\zeta_1|^2+|\zeta_2|^2-\bar\zeta_1 z_1-\bar\zeta_2 z_2)(1-\zeta_1\bar z_1-\zeta_2\bar z_2)}\, d\mu(z_1, z_2)$$

is a holomorphic extension of K_μ (denote it by K_μ^E) to a small neighborhood of T in $\mathbf{C}^2$.

Proof. The function $(|\zeta_1|^2+|\zeta_2|^2-\bar\zeta_1 z_1-\bar\zeta_2 z_2)(1-\zeta_1\bar z_1-\zeta_2\bar z_2)$ is not equal to zero on $T\times X$. Therefore, there exist neighborhoods U_T, U_X of T and X respectively in $\mathbf{C}^2$ such that $(|\zeta_1|^2+|\zeta_2|^2-\bar\zeta_1 z_1-\bar\zeta_2 z_2)(1-\zeta_1\bar z_1-\zeta_2\bar z_2) \neq 0$ on $U_T\times U_X$. So the function

$$\frac{\bar\zeta_1\bar z_2 - \bar\zeta_2\bar z_1}{(|\zeta_1|^2+|\zeta_2|^2-\bar\zeta_1 z_1-\bar\zeta_2 z_2)(1-\zeta_1\bar z_1-\zeta_2\bar z_2)}$$

is well defined and smooth on $U_T \times U_X$. Also,

$$\frac{\partial K_\mu^E}{\partial \bar{\zeta}_1}(\zeta_1, \zeta_2)$$

$$= \int \frac{\partial}{\partial \bar{\zeta}_1} \frac{\bar{\zeta}_1 \bar{z}_2 - \bar{\zeta}_2 \bar{z}_1}{(|\zeta_1|^2 + |\zeta_2|^2 - \bar{\zeta}_1 z_1 - \bar{\zeta}_2 z_2)(1 - \zeta_1 \bar{z}_1 - \zeta_2 \bar{z}_2)} d\mu(z_1, z_2)$$

$$= \int \frac{\bar{z}_2(|\zeta_1|^2 + |\zeta_2|^2 - \bar{\zeta}_1 z_1 - \bar{\zeta}_2 z_2) - (\bar{\zeta}_1 \bar{z}_2 - \bar{\zeta}_2 \bar{z}_1)(\zeta_1 - z_1)}{(|\zeta_1|^2 + |\zeta_2|^2 - \bar{\zeta}_1 z_1 - \bar{\zeta}_2 z_2)^2 (1 - \zeta_1 \bar{z}_1 - \zeta_2 \bar{z}_2)} d\mu(z_1, z_2)$$

$$= \int \frac{|\zeta_1|^2 \bar{z}_2 + |\zeta_2|^2 \bar{z}_2 - \bar{\zeta}_1 z_1 \bar{z}_2 - \bar{\zeta}_2 |z_2|^2 - |\zeta_1|^2 \bar{z}_2 + \bar{\zeta}_2 \zeta_1 \bar{z}_1 + \bar{\zeta}_1 \bar{z}_2 z_1 - \bar{\zeta}_2 |z_1|^2}{(|\zeta_1|^2 + |\zeta_2|^2 - \bar{\zeta}_1 z_1 - \bar{\zeta}_2 z_2)^2 (1 - \zeta_1 \bar{z}_1 - \zeta_2 \bar{z}_2)} d\mu(z_1, z_2)$$

$$= \int \frac{\bar{\zeta}_2(\zeta_2 \bar{z}_2 + \zeta_1 \bar{z}_1 - 1)}{(|\zeta_1|^2 + |\zeta_2|^2 - \bar{\zeta}_1 z_1 - \bar{\zeta}_2 z_2)^2 (1 - \zeta_1 \bar{z}_1 - \zeta_2 \bar{z}_2)} d\mu(z_1, z_2)$$

$$= 0$$

since $\mu \perp R(X)$. Similarly, $\dfrac{\partial K_\mu^E}{\partial \bar{\zeta}_2} = 0$. Clearly,

$$K_\mu^E(\zeta_1, \zeta_2) = \int \frac{\bar{\zeta}_1 \bar{z}_2 - \bar{\zeta}_2 \bar{z}_1}{(1 - \bar{\zeta}_1 z_1 - \bar{\zeta}_2 z_2)(1 - \zeta_1 \bar{z}_1 - \zeta_2 \bar{z}_2)} d\mu(z_1, z_2)$$

$$= K_\mu(\zeta_1, \zeta_2)$$

for $(\zeta_1, \zeta_2) \in \partial B$. So K_μ^E is a holomorphic extension of K_μ to U_T. For the sake of simplicity we denote it again by K_μ. Since T was arbitrary, we see that K_μ extends to a neighborhood U of $\partial B \setminus X$ in $\mathbb{C}^2$ by formula (4) above.

II. *Extendability*

Fix (a, b) in B. By the rational convexity of X, there exists a polynomial P, which we may suppose to be irreducible, such that P(a, b) = 0, yet P does not have any zero on X. By properly choosing P we may assume that the variety $\{P=0\}$ is non-singular everywhere in $\mathbb{C}^2$. (For a justification of this, see appendix.) Let Ω be a neighborhood of $\overline{B}$ in $\mathbb{C}^2$ and let Y be the connected component of $\{P=0\} \cap \Omega$ which contains the point (a, b). Let U be as above, so that U is a neighborhood of $\partial B \setminus X$ in $\mathbb{C}^2$ on which K_μ is analytic. Choose a connected two-dimensional submanifold Σ of Y such that $(a, b) \in \Sigma$, $\Gamma = \partial\Sigma \subset U \cap \overline{B}$ and Γ is a disjoint union of real analytic curves. Thus Σ is a finite Riemann surface with smooth boundary Γ. Since $\Gamma \subset U$, K_μ is defined on Γ.

Let ω be any holomorphic (1,0)-form on $\overline{\Sigma}$. In view of the fact that

$$\frac{\bar{\zeta}_1\bar{z}_2 - \bar{\zeta}_2\bar{z}_1}{(|\zeta_1|^2+|\zeta_2|^2-\bar{\zeta}_1 z_1-\bar{\zeta}_2 z_2)(1-\zeta_1\bar{z}_1-\zeta_2\bar{z}_2)}$$

is regular for $(\underline{\zeta}, \underline{z}) \in \Gamma \times X$, where $\underline{\zeta} = (\zeta_1, \zeta_2)$, $\underline{z} = (z_1, z_2)$, we get

$$(5) \quad \int_\Gamma K_\mu \omega = \int_\Gamma \left(\int_X \frac{\bar{\zeta}_1\bar{z}_2 - \bar{\zeta}_2\bar{z}_1}{(|\zeta_1|^2+|\zeta_2|^2-\bar{\zeta}_1 z_1-\bar{\zeta}_2 z_2)(1-\zeta_1\bar{z}_1-\zeta_2\bar{z}_2)} d\mu(z_1,z_2) \right) \omega(\zeta_1,\zeta_2)$$

$$= \int_X d\mu(z_1,z_2) \int_\Gamma \frac{\bar{\zeta}_1\bar{z}_2 - \bar{\zeta}_2\bar{z}_1}{(1-\zeta_1\bar{z}_1-\zeta_2\bar{z}_2)(|\zeta_1|^2+|\zeta_2|^2-\bar{\zeta}_1 z_1-\bar{\zeta}_2 z_2)} \omega(\zeta_1,\zeta_2) .$$

Let $$F(z_1,z_2) = \int_\Gamma \frac{\bar{\zeta}_1\bar{z}_2 - \bar{\zeta}_2\bar{z}_1}{(|z_1|^2+|z_2|^2-\zeta_1\bar{z}_1-\zeta_2\bar{z}_2)(|\zeta_1|^2+|\zeta_2|^2-\bar{\zeta}_1 z_1-\bar{\zeta}_2 z_2)} \omega(\zeta_1,\zeta_2).$$

We see that for (z_1, z_2) in X, $F(z_1, z_2)$ is the integrand in the right-hand side of relation (5). By our argument in (iv) of part I, $F(z_1, z_2)$ is well defined and smooth in a neighborhood U_X of X in $\mathbf{C}^2$. We will show that F is holomorphic there.

$$\frac{\partial F}{\partial \bar z_1} = \int_\Gamma \frac{\partial}{\partial \bar z_1}\left\{\frac{\bar\zeta_1\bar z_2 - \bar\zeta_2\bar z_1}{(|z_1|^2+|z_2|^2-\zeta_1\bar z_1-\zeta_2\bar z_2)(|\zeta_1|^2+|\zeta_2|^2-\bar\zeta_1 z_1-\bar\zeta_2 z_2)}\right\}\omega(\zeta_1,\zeta_2)$$

$$= \int_\Gamma \frac{-\bar\zeta_2(|z_1|^2+|z_2|^2-\zeta_1\bar z_1-\zeta_2\bar z_2) - (z_1-\zeta_1)(\bar\zeta_1\bar z_2-\bar\zeta_2\bar z_1)}{(|z_1|^2+|z_2|^2-\zeta_1\bar z_1-\zeta_2\bar z_2)^2(|\zeta_1|^2+|\zeta_2|^2-\bar\zeta_1 z_1-\bar\zeta_2 z_2)}\,\omega(\zeta_1,\zeta_2)$$

$$= \int_\Gamma \frac{-\bar\zeta_2|z_1|^2-\bar\zeta_2|z_2|^2+\bar\zeta_2\zeta_1\bar z_1+|\zeta_2|^2\bar z_2-z_1\bar\zeta_1\bar z_2+|\zeta_1|^2\bar z_2+\bar\zeta_2|z_1|^2-\zeta_1\bar\zeta_2\bar z_1}{(|z_1|^2+|z_2|^2-\zeta_1\bar z_1-\zeta_2\bar z_2)^2(|\zeta_1|^2+|\zeta_2|^2-\bar\zeta_1 z_1-\bar\zeta_2 z_2)}\,\omega($$

$$= \int_\Gamma \frac{\bar z_2}{(|z_1|^2+|z_2|^2-\zeta_1\bar z_1-\zeta_2\bar z_2)^2}\,\omega(\zeta_1,\zeta_2)$$

We observe that for (z_1, z_2) sufficiently close to X and for (ζ_1, ζ_2) in Σ, we have $|z_1|^2 + |z_2|^2 - \zeta_1\bar z_1 - \zeta_2\bar z_2 \neq 0$. So, by shrinking U_X, if necessary, the function

$$\frac{\bar z_2}{(|z_1|^2 + |z_2|^2 - \zeta_1\bar z_1 - \zeta_2\bar z_2)^2}$$

is defined for all (ζ_1, ζ_2) in $\bar\Sigma$, and is in fact analytic there. Stokes' formula then gives

$$\frac{\partial F}{\partial \bar z_1} = \int_\Gamma \frac{\bar z_2}{(|z_1|^2+|z_2|^2-\zeta_1\bar z_1-\zeta_2\bar z_2)^2}\,\omega(\zeta_1,\zeta_2)$$

$$= \int_\Sigma d_\zeta\left(\frac{\bar z_2}{(|z_1|^2+|z_2|^2-\zeta_1\bar z_1-\zeta_2\bar z_2)^2}\,\omega(\zeta_1,\zeta_2)\right)$$

$$= 0$$

since ω is a holomorphic 1-form and

$$\frac{\bar{z}_2}{(|z_1|^2+|z_2|^2-\zeta_1\bar{z}_1-\zeta_2\bar{z}_2)^2}$$

is holomorphic in ζ_1, ζ_2. Similarly, $\frac{\partial F}{\partial \bar{z}_2} = 0$ in U_X. It follows that F is holomorphic in U_X as claimed, and so the restriction of F to X lies in R(X), since X is rationally convex. Hence

$$\int F\, d\mu = 0 \quad \text{and consequently} \quad \int K_\mu \omega = 0 .$$

Since this holds for all ω, by a theorem of Royden, [4], K_μ can be extended on Σ from Γ to a holomorphic function of $\dot{\Sigma}$. For a fixed choice of P, the value $K_\mu^P(a, b)$ obtained in this way clearly does not depend on our choice of Σ. Also, if $(a, b) \in U$, $K_\mu^P(a, b) = K_\mu(a, b)$. We denote the value of this extension at (a, b) by $K_\mu^P(a, b)$.

III. *Uniqueness*

Fix (a, b) in B. Choose P, Σ, Γ as in II.

ASSERTION. For every choice of P, Σ, the value $K_\mu^P(a, b)$, obtained by the above extension, is the same.

Proof. Consider two tuples (P, Σ) and $(\tilde{P}, \tilde{\Sigma})$. Let g be a function holomorphic on Σ, continuous on $\bar{\Sigma}$, such that g has a simple zero at (a, b) and no other zeros. By multiplying by a constant we may assume that the residue of the differential $g^{-1}d\zeta_1$ at (a, b) is 1. By the residue theorem on Σ, then

$$K_\mu^P(a, b) = \frac{1}{2\pi i}\int_\Gamma K_\mu\, g^{-1} d\zeta_1 .$$

Let $\tilde{g}$ be the analogous function for $\tilde{\Sigma}$. Then

$$K_\mu^{\tilde{P}}(a,b) = \frac{1}{2\pi i}\int_{\tilde{\Gamma}} K_\mu\, \tilde{g}^{-1} d\zeta_1\,, \qquad \tilde{\Gamma} = \partial\tilde{\Sigma}\,.$$

We thus need to show that

$$\int_\Gamma K_\mu\, g^{-1} d\zeta_1 - \int_{\tilde{\Gamma}} K_\mu\, \tilde{g}^{-1} d\zeta_1 = 0\,.$$

Interchanging the order of integration, the left side above can be written as

$$\int_X d\mu(z_1,z_2)\left(\int_\Gamma \frac{\bar{\zeta}_1\bar{z}_2 - \bar{\zeta}_2\bar{z}_1}{(1-\zeta_1\bar{z}_1-\zeta_2\bar{z}_2)(|\zeta_1|^2+|\zeta_2|^2-\bar{\zeta}_1 z_1-\bar{\zeta}_2 z_2)}\, g^{-1}d\zeta_1 \right.$$

$$\left. - \int_{\tilde{\Gamma}} \frac{\bar{\zeta}_1\bar{z}_2 - \bar{\zeta}_2\bar{z}_1}{(1-\zeta_1\bar{z}_1-\zeta_2\bar{z}_2)(|\zeta_1|^2+|\zeta_2|^2-\bar{\zeta}_1 z_1-\bar{\zeta}_2 z_2)}\, \tilde{g}^{-1}d\zeta_1\right)$$

Let $F(z_1,z_2) = \displaystyle\int_\Gamma \frac{\bar{\zeta}_1\bar{z}_2 - \bar{\zeta}_2\bar{z}_1}{(|z_1|^2+|z_2|^2-\zeta_1\bar{z}_1-\zeta_2\bar{z}_2)(|\zeta_1|^2+|\zeta_2|^2-\bar{\zeta}_1 z_1-\bar{\zeta}_2 z_2)}\, g^{-1}d\zeta_1$

$$- \int_{\tilde{\Gamma}} \frac{\bar{\zeta}_1\bar{z}_2 - \bar{\zeta}_2\bar{z}_1}{(|z_1|^2+|z_2|^2-\zeta_1\bar{z}_1-\zeta_2\bar{z}_2)(|\zeta_1|^2+|\zeta_2|^2-\bar{\zeta}_1 z_1-\bar{\zeta}_2 z_2)}\, \tilde{g}^{-1}d\zeta_1$$

We claim that F is holomorphic provided (z_1,z_2) is in a small neighborhood of X in $\mathbf{C}^2$. We calculate the partial derivatives $\frac{\partial F}{\partial \bar{z}_1}$ as above and get

$$\frac{\partial F}{\partial \bar{z}_1} = \int_\Gamma \frac{\bar{z}_2}{(|z_1|^2+|z_2|^2-\zeta_1\bar{z}_1-\zeta_2\bar{z}_2)^2}\, g^{-1}d\zeta_1 - \int_{\tilde{\Gamma}} \frac{\bar{z}_2}{(|z_1|^2+|z_2|^2-\zeta_1\bar{z}_1-\zeta_2\bar{z}_2)^2}\, \tilde{g}^{-1}$$

$$= 2\pi i\left(\frac{\bar{z}_2}{(|z_1|^2+|z_2|^2-a\bar{z}_1-b\bar{z}_2)^2} - \frac{\bar{z}_2}{(|z_1|^2+|z_2|^2-a\bar{z}_1-b\bar{z}_2)^2}\right)$$

$$= 0$$

by the residue theorem and the fact that

$$\frac{1}{(|z_1|^2+|z_2|^2-\zeta_1\bar{z}_1-\zeta_2\bar{z}_2)^2}$$

is a holomorphic function on $\Sigma \cup \tilde{\Sigma}$ for fixed (z_1, z_2). Similarly $\frac{\partial F}{\partial \bar{z}_2} = 0$. Hence F is holomorphic as claimed and so $\int_X F\, d\mu = 0$. Thus $K_\mu^P(a,b) = K_\mu^{\tilde{P}}(a,b)$ as desired.

We have thus obtained a unique extension of K_μ from $\partial B \setminus X$ to B. We denote it again by K_μ.

IV. *Analyticity*

Fix (a, b) in B. We shall show that K_μ is analytic in some neighborhood of (a, b) in $\mathbb{C}^2$. Let P, Σ, Γ be as in II, III, and recall that Ω is a neighborhood of $\bar{B}$ in $\mathbb{C}^2$. We may assume that $\frac{\partial P}{\partial z_1} \neq 0$ at (a,b). Let δ be small enough so that $\{P=u\} \cap X = \phi$ for all u, $|u| < \delta$. Also, we can choose ε small enough so that if we put $Q(z_1, z_2) = P(z_1, z_2) + \varepsilon(z_2 - b)$ then $\{Q=v\} \cap X = \phi$ and $\{Q=v\} \cap \Omega$ is non-singular for all v, $|v| < \delta$. Therefore, all the previous arguments in II, III are applicable for the Riemann surfaces $\{P=u\} \cap \Omega$, $\{Q=v\} \cap \Omega$, with $|u|$, $|v| < \delta$, and we have that the restriction of K_μ is analytic on $\{P=u\} \cap \Omega$, $\{Q=v\} \cap \Omega$, for each u, v. Clearly, the set $W = \{(z_1, z_2)$ in B, $P(z_1, z_2) = u$, $Q(z_1, z_2) = v$, $|u|, |v| < \delta\}$ is an open neighborhood of (a, b). Let ϕ be the map from W to $\mathbb{C}^2$ such that $\phi(z_1, z_2) = (P(z_1, z_2), Q(z_1, z_2))$. Then, since the Jacobian determinant $J\phi(a, b) = \varepsilon \frac{\partial P}{\partial z_1}$, is non-zero by our choice of P, ϕ is a biholomorphic map from a neighborhood of (a,b) onto a neighborhood of (0, 0) in $\mathbb{C}^2$. $K_\mu \circ \phi^{-1}$ is holomorphic along each coordinate axis $u = u_o$ and also along each axis $v = v_o$. Hartogs' Theorem now gives that $K_\mu \circ \phi^{-1}$ is holomorphic in a neighborhood of (0, 0). So, K_μ is holomorphic in a neighborhood of (a,b). This establishes the analyticity of K_μ.

ACKNOWLEDGEMENT

Our original proof for part IV is a much more complicated one. The present one is due to Brian Cole, to whom we are most grateful. We also thank Alan Landman for his very helpful remarks.

A special case of this theorem is contained in [3].

Appendix

Let Q be an irreducible polynomial which vanishes at (a, b) and has no zeros on X. By adding $\varepsilon(z_1-a)+\varepsilon(z_2-b)$ to Q for ε sufficiently small, we may assume that $\frac{\partial Q}{\partial z_1}, \frac{\partial Q}{\partial z_2} \neq 0$ at (a, b). The singular points of $\{Q=0\}$, if any, are the common zeros of Q, $\frac{\partial Q}{\partial z_1}, \frac{\partial Q}{\partial z_2}$. There are only finitely many of them. Hence, if ε_0 is small enough, the sets $\{Q=\varepsilon\}$ will miss X and $\{Q=\varepsilon\}$ consists of non-singular points for all ε, $0<|\varepsilon|<\varepsilon_0$. Choose c_1, c_2 small enough so that $Q(a+c_1, b+c_2)=c$ with $|c|<\varepsilon_0$, and the translation set $\{(z_1-c_1, z_2-c_2); (z_1, z_2)$ in $\{Q=c\}\}$ still misses X. Let $P(z_1, z_2)=Q(z_1+c_1, z_2+c_2)-c$. Then $P(a,b)=0$ and $\{P=0\}$ misses X. Moreover, $\{P=0\}$ is non-singular, for,

$$\{P=0\} \cap \left\{\frac{\partial P}{\partial z_1}=0\right\} \cap \left\{\frac{\partial P}{\partial z_2}=0\right\}$$

$$=\left\{(z_1-c_1, z_2-c_2) : (z_1, z_2) \in \{Q=c\} \cap \left\{\frac{\partial Q}{\partial z_1}=0\right\} \cap \left\{\frac{\partial Q}{\partial z_2}=0\right\}\right\}=\phi .$$

BROWN UNIVERSITY

REFERENCES

[1] R. Basener, On rationally convex hulls, Trans. Amer. Math. Soc. 182 (1973), 353-381.

[2] G. M. Henkin, The Lewy equation and analysis on pseudoconvex manifolds, Russian Math. Surveys, 32:3 (1977), 59-130, Uspehi Mat. Nauk 32:3 (1977), 57-118.

[3] H. P. Lee, Thesis, Brown University (1979).

[4] H. L. Royden, The boundary values of analytic and harmonic functions, Math. Zeitschrift 78 (1962), 1-24.

[5] H. Skoda, Valeurs au bord pour les solutions de l'opérateur d'' dans les ouverts strictement pseudoconvexes, C. R. Acad. Sci. Paris 280 (1975), A633-A636.

[6] N. Th. Varopoulos, BMO functions and the $\bar{\partial}$-equation, Pac. J. Math. Vol. 71, No. 1 (1977), 221-273.

[7] J. Wermer, Banach Algebras and Several Complex Variables, 2nd ed., New York, Springer-Verlag, 1976.

A HOLOMORPHICALLY CONVEX ANALOGUE OF CARTAN'S THEOREM B

Andrew Markoe

In the conference at Princeton last spring I presented four topics on the analogy between holomorphically convex spaces and Stein spaces.

Sometimes the analogy is close. This is the case in the first topic which reflects the title by presenting a characterization of holomorphic convexity in terms of the cohomology groups analogous to Cartan's Theorem B characterization of Stein spaces.

Actually this analogy is not complete since it involves an auxiliary condition which is not cohomological. The second topic shows that this extra condition is necessary by proving an extension of part of Ramis's generalization of the Andreotti-Grauert finiteness theorem.

Sometimes the analogy does not hold. Narasimhan showed that holomorphic completeness is invariant under proper maps [8], but I showed [6] that holomorphic convexity is not invariant under proper maps. The third topic presents a positive aspect of this analogy for analytic fiber bundles.

Every one-dimensional analytic space without compact subvarieties of positive dimension is a Stein space. The analogy for the holomorphically convex case would be that every one-dimensional analytic space is holomorphically convex. The last topic is concerned with this analogy, which happens to be false.

Recent Developments in Several Complex Variables
0-691-08285-5/81/000291-08 $00.50/0 (cloth)
0-691-08281-2/81/000291-08 $00.50/0 (paperback)

1. *A Cartan's Theorem B for holomorphically convex spaces*

One aspect of Cartan's Theorem B is that the vanishing of $H^1(X, \mathcal{J})$ for all ideal sheaves $\mathcal{J}$ characterizes Stein spaces. The analogy continues to strongly pseudoconvex spaces by replacing the vanishing by finite dimensionality as shown by Narasimhan [7].

The next step in the hierarchy are the holomorphically convex spaces. The first result gives the analogous characterization in terms of a topological separation property of the cohomology groups. By computing cohomology from the Cech complex over a countable Stein cover, the cohomology groups of a coherent sheaf inherit a topology. This topology is an invariant of the Stein cover chosen and is a QFS topology; quotient of Frechet spaces. However only the numerator is complete, the denominator may not be. Hence the topology may not be Hausdorff separated. The term *separated* will always be taken to mean Hausdorff separated. Also all complex spaces are assumed to be second countable.

Before proceeding to the theorem two comments are necessary. First there can be no purely cohomological characterization of holomorphically convex spaces (cf. section 2). The extra condition involves sets of the form $L_x = \{y \in X : f(y) = f(x) \text{ for all } f \in \mathcal{O}(X)\}$; the *level set* of X through x.

Second Prill [9] proved and Ramis [10] announced that $H^p(X, \mathcal{F})$ is separated for any coherent sheaf $\mathcal{F}$ and any p. This aspect of the theorem will not be repeated here.

THEOREM 1. *For a complex space* X *to be holomorphically convex it is necessary and sufficient that*

1. $H^1(X, \mathcal{J})$ *is separated for all ideal sheaves* $\mathcal{J}$ *and*
2. *All level sets of* X *are compact.*

Proof. The necessity of *1* follows from the Prill-Ramis theorem.

The necessity of *2* follows directly from the definition of holomorphic convexity; not all holomorphic functions can be constant on a discrete set of a holomorphically convex space.

To prove sufficiency, let $\{x_1, \cdots, x_n, \cdots\}$ be an infinite discrete set in X. I will show that there is a holomorphic function f with $\limsup_{n\to\infty} |f(x_n)| = \infty$. By *2* only finitely many x_n can be in any level set. Therefore by discarding a subsequence, if necessary, it may be assumed that $L_{x_n} \cap L_{x_m} = \emptyset$ if $m \neq n$.

Let $\mathcal{I}$ be the ideal sheaf of $\{x_1, \cdots\}$.

The assumption that $L_{x_n} \cap L_{x_m} = \emptyset$ for $m \neq n$ together with standard interpolation methods show that for every n the values of a holomorphic function may be specified on $\{x_1, \cdots, x_n\}$. The quotient map $\pi: \mathcal{O}(X) \to (\mathcal{O}/\mathcal{I})(X)$ assigns the sequence $(f(x_1), \cdots, f(x_n), \cdots)$ to $f \in \mathcal{O}(X)$ once $(\mathcal{O}/\mathcal{I})(X)$ is identified with the Frechet space $C^{\{x_1, \cdots\}}$. The ability to specify arbitrary values of a holomorphic function to every $\{x_1, \cdots, x_n\}$ implies that the image of $\mathcal{O}(X)$ under π is dense in $(\mathcal{O}/\mathcal{I})(X)$.

But the exact sequence $\mathcal{O}(X) \xrightarrow{\pi} (\mathcal{O}/\mathcal{I})(X) \xrightarrow{\partial} H^1(X, \mathcal{I})$ and the assumption *1* that $H^1(X, \mathcal{I})$ is separated imply that $\ker \partial = \operatorname{im} \pi$ is closed in $(\mathcal{O}/\mathcal{I})(X)$. A closed dense set must equal the whole space so π is surjective and it is easy to find an $f \in \mathcal{O}(X)$ which blows up along $\{x_1, \cdots\}$.

2. *A condition for the separation of certain cohomology groups*

The Andreotti-Grauert finiteness theorem for strongly pseudoconcave spaces as generalized by Ramis [10] and Andreotti-Kas [2] states that $H^p(X, \mathcal{F})$ is separated for $p \leqq \operatorname{codh} \mathcal{F} - q$ and is finite dimensional for $p < \operatorname{codh} \mathcal{F} - q$ provided that $\mathcal{F}$ is a coherent sheaf on the strongly q-concave analytic space X.

Unfortunately, this result says nothing about sheaves of lower homological codimension. The next result is a first step in obtaining some separation information for the cohomology of sheaves of lower codimension. An application of this result shows that the level set condition in Theorem 1 is necessary.

THEOREM 2. *Let* $0 \to \mathfrak{F}' \to \mathfrak{F}$ *be an exact sequence of coherent analytic sheaves on* X *such that* $H^1(X, \mathfrak{F})$ *is separated and* $\dim H^0(X, \mathfrak{F})$ *is finite.*

Then $H^1(X, \mathfrak{F}')$ *is separated.*

Proof. First it must be observed that $\mathfrak{F}'$ may be treated entirely like a subsheaf of $\mathfrak{F}$. This is entirely obvious from the algebraic standpoint, but one must be careful to check that the relative topologies on cochains, cocycles and coboundaries with coefficients in $\mathfrak{F}'$ (considered as subgroups of the corresponding groups with coefficients in $\mathfrak{F}$) coincide with the intrinsic topologies on these groups. This is not hard, but it is boring and I have taken the liberty of omitting the argument.

Let $\mathfrak{U}$ be a countable Stein cover of X and let $C^0(\mathfrak{F}) = C^0(\mathfrak{U}, \mathfrak{F})$, with similar notations for the other Cech groups.

By the remarks on the relative topologies, $C^0(\mathfrak{F}')$ is a closed subspace of $C^0(\mathfrak{F})$. Hence its sum with a finite dimensional subspace is also closed. Thus $(C^0(\mathfrak{F}') + Z^0(\mathfrak{F}))/Z^0(\mathfrak{F})$ is a Frechet subspace of $C^0(\mathfrak{F})/Z^0(\mathfrak{F})$.

But the open mapping theorem implies that the algebraic and continuous isomorphism $\delta: C^0(\mathfrak{F})/Z^0(\mathfrak{F}) \to B^1(\mathfrak{F})$ is actually a topological isomorphism. Here δ is the coboundary operator and the separation of $H^1(X, \mathfrak{F})$ was used to show that $B^1(\mathfrak{F})$ is closed in $C^1(\mathfrak{F})$.

Hence $\delta((C^0(\mathfrak{F}') + Z^0(\mathfrak{F}))/Z^0(\mathfrak{F}))$ is closed in $B^1(\mathfrak{F})$. The proof is completed by observing that $B^1(\mathfrak{F}') = \delta((C^0(\mathfrak{F}') + Z^0(\mathfrak{F})/Z^0(\mathfrak{F}))$. A QFS-space with closed denominator is a Frechet space, hence separated.

COROLLARY. *If* X *is strongly* q-*concave of homological codimension at least* q+1, *then* $H^1(X, \mathfrak{I})$ *is separated for all ideal sheaves* $\mathfrak{I}$.

Proof. By the hypothesis that $\operatorname{codh} \mathcal{O} \geq q+1$ and by the theorem of Ramis [10] it follows that $H^0(X, \mathcal{O})$ is finite dimensional and that $H^1(X, \mathcal{O})$ is separated. Now apply the theorem.

EXAMPLE. If X is a compact complex manifold of dimension 2 with a small ball excised, then $H^1(X, \mathcal{J})$ is separated for every ideal sheaf $\mathcal{J}$, but X is not holomorphically convex. Thus condition *2* of Theorem 1 is necessary.

To verify the example, just observe that such a manifold obeys the hypotheses of the corollary to give the separation. But all holomorphic functions are constant so it cannot be holomorphically convex.

3. *Proper maps of fiber bundles*

Holomorphic convexity is not invariant under proper maps although holomorphic completeness is. However certain proper maps do preserve holomorphic convexity as shown by the next result.

THEOREM 3. *If $\pi: X^* \to X$ is an analytic fiber bundle with compact fibers such that the total space is holomorphically convex, then the base space is also holomorphically convex.*

Proof. Let K be a model for the fiber of π. This means that every fiber of π is biholomorphic to K. In particular K can be decomposed into a finite number of connected compact components K_α; $\alpha = 1, \cdots, k$.

By definition of an analytic fiber bundle, there is a countable open cover $\mathfrak{U}$ of X such that

1. Every $U_i \in \mathfrak{U}$ is relatively compact, Stein and connected,

2. There are biholomorphic maps $T_i : U_i \times K \to \pi^{-1}(U_i)$ which respect fibers: $\pi(T_i(x, y)) = y$.

Let y_α be some point chosen in K_α.

Let $\{x_n\}$ be an infinite discrete set in X. For each n choose an index $i(n)$ such that $x_n \in U_{i(n)}$. Let $x^*_{n,\alpha} = T_{i(n)}(x_n, y_\alpha) \in \pi^{-1}(U_{i(n)})$. By taking a subsequence if necessary, it may be assumed that $x^*_{n,\alpha}$ and $x^*_{m,\beta}$ are in different elements of the Stein decomposition of X^*. Hence the holomorphic convexity of X^* implies that there is a holomorphic function $f^* \in \mathcal{O}(X^*)$ such that $f^*(x^*_{n,\alpha}) = n$ for all n, α.

Define $f_i : U_i \to \mathbf{C}$ by $f_i(x) = \sum_{\alpha=1}^{k} f^*(T_i(x, y))$.

Clearly (f_i) defines a holomorphic co-chain. It must in fact be proved that this is a holomorphic function.

Let $x \in U_i \cap U_j$. If K has some biholomorphic components, there is no guarantee that $T_i(x, y_\alpha) = T_j(x, y_\alpha)$. In fact, even if $T_i(x, y_\alpha)$ and $T_j(x, y_\alpha)$ are in the same component of K, there is no reason to suppose that they will be the same point. But, for fixed x, $T_i(x, \cdot)$ and $T_j(x, \cdot)$ are both biholomorphic maps onto $\pi^{-1}(x)$. Hence the sets $\{f^* T_i(x, y_\alpha)\}$ and $\{f^* T_j(x, y_\alpha)\}$ are identical since a holomorphic function is constant on a connected compact subvariety.

Therefore $f_i(x) = f_j(x)$, proving that the definition $f(x) = f_i(x)$ for $x \in U_i$ gives a holomorphic function.

Now, $f(x_n) = f_{i(n)}(x_n) = \sum_{\alpha=1}^{k} f^*(T_{i(n)}(x_n, y_\alpha)) = \sum_{\alpha=1}^{k} f^*(x^*_{n,\alpha}) = kn$.

Thus a function which blows up along $\{x_n\}$ has been found and X is thus holomorphically convex.

4. *On the holomorphic convexity of* 1*-dimensional complex spaces*

When I was studying complex analysis I read Theorem IX C 10 of Gunning and Rossi's book [4], but I did not believe the statement that every one-dimensional analytic space is a Stein space. However, I did believe that the basic technique was right and could prove that every one-dimensional analytic space is holomorphically convex.

But then I read an announcement of an example of H. W. Schuster in Kramm's paper [5], which showed that not every one-dimensional analytic space was holomorphically convex.

The following is a slight modification of Schuster's "necklace" suggested to me by R. Narasimhan. Take an infinite union of spheres each having one point in common with the next. This is a one-dimensional space on which all holomorphic functions are constant. Yet it is not compact and therefore not holomorphically convex.

DEFINITION. Let K_X be the union of the compact irreducible branches of the analytic space X. A *necklace* in X is a connected component of K_X in the relative topology induced by X. A *finite necklace* is one which is the union of finitely many compact irreducible branches, while an *infinite necklace* is the remaining case.

The next theorem shows that one-dimensional spaces are holomorphically convex if and only if there are no infinite necklaces. Thus Schuster's example is typical of what can go wrong.

THEOREM 4. *A one-dimensional analytic space is holomorphically convex if and only if every necklace has only finitely many irreducible components.*

Proof. Since holomorphic functions are constant on connected compact subvarieties, the existence of an infinite necklace shows that there will be a discrete sequence along which every holomorphic function is constant. Thus the condition is necessary.

For the other direction, let $\pi: X \to X$ be the normalization of a one-dimensional analytic space with only finite necklaces. Also let $\{x_n\}$ be an infinite discrete set in X.

Eliminate any members of $\{x_n\}$ that are in the same necklace as x_1. This necklace is compact by hypothesis so the new $\{x_n\}$ is still infinite and discrete. Continue until an infinite discrete subset $\{x_n\}$ of the original set is reached and such that x_n and x_m are in different necklaces if $n \neq m$. Also there is nothing lost by assuming that all necklaces have a representative x_n.

There is no loss in generality in assuming that X is connected. Hence every necklace must have a non-empty intersection with a non-compact irreducible component of X. Let $X_0, X_1, \cdots$ be the non-compact irreducible components of X and let K_1 be the necklace containing x_1, etc.

Define a holomorphic function f on K_n by setting f to be n on K_n. Let $\{y_n\}$ be points in $\cup X_i$ such that $y_n \in K_n$. Since the normalization of a 1-dimensional space is just the union of irreducible components it follows that any singularities of any X_i must lie in a codimension 2 subvariety. In other words, every X_i is a Riemann surface. As shown for example by Forster [3], every Riemann surface is Stein. Use the Mittag-Leffler property of Stein spaces to extend f from $\cup K_n$ to $\cup K_n \cup X_0$ by taking a holomorphic function on X_0 with the prescribed value n at any y_n that happens to lie in X_0. Then, using the fact that X_1 is Stein, extend f to $\cup K_n \cup X_0 \cup X_1$ by taking an extension to X_1 with the value n at any y_n which lies in X_1 but not in X_0. By continuing in this way a holomorphic function is constructed on X which blows up along $\{x_n\}$.

RIDER COLLEGE
TRENTON, NEW JERSEY 08602

REFERENCES

[1] A. Andreotti and H. Grauert, Théorèmes de finitude pour la cohomologie des espaces complexes, Bull. Soc. Math. France 90, (1962), 193-259.

[2] A. Andreotti and A. Kas, Duality on complex spaces, Ann. d. Sc. Norm. Sup. di Pisa 27 (1973), 187-263.

[3] O. Forster, Riemannsche-Flächen, Springer-Verlag, Berlin, New York, 1977.

[4] R. C. Gunning and H. Rossi, Analytic functions of several complex variables, Prentice-Hall, Englewood Cliffs, 1965.

[5] B. Kramm, Eine Funktionalanalytische Charakterisierung der Steinschen Algebren, Preprint, Universität Bayreuth, 1976.

[6] A. Markoe, Invariance of Holomorphic Convexity Under Proper Mappings, Math. Ann. 217 (1975), 267-270.

[7] R. Narasimhan, Levi Problem for Complex Spaces, II, Math. Ann. 146 (1962), 195-216.

[8] ———, A note on Stein spaces and their normalization, Ann. d. Scuola Norm. di Pisa 16 (1962), 327-333.

[9] D. Prill, The divisor class groups of some rings of holomorphic functions, Math. Z. 121 (1971), 58-80.

[10] J. P. Ramis, Théorèmes de séparation et de finitude pour l'homologie et la cohomologie des espaces (p,q)-convexes-concaves, Ann. d. Sc. Norm. di Pisa, (1973), 933-997.

SOME GENERAL RESULTS ON EQUIVALENCE OF EMBEDDINGS

J. Morrow and H. Rossi[1]

1. *Introduction*

In the late 1950's, L. Nirenberg and D. C. Spencer wrote a paper (unpublished, but for the summary [15]) in which they set down the formal machinery involved in extending finite, or formal equivalences to actual equivalence of embeddings. They considered the case of positive embeddings and were able to show that a finite equivalence of high enough order extends uniquely to a formal equivalence, but they were unable to show convergence (in these proceedings, Grauert and Commichau [8] succeeds in completing this program, showing the convergence under substantially weaker hypotheses). In 1962, in [7] Grauert solved this finite equivalence problem (affirmatively) in the case of negative embeddings.

In this paper, we return to the consideration of the cohomological obstructions to extending finite equivalences, and compute them in certain cases. We shall consider in detail (in this and a subsequent paper [14]) the embeddings

(1.1) $z_T : K \to T(K)$ of K as the zero section of its tangent bundle,

(1.2) $\Delta : K \to K \times K$ of K as the diagonal in the Cartesian product.

[1]This research was partially supported by the National Science Foundation under grants MCS78-02139 (Morrow), MCS78-08126 (Rossi).

Recent Developments in Several Complex Variables
0-691-08285-5/81/000299-27 $01.35/0 (cloth)
0-691-08281-2/81/000299-27 $01.35/0 (paperback)

These two embeddings are first-order equivalent, and the obstruction to extending this to a second-order equivalence is the total Chern class of K. We have not determined (except for curves) whether or not there are more obstructions to higher order equivalence.

We consider especially embeddings $K \to X$ which are negative and of codimension one. In this case K blows down (see [7]) in both X and as the zero section of the normal bundle N. Let $\tilde{X}$ and $\tilde{N}$ be the blown down varieties. The blowdown is monoidal (or quadratic) only in special circumstances, and even more special are the circumstances where $\tilde{N}$ is the tangent cone to $\tilde{X}$. We show

1.3. THEOREM. $\tilde{N}$ *is the tangent cone to* $\tilde{X}$ *if and only if*

a) $H^0(K, N^{-1})$ *projectively embeds* K,

b) $H^0(K, \mathcal{I}_K) \to H^0(K, N^{-1})$ *is surjective, where* $\mathcal{I}_K$ *is the ideal sheaf of* K,

c) $H^0(K, N^{-1})$ *generates* $\bigoplus_{\nu>0} H^0(K, N^{-\nu})$ *as a ring.*

In particular N, X *are the quadratic transforms of* $\tilde{N}, \tilde{X}$ *respectively.*

We would like to thank Harry Corson for asking us the right questions which led to this work, and for his continuing interest; also Nicholas Coleff for checking our calculations and many helpful and insightful conversations; also Alan Norton for many useful discussions involving this work and his own interest in this topic.

2. *Obstructions to extending equivalences*

Throughout this article X is a reduced analytic space with structure sheaf ${}_X\mathcal{O}$, and K is a reduced compact analytic subvariety of X with ideal sheaf $\mathcal{I}_K$. Most of the time K and X shall be nonsingular. The following definitions are standard, see [7, 10, 15].

2.1. DEFINITION. ${}_K\mathcal{O}^{(\nu)} = {}_X\mathcal{O}/\mathcal{I}_K^{\nu+1}$ is the structure sheaf of *the* ν*th neighborhood* $K^{(\nu)}$ of K in X.

$$_K\hat{\mathcal{O}} = \varprojlim_{\nu\to\infty} {}_K\mathcal{O}^{(\nu)}$$

is the structure sheaf of *the formal neighborhood* $\hat{K}$ of K in X.

2.2. DEFINITION. Let K be embedded in X and X'. A *νth order* {*formal, actual*} equivalence of these embeddings is a bimorphism of ringed spaces $(K, {}_K\mathcal{O}^{(\nu)}) \simeq (K, {}_K\mathcal{O}'^{(\nu)})\{(K, {}_K\hat{\mathcal{O}}) \simeq (K, {}_K\hat{\mathcal{O}}'), (K, {}_X\mathcal{O}|_K) \simeq (K, {}_{X'}\mathcal{O}|_K)\}$, which induces the identity on K.

${}_X\mathcal{O}|_K$ is the "physical" restriction of ${}_X\mathcal{O}$ to K. It is trivial to show, and we shall take it for granted, that an actual equivalence is induced by a biholomorphic map of a neighborhood of K in X with a neighborhood of K in X'.

2.3. DEFINITION. $\mathfrak{N}^* = (\mathcal{I}_K/\mathcal{I}_K^2) \otimes_{{}_X\mathcal{O}_K} \mathcal{O}$ is the *conormal sheaf* of the embedding $K \to X$.

If K is a local complete intersection then $\mathfrak{N}^*$ is the sheaf of sections on K, of a vector bundle N^*. N is called the *normal bundle* of $K \to X$. In the nonsingular case N also arises in the sequence

$$0 \to {}_K\Theta \to {}_X\Theta \to \mathfrak{N} \to 0 \tag{2.4}$$

where ${}_Y\Theta$ is the sheaf of germs of holomorphic vector fields on Y. If K has codimension one, $\mathfrak{N}^* = \mathcal{I}_K \otimes_{{}_X\mathcal{O}_X} \mathcal{O}$.

We shall be concerned with the following result of Nirenberg and Spencer [15]:

2.5. THEOREM. *Suppose that* K, X *and* X' *are all manifolds,* K *compact, and* $i: K \to X$, $j: K \to X'$ *are embeddings. Assume that the normal bundle of both embeddings are the same,* N. *The obstruction to extending a given* ν*-equivalence to a* $(\nu+1)$*-equivalence lies in*

$$(\nu = 0):\ H^1(K, {}_K\Theta \otimes N^*)\,,$$

$$(\nu \geq 0):\ H^1(K, {}_X\Theta \otimes \mathcal{I}_K^{\nu+2}) \cong H^1(K, {}_X\Theta \otimes S^{\nu+1}(N^*))\,,$$

where $S^{\nu}(N^*)$ *is the* ν*th symmetric product of* N^*.

We now examine an embedding in terms of local charts. Let $K \to X$ be an embedding of manifolds. Let $\{U_\alpha\}$ be a coordinate cover of K with coordinate z_α and $U_\alpha \times V_\alpha$ a corresponding coordinate cover of K in X, where V_α is a polydisc with coordinate w_α so that $K \cap (U_\alpha \times V_\alpha) = \{w_\alpha = 0\}$. Then a neighborhood of K in X is given by the $U_\alpha \times V_\alpha$ patched together by certain transition functions

(2.6) $$z_\beta = G_\beta^\alpha(z_\alpha, w_\alpha), \quad w_\beta = F_\beta^\alpha(z_\alpha, w_\alpha)\,.$$

Expand the functions G, F in series of homogeneous polynomials in w_α:

(2.7) $$z_\beta = \sum_{n \geq 0} G_\beta^{\alpha(n)}(z_\alpha), \quad w_\beta = \sum_{n > 0} F_\beta^{\alpha(n)}(z_\alpha)\,,$$

where $G_\beta^{\alpha(n)}$, $F_\beta^{\alpha(n)}$ are nth order homogeneous polynomials in w_α. If K has codimension 1, (2.7) are power series

(2.7)′ $$z_\beta = \sum_{n \geq 0} G_\beta^{\alpha(n)}(z_\alpha)\, w_\alpha^n, \quad w_\beta = \sum_{n > 0} F_\beta^{\alpha(n)}(z_\alpha)\, w_\alpha^n\,.$$

REMARKS 1. Two embeddings are ν-equivalent if we can find coordinatizations (2.6) for both of them so that the expansions (2.7) agree to νth order. In this case the difference of the terms of order $\nu + 1$ determines the obstruction class of Theorem (2.5).

2. Given the embeddings described by (2.6-7), the normal bundle has the coordinate cover $\{U_\alpha \times \mathbb{C}^d\}$ with coordinates z_α, ξ_α related by

(2.8) $$z_\beta = G_\beta^{\alpha(0)}(z_\alpha), \quad \xi_\beta = F_\beta^{\alpha(1)}(z_\alpha) \cdot \xi_\alpha\,.$$

We shall let $z_N : K \to N$ represent the embedding of K as the zero section of N. The given embedding (2.6-7) is first order equivalent to z_N if and only if we can make the first order terms of (2.7) and (2.8) agree. In the second relation they obviously do; thus what has to be done is to find a substitution $z_\alpha = \phi_\alpha(z'_\alpha)$ which will kill the term $\{G_\beta^{\alpha(1)}\}$ of (2.7). Making the necessary computation, we find that $\{[F_\beta^{\alpha(1)}]^{-1} \cdot G_\beta^{\alpha(1)}\}$ defines a class in $H^1(K, {}_K\Theta \otimes N^*)$, and we are asking precisely that this class vanish. A comparison of this computation with a similar one in [5] shows that this class is precisely the obstruction to splitting the normal bundle sequence (2.4) (i.e., writing ${}_X\Theta = {}_K\Theta \oplus \mathcal{N}$).

2.9. PROPOSITION. *An embedding* $K \to X$ *is first-order equivalent to the embedding* z_N *if and only if the normal bundle sequence (2.4) splits.*

2.10. COROLLARY (see [11]). *If* $X_0 \to \mathcal{X}$ *is the embedding as the fiber over* $0 \in S$ *in a family of deformations* $\mathcal{X} \to S$, *then the obstruction to first order equivalence with the trivial embedding* $X_0 \to X_0 \times S$ *is the Kodaira-Spencer class of the family at* 0.

In [13] we showed that for embeddings in $\mathbf{P}^n$ the normal bundle sequence splits if and only if everything is linear. Thus

2.11. COROLLARY. $K \to \mathbf{P}^n$ *is first order equivalent with* z_N *if and only if* K *is a linear subvariety of* $\mathbf{P}^n$.

REMARK. In [13] we called a submanifold $K \subset X$ and *HNR* if there is a holomorphic contraction of a neighborhood of K onto K. In this case we may choose local coordinates (2.6-7) so that the functions G_β^α are independent of w_β. In this case ${}_X\Theta = {}_K\Theta \oplus \mathcal{N}$, and the ${}_K\Theta$ part of the obstruction to extending a ν-equivalence vanishes, so that obstruction of Theorem 2.5 lies, in the case of an HNR, in $H^1(K, \mathcal{N} \otimes S^{\nu+1}(\mathcal{N}^*))$. We shall capitalize on this simplification later.

3. *Second-order obstructions*

Now we shall explicitly compute the second order obstructions to equivalence of an embedding $K \to X$ with the embedding z_N as the zero section of the normal bundle N. We shall assume that these embeddings are first order equivalent; in fact that K is an HNR in X. Take a coordinatization (2.6-7) where the G^{α}_{β} are independent of w_{α}. We shall assume (to make the computation easier) that if $U_i \cap U_j \cap U_k$ is non-empty, then $U_i \cup U_j \cup U_k$ is contained in a single coordinate neighborhood on which N is trivial. This allows us to take $F^{\alpha(1)}_{\beta}(z_{\alpha}) \equiv I$ for (α, β) any of the pairs (i, j), (j, k), (k, i). Now use the (j, k) equation (2.7) to substitute for w_j in the (i, j) equation, obtaining an expression for w_i in terms of w_k:

$$(3.1) \qquad w_{\alpha} = w_{\beta} + F^{\alpha(2)}_{\beta} + F^{\alpha(3)}_{\beta} + F^{\beta(2)}_{\alpha} \vee F^{\gamma(2)}_{\beta} + F^{\beta(3)}_{\alpha} + \cdots$$

where the ellipsis refers to terms of order 4 or more. The wedge and the cup product used below are as defined by Douady in [6]. Comparing (3.1) with the (i, k) equation (2.7), we must have

$$(3.2) \qquad F^{k(2)}_{j} + F^{j(2)}_{i} = F^{k(2)}_{i} .$$

Thus is a cocycle condition for $S^2(\mathfrak{N}^*)$ (if we did not assume N trivial on $U_i \cup U_j \cup U_k$, the transition functions for $\mathfrak{N} \otimes S^2(\mathfrak{N}^*)$ would appear. This cocycle, made of the second order terms of the transition equations (2.7) shall be denoted $\sigma^{(2)}_{K,X}$.

3.3. PROPOSITION. *In order for equations (2.7) to be the transition equations for an HNR embedding it is necessary that the second order terms define a cocycle* $\sigma^{(2)}_{K,X} \in H^1(K, \mathfrak{N} \otimes S^2(\mathfrak{N}^*))$. *For the embedding to be second order equivalent to the normal bundle embedding it is necessary and sufficient that* $\sigma^{(2)}_{K,X}$ *vanish.*

The third order terms in (3.1) can be used to answer another question.

3.4. PROPOSITION. *Let* $N \to K$ *be a vector bundle and* $\sigma \in H^1(K, \mathfrak{N} \otimes S^2(\mathfrak{N}^*))$. *In order for there to be an embedding* $K \to X$ *with normal bundle* N *and second order obstruction* σ *it is necessary that* $\sigma \cup \sigma \in H^2(K, \mathfrak{N} \otimes S^3(\mathfrak{N}^*))$ *vanish. If* N *is a line bundle this condition is automatically satisfied.*

Proof. Comparing third order terms, we have

$$F_j^{k(3)} + F_i^{j(3)} + F_i^{j(2)} \vee F_j^{k(2)} = F_j^{k(3)} . \tag{3.5}$$

But $\{F_i^{j(2)} \vee F_j^{k(2)}\}$ represents $\sigma_{K,X}^{(2)} \cup \sigma_{K,X}^{(2)}$ on $U_i \cap U_j \cap U_k$ and (3.5) just reads that this class is a coboundary.

In case N is a line bundle, equations (2.7) are replaced by (2.7)′ and (3.5) becomes the scalar equation

$$F_j^{k(3)} + F_i^{j(3)} + 2F_i^{j(2)}F_j^{k(2)} = F_i^{k(3)} . \tag{3.5$'$}$$

Conversely, given $\sigma \in H^1(K, \mathfrak{N}^*)$ (note $\mathfrak{N} \otimes S^2(\mathfrak{N}^*) = \mathfrak{N}^*$), represented by $\{f_i^j\}$, then $\sigma \cup \sigma$ is realized by $2f_i^j f_j^k$ on $U_i \cap U_j \cap U_k$. We compute that this is the coboundary of the cochain $\{-(f_i^j)^2\}$:

$$\begin{aligned}(f_j^k)^2 + (f_i^j)^2 - (f_i^k)^2 &= (f_j^k)^2 + (f_i^j - f_i^k)(f_i^j + f_i^k) \\ &= (f_j^k)^2 - f_j^k(f_i^j + f_i^k) \\ &= f_j^k(f_j^k - f_i^j - f_i^k) = -2f_j^k f_i^j\end{aligned}$$

as desired.

REMARK. N. Coleff demonstrated to us that the codimension one case can always be realized. Let $L \to K$ be a line bundle, and take $\sigma \in H^1(K, \mathfrak{L}^*)$. Let $\{U_\alpha\}$ be coordinate neighborhoods for K trivializing L. Let $\{f_{\alpha\beta}\}$ be the transition functions for L and let σ be represented by the Cech cocycle $\{\sigma_{\alpha\beta}\}$. Let Δ_α be a disc in C with coordinate w_α. Patch $U_\alpha \times \Delta_\alpha$ to $U_\beta \times \Delta_\beta$ (over $U_\alpha \cap U_\beta$) using the transition function

$$w_\alpha = \frac{f_{\alpha\beta}(z)\, w_\beta}{1 - a_{\alpha\beta}(z)\, w_\beta}. \tag{3.6}$$

One easily checks the consistency criteria for these patchings, so we do obtain an embedding of K (as $\{w_\alpha = 0\}$ in U_α) in a manifold X. Furthermore, it is easily checked that σ is the obstruction to second order equivalence of X with L along K.

4. *Second order obstructions as curvature classes*

In [2], Atiyah calculated the obstructions to finding holomorphic connections for principal bundles, and found them to be representatives of the total Chern class. These calculations are relevant to the study of embeddings; we summarize the calculations of Atiyah which are relevant here.

4.1. PROPOSITION. *Let* $P \to X$ *be a principal* G-*bundle,* G *a lie group. Let* $\mathrm{ad}P \to X$ *be the* $\mathfrak{g}$-*bundle associated to* G *by the adjoint presentation of* G *on its Lie algebra* $\mathfrak{g}$. *If the matrices* $\{f_{\alpha\beta}\}$ *are the transition functions for* P, $\{f_{\alpha\beta}^{-1} df_{\alpha\beta}\}$ *defines an element of* $H^1(X, \Omega^1 \otimes \mathrm{ad}P)$, *called the curvature class of* P.

(*Note: here* Ω^1 *is the sheaf of holomorphic* 1-*forms on* X, *and the cochain* $\{f_{\alpha\beta}^{-1} df_{\alpha\beta}\}$ *is considered as explicitly expressed in the coordinatization of the second index.*)

Proof. Since $f_{\alpha\beta} f_{\beta\gamma} f_{\gamma\alpha}$ = identity, we have

$$df_{\alpha\beta} \cdot f_{\beta\gamma} \cdot f_{\gamma\alpha} + f_{\alpha\beta} \cdot df_{\beta\gamma} \cdot f_{\gamma\alpha} + f_{\alpha\beta} \cdot f_{\beta\gamma} \cdot df_{\gamma\alpha} = 0$$

or

$$(f_{\alpha\beta} f_{\beta\alpha}) \cdot df_{\alpha\beta} \cdot f_{\gamma\alpha} + f_{\alpha\beta} \cdot df_{\beta\gamma} \cdot df_{\gamma\alpha} + f_{\alpha\beta} f_{\beta\gamma} \cdot df_{\gamma\alpha} = 0$$

or

$$f_{\beta a}^{-1}[f_{a\beta}^{-1}df_{a\beta}]f_{\beta a} + f_{\gamma a}^{-1}[f_{\beta\gamma}^{-1}df_{\beta\gamma}]f_{\gamma a} + f_{\gamma a}^{-1}df_{\gamma a} = 0 ,$$

which is the cocycle condition, computed in the coordinates of this covering, for $H^1(X, \Omega^1 \otimes \mathrm{adP})$.

REMARK. Under the Dolbeault isomorphism $H^1(X, \Omega^1 \otimes \mathrm{adP}) \simeq H^{1,1}(X, \mathrm{adP})$; the curvature class is the class of the curvature of any (1,0)-connection for P.

Now, let K be a compact manifold, and $z_T : K \to T(K)$ the embedding of K as the zero section of the holomorphic tangent bundle to K. Let $\Delta : K \to K \times K$ be the embedding of K as the diagonal.

4.2. THEOREM. *z_T and Δ are first-order equivalent. The obstruction to extending this to a second-order equivalence is the curvature class (of the tangent bundle) of K.*

Proof. Let $\{U_a\}$ be a coordinate cover of K, with local coordinate z_a. Let

$$z_a = f_{a\beta}(z_\beta) \tag{4.3}$$

be the transition functions. Then $U_a \times U_{a'}$ is a coordinate cover of $K \times K$ with local coordinates z_a, w_a and transition functions

$$z_a = f_{a\beta}(z_\beta), \quad w_a = f_{a'\beta'}(w_\beta) . \tag{4.4}$$

Since we are interested only in the diagonal, we may consider only the covering $\{U_a \times U_a\}$. Introduce the variable $\xi_a = w_a - z_a$, whose zero locus is the diagonal. Then we may take z_a, ξ_a as local coordinates in a neighborhood of the diagonal. The transition functions in terms of these coordinates are

$$\begin{cases} z_a = f_{a\beta}(z_\beta) \\ \xi_a = f_{a\beta}(z_\beta + \xi_\beta) - f_{a\beta}(z_\beta) . \end{cases} \tag{4.5}$$

(Δ is an HNR embedding.) We expand the second equation in a Taylor series. Let $f^{(n)}_{\alpha\beta}(z_\beta)$ be the homogeneous polynomial of degree n in the Taylor series for $f_{\alpha\beta}$. Then (4.5) becomes

$$\xi_\alpha = \sum_{n>0} f^{(n)}_{\alpha\beta}(z_\beta)(\xi_\beta)\,. \tag{4.6}$$

Now a homogeneous polynomial of degree n on the vector space V is the same as an n-linear symmetric function restricted to the diagonal. We shall use these notations interchangeably. In particular, a second degree homogeneous polynomial on a vector space V, taking values in V is in $(V^* \otimes V^*) \otimes V$, and thus determines an element of $V^* \otimes \mathrm{Hom}\,(V, V) = V^* \otimes \mathrm{ad}\ GL(V)$.

Let $g_{\alpha\beta} = f^{(1)}_{\alpha\beta}$. Then

$$f^{(2)}_{\alpha\beta}(z_\beta)(\xi_\beta) = dg_{\alpha\beta}(z_\beta)(\xi_\beta, \xi_\beta) \tag{4.7}$$

(as a 2-linear symmetric function). The $g_{\alpha\beta}$ are transition functions for the tangent bundle: if ζ_α is the fiber coordinate on T(X) over U_α, we have

$$\zeta_\alpha = g_{\alpha\beta}(z_\beta) \cdot \zeta_\beta\,. \tag{4.8}$$

Now, suppose the embeddings z_T and Δ are second order equivalent. Then we can write

$$\xi_\alpha = \zeta_\alpha + \phi_\alpha(z_\zeta)(\zeta_\alpha) + \cdots \tag{4.9}$$

(where $\phi_\alpha(z_\alpha)$ is a homogeneous polynomial of second order and the ellipsis refers to higher order terms) so that the substitution of (4.9) into (4.6) coincides with (4.8) through second order. Substituting, we obtain

$$\zeta_\alpha + \phi_\alpha(z_\alpha)(\zeta_\alpha) + \cdots =$$
$$= g_{\alpha\beta}(z_\beta)(\zeta_\beta + \phi_\beta(z_\beta)(\zeta_\beta) + \cdots) + dg_{\alpha\beta}(z_\beta)(\zeta_\beta + \cdots) + \cdots$$

or

$$g_{\alpha\beta}\cdot\zeta_\beta + \phi_\alpha(g_{\alpha\beta}\cdot\zeta_\beta) = g_{\alpha\beta}\cdot\zeta_\beta + g_{\alpha\beta}\cdot\phi_\beta(\zeta_\beta) + dg_{\alpha\beta}(\zeta_\beta) + \cdots .$$

Equating second order terms and multiplying by $g_{\alpha\beta}^{-1}$ we obtain

$$g_{\alpha\beta}^{-1}dg_{\alpha\beta}(\zeta_\beta) = g_{\alpha\beta}^{-1}\phi_\alpha(g_{\alpha\beta}\cdot\zeta_\beta) - \phi_\beta(\zeta_\beta) . \tag{4.10}$$

This is just the coboundary equation, in these coordinates, for $H^1(K, \Omega^1 \otimes \mathrm{Hom}(T, T))$, and the left hand side of (4.10) is the curvature class of $T(K)$. Thus the vanishing of the curvature class is necessary (and sufficient, for the argument easily reverses) for a second-order equivalence.

4.11. COROLLARY. *For a compact Riemann surface* K, *the embeddings* z_T *and* Δ *are equivalent if and only if* K *has genus* 1.

Proof. It is easy to check that if K is a torus, both embeddings are equivalent to the embedding as a fiber in the product (for example see Proposition 6.1 of [14]). On the other hand, if genus $(K) \neq 1$, the Chern class is not zero, so these embeddings are not even second-order equivalent.

Let us recalculate this second-order obstruction for a general HNR embedding. Let $K \to X$ be such an embedding. We can find local coordinates $\{z_\alpha, w_\alpha\}$ with transition functions

$$z_\alpha = g_{\alpha\beta}(z_\beta), \quad w_\alpha = \sum_{n>0} f_{\alpha\beta}^{(n)}(z_\beta)(w_\beta) ,$$

where $f_{\alpha\beta}^{(n)}(z_\beta)$, is an nth degree homogeneous polynomial. The normal bundle of this embedding has coordinates $\{z_\alpha, u_\alpha\}$ with transition functions

$$z_\alpha = g_{\alpha\beta}(z_\beta), \quad u_\alpha = f_{\alpha\beta}^{(1)}(z_\beta)(u_\beta) .$$

The obstruction to a second-order equivalence is given by the cocycle $\{f_{\alpha\beta}^{(1)}(z_\beta)^{-1} f_{\alpha\beta}^{(2)}\}$ as an element of $H^1(K, \mathfrak{N}^{-1} \otimes \mathrm{ad}P_N)$, where P_N is the principal bundle of N.

Higher order obstructions will lie in $H^1(K, \mathfrak{N}^{-v} \otimes adP_N)$. In the case of a codimension 1 HNR embedding, $adP_N = \mathcal{O}$, so the obstructions all lie in $H^1(K, \mathfrak{N}^{-v})$.

We can generalize the result on the diagonal embedding in another way. Let $h: K \to S$ be a holomorphic mapping of compact manifolds. We consider the embedding $\Gamma: K \to K \times S$ as the graph of h. The normal bundle to the graph is $h^*(T(S))$, the pullback of the tangent bundle of S, and the criterion for second order equivalence of these two embeddings can be computed.

4.12. THEOREM. *Let K, S be compact manifolds and $h: K \to S$ a holomorphic mapping. The embedding Γ of K as the graph of h is first-order equivalent with the zero section of its normal bundle. Those embeddings are second-order equivalent if and only if the pullback via h of the curvature class of S vanishes.*

Proof. Let $\{U_\alpha \times V_\alpha\}$ be a coordinate covering of $K \times S$ with coordinates $\{z_\alpha, w_\alpha\}$ and transition functions

$$z_\alpha = f_{\alpha\beta}(z_\beta), \quad w_\alpha = g_{\alpha\beta}(w_\beta).$$

Let the local representation of h be given by equations: $w_\alpha = h_\alpha(z_\alpha)$. We must have

$$h_\alpha(z_\alpha) = g_{\alpha\beta}(h_\beta(z_\beta)).$$

We introduce the normal (to the graph) coordinate $u_\alpha = w_\alpha - h_\alpha(z_\alpha)$. Then in the coordinates z_α, u_α we have the transition functions

$$\begin{aligned} u_\alpha &= g_{\alpha\beta}(w_\beta) - h_\alpha(f_{\alpha\beta}(z_\beta)) \\ &= g_{\alpha\beta}(u_\beta + h_\beta(z_\beta)) - g_{\alpha\beta}(h_\beta(z_\beta)) \\ &= \sum_{n>0} g_{\alpha\beta}^{(n)}(h_\beta(z_\beta))(u_\beta). \end{aligned}$$

Thus the normal bundle has transition functions $g^{(1)}_{\alpha\beta} \circ h_\beta$, and thus the normal bundle is $h^*(T(S))$. Since the graph is HNR, these embeddings are first-order equivalent, and the obstruction to a second-order equivalence is (by the above remarks) seen to be the curvature class of $h^*(T(S))$. But by functoriality of characteristic classes, the pullback of the curvature class of S represents the pullback of the curvature class of T(S) which represents the curvature class of the pullback of T(S). Thus the theorem is proven.

4.13. THEOREM. *Let* $f: K \to P^1$ *be a holomorphic map. The embedding of* K *as the graph of* f *is second-order equivalent to the zero section of its normal bundle if and only if* f *is constant.*

Proof. If f is constant, clearly both embeddings are equivalent to the embeddings in a product.

Let V_0, V_1 be coordinate neighborhoods of P^1 with coordinates w_0, w_1, and transition function $w_0 = w_1^{-1}$. Let $U_i = f^{-1}(V_i)$. Then in the covering $\{U_i \times V_i\}$, the graph is given by the equations $w_i = f_i(x)$. Let u_0, u_1 be the normal coordinates $u_i = w_i - f_i(x)$. Then

$$u_0 + f_0(x) = (u_1 + f_1(x))^{-1} = \sum_{n \geq 0} (-u_1)^n (f_1(x))^{-(n+1)}$$

so that

$$u_0 = \sum_{n>0} \frac{(-u_1)^n}{f_1(x)^{n+1}} .$$

Comparing this with (4.10), via (4.7) we find that the condition for second-order equivalence is the existence of functions ϕ_i holomorphic in U_i so that

$$(-f_1^2)(f_1^{-3}) = (\phi_0)(-f_1^{-2}) - \phi_1 ,$$

or

$$-\phi_0 = \phi_1 f_1^2 - f_1 \quad \text{in} \quad U_0 \cap U_1 .$$

Thus $\phi_1 f_1^2 - f_1$ is a global holomorphic function on K, so is constant. If f_1 never takes the value zero, then f is a holomorphic $\mathbb{C}$-valued map in K, so is constant. Thus, we may assume that

$$(\phi_1 f_1 - 1) f_1 \equiv 0 .$$

If $f_1 \equiv 0$, then again f is constant, so we may assume that $\phi_1 f_1 - 1 \equiv 0$, from which we conclude after all that f_1 is never zero and thus f is constant.

5. *Negative embeddings and blowdowns*

Let K be a compact subvariety of X. Let X be the topological space obtained by identifying K to a point, denoted 0. Let $\pi: X \to \tilde{X}$ be the quotient map. We make $\tilde{X}$ a ringed space by taking the direct image of ${}_X\mathcal{O}$ as its structure sheaf ${}_{\tilde{X}}\mathcal{O}$.

5.1. DEFINITION [7]. K is *exceptional* in X if $(\tilde{X}, {}_{\tilde{X}}\mathcal{O})$ is an analytic space.

When K is exceptional, we shall call $\tilde{X}$ the *Grauert blowdown* of X at K. Notice that if X is normal, so also is $\tilde{X}$.

5.2. DEFINITION [7]. Let K be a compact subvariety of X whose ideal sheaf is invertible. Let $N \to K$ be the normal bundle of K in X. We say that K is *negatively embedded* if the zero section of N is exceptional.

Grauert [7] proved that a negatively embedded variety is exceptional.

5.3. DEFINITION [10]. Let $\tilde{X}$ be an analytic space, $0 \in \tilde{X}$, and I an ideal supported at 0. Let $f_0, \cdots, f_N$ generate I and consider $[f_0, \cdots, f_N]: \tilde{X} - \{0\} \to P^N$. This map is meromorphic; the closure X of its graph is an analytic space, and for $\pi: X \to \tilde{X}$ the projection onto the first factor, $K = \pi^{-1}(0)$ is exceptional in X. We call X the *monoidal transform* of $\tilde{X}$ *with center* I. If $I = m_0$, the maximal ideal at 0, we call X the *quadratic transform*.

Throughout this section K shall be a compact manifold, X a manifold of one more dimension and $K \to X$ an embedding with negative normal bundle. First we shall ask: under what conditions is $K \to X$ the result of a monoidal (or quadratic) transformation? To consider this we need to introduce a little more local information. Let $N \to K$ be the normal bundle. For $f \in H^0(X, \mathcal{I}_K)$, we expand f locally (about K) in a Taylor series

$$f = \sum_{\nu \geq \nu_0} f_\nu, \quad f_\nu \in H^0(K, \mathcal{N}^{-\nu}), \quad f_{\nu_0} \neq 0 .$$

We shall call $\nu_0 = \nu(f)$ the *order of vanishing of* f and $f_{\nu_0} = \sigma(f)$ the *initial form of* f.

Let $\sigma(X)$ be the ring of initial forms in $\mathcal{K}(X) = \bigoplus_{\nu \geq 0} H^0(K, \mathcal{N}^{-\nu})$.

This latter ring is (in the algebraic sense) the ring of the Grauert blow-down $\tilde{N}$ of N. It is the normalization of $\sigma(X)$.

This can be seen as follows. By [1], there is a $\nu_0 > 0$ such that $H^1(K, \mathcal{I}_K^{\nu+1}) = 0$ for $\nu \geq \nu_0$, so that

$$H^0(K, \mathcal{I}_K^\nu) \to H^0(K, \mathcal{N}^{-\nu}) \tag{5.4}$$

is surjective for $\nu \geq \nu_0$. Thus the graded ring $\sigma(X)$ is of finite codimension in $\mathcal{K}(X)$, and in fact they agree in degrees greater than ν_0. It is now easy to show that $\mathcal{K}(X)$ is algebraic over $\sigma(X)$ and contained in its field of quotients. Since $\mathcal{K}(X)$ is normal, it is the normalization of $\sigma(X)$.

We can see this geometrically. Because (5.4) is surjective, and for ν large enough, $H^0(K, N^{-\nu})$ blows down the zero section, we can blow down the zero section using only initial forms. The image is a variety N_0 whose local ring at 0 is (algebraically) $\sigma(X)$. Also N_0 is the image of $\tilde{N}$ under an injective holomorphic map. Thus $\tilde{N}$, being normal, is the normalization of N_0.

Let m_0 be the maximal ideal at the singularity at $\tilde{X}$. Then

$$\bigoplus_{\nu \geq 0} m_0^\nu / m_0^{\nu+1}$$

is algebraically the ring of the tangent cone $C_0(\tilde{X})$ of $\tilde{X}$ at 0. The differential of the map $\pi: X \to \tilde{X}$ kills all vectors tangent to K, so descends to a map of the normal bundle N into $C_0(\tilde{X})$. Since $d\pi$ kills the zero vector, it induces a map $d\tilde{\pi}: \tilde{N} \to C_0(\tilde{X})$. We shall see that it is rare for this to be an isomorphism.

First, we show that if K is negatively embedded in X, and of codimension 1, the Grauert blowdown is always monoidal.

5.5. THEOREM. *Let* K *be negatively embedded in* X, X *a manifold. Let* $\pi: X \to \tilde{X}$ *be the Grauert blowdown. There is an integer* $s > 0$ *and an ideal* I *supported at* 0 *in* $\tilde{X}$ *such that*

(a) $\mathcal{I}_k^s = \pi^{-1}(I) \cdot {}_X\mathcal{O}$,

(b) X *is the monoidal transform of* $\tilde{X}$ *centered at* I.

Proof. First we note that a monoidal transform is trivially negatively embedded since some power of N^* is ample [7]. Now, under the hypotheses given, $\mathcal{I}_K$ is an invertible sheaf, i.e., the sheaf of sections of a line bundle L such that $L|K < 0$. By [1, 20] there is an integer $s > 0$ and $f_0, \cdots, f_N \in H^0(X, \mathcal{I}_k^s)$ which gives a projective embedding of X (considered as sections of L). By normality of $\tilde{X}$, there are $g_j \in m_0$ such that $f_j = g_j \circ \pi$. Then (a) is true, with I the ideal generated by $g_0, \cdots, g_N$.

Now consider $\Phi: X \to X \times P^N$, $\Phi = (\pi, \Phi_0)$ where $\Phi_0 = [f_0, \cdots, f_N]$. Since Φ_0 is an embedding, Φ also is an embedding, i.e., a biholomorphic map of X onto its image. But $\Phi(X-K)$ is clearly the graph of $[g_0, \cdots, g_N]$, so its closure $\Phi(X)$ is the monoidal transform of I.

5.6. THEOREM. *Let* K *be negatively embedded in* X *with normal bundle* $\mathfrak{N}$, *both* K *and* X *nonsingular.* X *is obtained as the quadratic transform of its Grauert blowdown* $\tilde{X}$ *if and only if*

(a) $H^0(K, \mathfrak{N}^*)$ *gives a projective embedding of* K,

(b) $H^0(K, \mathcal{I}_K) \to H^0(K, \mathfrak{N}^*)$ *is surjective.*

Proof. Let $z_0, \cdots, z_N$ generate the maximal ideal m_0 of 0 in $\tilde{X}$. The quadratic transform $Q\tilde{X}$ is defined as the closure of the graph of the map

$\Phi: [z_0, \cdots, z_N]: \tilde{X} - \{0\} \to P^N$ in the variety $\{z_i w_j = w_i z_j;\ 0 \le i, j \le N\}$, where $[w_0, \cdots, w_N]$ are homogeneous coordinates for P^N. Let K_0 be the blowup of 0. Then $w_j | K_0 \in H^0(K_0, \mathfrak{N}^*)$ and $w_j = \sigma(z_j)$, $0 \le j \le N$. Thus (a), (b) hold on $Q\tilde{X}$.

Now, we prove the sufficiency. Let $\pi: X \to \tilde{X}$ be the Grauert blowdown. Let $f_j = z_j \circ \pi$, $0 \le j \le N$, where the $\{z_j\}$ generate m_0. Then $f_0, \cdots, f_N$ generate $H^0(K, \mathfrak{I}_K)$ as $H^0(K, \mathcal{O})$-module so, by (b), $\sigma(f_0), \cdots, \sigma(f_N)$ span $H^0(K, \mathfrak{N}^*)$. Because of (a), $\Phi = (\pi, [f_0, \cdots, f_N])$: $X \to X \times P^N$ maps X holomorphically and injectively onto $Q\tilde{X}$. We show $d\Phi$ has maximal rank points x_0 of K. Clearly $d\Phi|_{T_x(K)}$ has maximal rank so we need only show $d\Phi$ is nonzero on vectors normal to K. Let ξ be part of a coordinate system for X at x so that $K = \{\xi = 0\}$ near x. There is a j such that $\sigma(f_j)(x) \neq 0$, thus $f_j = q \cdot \xi$ with $q(x) \neq 0$. Then $\partial f_j / \partial \xi(x) = q(x) \neq 0$.

5.7. PROPOSITION. *Let* K *be negatively embedded in* X, *with* X *nonsingular. Let* $f_1, \cdots, f_N \in H^0(K, \mathfrak{I}_K)$ *be such that* $F = (f_1, \cdots, f_N)$ *maps* $X - K$ *biholomorphically into* $C^N - \{0\}$. *Then* $X_0 = F(X)$ *is a variety with an isolated singularity at* X_0. X_0 *is the Grauert blowdown of* X *if and only if* $\sigma(f_1), \cdots, \sigma(f_N)$ *generate* $\sigma(X)$ *(algebraically).*

Proof. Suppose X_0 is the Grauert blowdown. Let $\sigma = \sigma(f) \in \sigma(X)$. Since f is a holomorphic function of $f_1, \cdots, f_N$ we have a Taylor series:

$$f = \sum_{\alpha_1 \ge 0, \cdots, \alpha_N \ge 0} a_{\alpha_1, \cdots, \alpha_N} f_1^{\alpha_1}, \cdots, f_N^{\alpha_N}.$$

Let f_j have order ν_j. Then

$$\sigma(f) = \sum_{\alpha_1 \nu_1 + \cdots + \alpha_N \nu_N = \nu(f)} a_{\alpha_1 \cdots \alpha_N} \sigma(f_1)^{\alpha_1} \cdots \sigma(f_N)^{\alpha_N}.$$

Conversely we have to show, under the given hypothesis, that $f_1, \cdots, f_N$ generate $H^0(X, \mathfrak{I}_K)$ analytically. Let $\pi: X \to \tilde{X}$ be the

Grauert blowdown. Clearly $f_j = g_j \circ \pi$, $g_j \in {}_{\tilde{X}}\mathcal{O}$, and $(g_1, \cdots, g_N): \tilde{X} \to X_0$ holomorphically and injectively. We conclude that $\tilde{X}$ is the normalization of X_0. Then there is an integer s such that ${}_{\tilde{X}}m^s \subset m$. An easy argument using ([3], p. 113) shows that there is an integer μ such that

$$H^0(K, \mathcal{I}_K^M) \subset H^0(K, \mathcal{I}_K)^s = {}_{\tilde{X}}m^s .$$

Inductively, then, we need only show that for $f \in H^0(K, \mathcal{I}_K)$, if f has order ν, then there is a polynomial P such that $f - P(f_1, \cdots, f_N)$ has order $\nu + 1$. But, by the given hypothesis there is a polynomial P such that

$$\sigma(f) = P(\sigma(f_1), \cdots, \sigma(f_N)) .$$

Since $\sigma(P(f_1), \cdots, P(f_N)) = P(\sigma(f_1), \cdots, \sigma(f_N))$ (because the latter is a symbol) $f - P(f_1, \cdots, f_N)$ has order $\nu + 1$. The theorem is proven.

Once again, K is negatively embedded in X, and N is the normal bundle. Let $\pi: X \to \tilde{X}$, $\tau: N \to \tilde{N}$ be the Grauert blowdowns. $d\pi: T(X)|_K \to C_0(\tilde{X})$, where $C_0(\tilde{X})$ is the tangent cone of $\tilde{X}$ at the origin. Since $d\pi$ vanishes on vectors tangent to K we have an induced map $d\tilde{\pi}: \tilde{N} \to \tilde{C}_0(\tilde{X})$.

For example, let $L \to K$ be a negative bundle on a curve and $\pi: L \to \tilde{L}$ the blowdown of the zero section. Then $d\tilde{\pi}: \tilde{L} \to C_0(\tilde{L})$ is given by the first order part of $d\pi$; i.e., by a basis of $H^0(K, L^{-1})$. Since $H^0(K, L^{-1})$ is almost always of dimension 0 or 1, $d\tilde{\pi}$ is almost never even surjective. Specifically, let p_1 and p_2 be two distinct points on a Riemann surface R such that $H^0(R, (p_1 + p_2)) \cong \mathbf{C}$ (if R has genus 2 this is generically the case). Let σ generate $H^0(R, (p_1 + p_2))$ and q such that $\sigma(q) \neq 0$. Then $(p_1 + p_2 - q)$ is a positive divisor with no non-zero sections. For if ξ were such a section and τ the canonical section of (q), $\xi\tau$ would be a section of $(p_1 + p_2)$ vanishing at q and thus not colinear with σ. Now, if L is the line bundle associated to the divisor $(p_1 + p_2 - q)$, L^{-1} is negative so there is a Grauert blowdown $\pi: L^{-1} \to X$, and $d\tilde{\pi} = 0$ since π vanishes to second order along the zero section. The following result tells us that $d\tilde{\pi}$ is biholomorphic only in very special circumstances.

5.8. THEOREM. *$d\tilde{\pi}$ is biholomorphic, so that $\tilde{N}$ is (in a natural way) the tangent cone to $\tilde{X}$ if and only if*

(a) X is the quadratic transform of $\tilde{X}$,

(b) $H^0(K, \mathfrak{N}^{-1})$ generates $\bigoplus_{\nu>0} H^0(K, N^{-\nu})$ algebraically.

Proof. Let $r'\colon X_0 \to C_0(\tilde{X})$ be the quadratic transform of $C_0(\tilde{X})$ at the vertex. X_0 is a line bundle over some projective variety K'.

First, let us suppose that $d\tilde{\pi}$ is biholomorphic. Then $d\tilde{\pi}\colon \tilde{N} - \{0\} \to C_0(\tilde{X}) - \{0\}$ certainly commutes with the $\mathbb{C}^*$ action and thus defines an isomorphism of these considered as the principal bundles associated to $N \to K$ and $X_0 \to K'$. In particular, $d\tilde{\pi}$ (since it takes fibers one-one to fibers) defines a biholomorphic map $K \to K'$. So $N \to K$ and $X_0 \to K'$ coincide, i.e., $N \to K$ is the quadratic transform of $\tilde{N}$ $(= C_0(\tilde{X}))$. Thus (a) of Theorem 5.6 holds as does (b) above, by Proposition 5.7. To say that $\tilde{N}$ is in addition the tangent cone to $\tilde{X}$ is to assert that the symbol map

$$\bigoplus_{\nu>0} H^0(K, \mathcal{I}_K)^{\nu}/H^0(K, \mathcal{I}_K)^{\nu+1} \to \bigoplus_{\nu>0} H^0(K, \mathfrak{N}^{-\nu}) \tag{5.9}$$

is an isomorphism of (not necessarily graded) rings. Thus every element of $H^0(K, \mathcal{I}^{-1})$ is the symbol of some $f \in H^0(K, \mathcal{I}_K)^{\nu}$. Certainly we have to have $\nu = 1$, thus (b) of Theorem 5.6 is true. Thus X also is the quadratic transform of $\tilde{X}$.

Conversely, if (a) and (b) are true, then both (a) and (b) of Theorem 5.6 are true. By 5.6(a), we conclude that N is also the quadratic transform of $\tilde{N}$, (since 5.6(b) is trivial for a bundle). By (b) above we conclude that $\tilde{N}$ is a cone. Since the map (5.9) above is injective, we need only show that it is surjective, i.e., that $\sigma(X) = \bigoplus_{\nu>0} H^0(K, \mathfrak{N}^{-\nu})$. Let

$\sigma \in H^0(K, \mathcal{N}^{-\nu})$. By (b) we can write $\sigma = P(\sigma_1, \cdots, \sigma_N)$, $\sigma_1, \cdots, \sigma_N \in H^0(K, \mathcal{N}^{-1})$, and P a homogeneous polynomial. By 5.6 (b), $\sigma = \sigma(f_j)$ with $f_j \in H^0(K, \mathcal{I}_K)$. Then $\sigma = \sigma(P(f_1, \cdots, f_N))$.

REMARK. In the above theorem we can just as well replace X by N in (a). This is not to say that X is a quadratic transform if and only if N is; in fact N could easily be a quadratic transform while X is not. We just need $H^0(K, \mathcal{I}) \to H^0(K, \mathcal{N}^{-1})$ not surjective.

As a corollary of the proof we obtain the following result.

5.10. COROLLARY. *Let* $L \to K$ *be a negative line bundle.* L *is a cone if and only if* $H^0(K, \mathcal{L}^{-1})$ *projectively embeds* K *and generates* $\bigoplus_{\nu>0} H^0(K, \mathcal{L}^{-\nu})$ *as a ring.*

The latter condition is strong: if K is a curve and L the tangent bundle, it holds only if K is not hyperelliptic, and it is the assertion of Noether's theorem that the linear differentials generate all the differentials. We shall study these embeddings in detail in [14].

6. *Families of embeddings* (dim K = 1)

In order to solve the Kuranishi problem for embeddings (to construct the family of all embeddings of a given compact manifold with given normal bundle) it is necessary to understand the obstructions to existence (for example, recall Proposition 3.4). In dimension 1 this is easy, for there are no obstructions. Thus the "space of all embeddings" can be easily parametrized. We shall restrict attention here to HNR embeddings. First, we need the theorem of Docquier and Grauert:

6.1. PROPOSITION [5]. *Let* S *be a Stein manifold, and* $j: S \to X$ *an embedding with normal bundle* N. *Let* K *be a compact subset of* S. *Then, in some neighborhood* U *of* K, $j|U$ *is equivalent to the embedding of* U *in the zero section of* N.

In [5], X is Euclidean space, but since K has, for the general X, a Stein neighborhood in X, Proposition 1 easily follows.

Let $V = V_1 \oplus \cdots \oplus V_n$ be a direct sum decomposition of a vector space. Define the C^* action on V:

$$t \cdot (v_1, v_2, \cdots, v_n) = (tv_1, t^2v_2, \cdots, t^n v_n) . \tag{6.2}$$

Let E(V) be the space of orbits, $E: V \to E(V)$ the quotient map. E(V) is compact, but not Hausdorff: $E(V) = E(V - \{0\}) \cup E(0)$, and the only neighborhood of E(0) is E(V). Nevertheless, $E(V - \{0\})$ is a compact analytic space.

6.3. THEOREM. *Let R be a compact Riemann surface and $L \to R$ a negative line bundle. Then, there is a ν_0 such that*

$$E\Big(\bigoplus_{0<\nu<\nu_0} H^0(R, L^{-\nu})\Big)$$

parametrizes the space of HNR embeddings of R with normal bundle L.

Proof. Let U, V be a covering of R such that U is a coordinate disc with coordinate z and $U \cap V$ is an annulus. Since line bundles on 1-dimensional noncompact manifolds are analytically trivial, the preceding result tells us that we may choose U, V so that $j|U$, $j|V$ are the trivial embeddings. Thus, any HNR embedding is described by the transition function of the normal coordinate u, v in U, V respectively:

$$v = f(z,u) = \sum_{\nu>0} f_\nu(z) u^\nu , \tag{6.4}$$

with $f_\nu \in \mathcal{O}(U \cap V)$. We need only identify the equivalence relation induced on the space of such sequences $\{f_\nu\}$ by equivalence of embeddings.

First of all, if the embedding described by (6.4) has L as normal bundle, then f_1 is invertible in $\mathcal{O}(U \cap V)$ and is the transition function for L. Letting g represent this transition function we may write (6.4)

(6.5) $$v = g(z)\left[u + \sum_{\nu>0} f_\nu(z)u^{\nu+1}\right]$$

Now $\{U, V\}$ is a Leray cover for the sheaf $L^{-\nu}$, so $H^1(R, L^{-\nu})$ is represented by the Cech cohomology

$$H^1(R, L^{-\nu}) \cong \frac{\mathcal{O}(U \cap V)}{\delta(\mathcal{O}(U) \oplus \mathcal{O}(V))}$$

where $\delta(h_U, h_V) = h_U - g^\nu h_V$. Let $h_1^\nu, \cdots, h_d^\nu \in \mathcal{O}(U \cap V)$ descend to a basis for $H^1(R, L^{-\nu})$.

Now, since L is negative, there is a ν_0 such that $H^1(R, L^{-\nu}) = 0$ for $\nu \geq \nu_0$. By Grauert's theorem [7], any two embeddings which are ν_0-equivalent are actually equivalent. Let $V = \bigoplus_{0<\nu<\nu_0} V_\nu$, where $V_\nu = H^1(R, L^{-\nu})$. Let $\zeta \in V$ be represented, in terms of the basis $\{h_\mu^\nu\}$ by

(6.6) $$\bigoplus_{0<\nu<\nu_0} \sum_{\mu=1}^{d_\nu} c_\mu^\nu h_\mu^\nu .$$

Let $j_\zeta : R \to X_\zeta$ be the HNR embedding given by the transition function (for the normal coordinate)

(6.7) $$v = g(z)\left[u + \sum_{0<\nu<\nu_0} \left(\sum_{\mu=1}^{d_\nu} c_\mu^\nu h_\mu^\nu(z)\right) u^{\nu+1}\right].$$

We have to show

(A) Any embedding $j: R \to X$ is ν_0-equivalent to some j_ζ.

(B) j_ζ is equivalent to j_ξ if and only if $t \circ \zeta = \xi$ for some $t \neq 0$, where $t\circ$ is defined by (6.2).

(A) follows directly from the following proposition which we prove by induction.

6.7. PROPOSITION. *Let* $j: R \to X$ *be an HNR embedding with normal bundle* L. *Let* η *be an integer,* $0 \leq \eta \leq \nu_0$. *We can write the transition function of the embedding in the form*

$$v = g(z)\left[u + \sum_{0<\nu<\eta} \left(\sum c^{\nu}_{\mu} h^{\nu}_{\mu}(z)\right) u^{\nu+1} + o(u^{\eta})\right].$$

Proof. The case $\eta = 0$ is obvious. We consider the induction from η to $\eta + 1$. Assume then that the embedding is given by the transition function

$$v = g(z)\left[u + \sum_{0<\nu<\eta} h^{\nu}_{0}(z) u^{\nu+1} + h^{\eta} u^{\nu+1} + o(u^{\eta+1})\right] \tag{6.8}$$

where the h^{ν}_{0} are known to be of the desired form. Now $h^{\eta} \epsilon\, \mathcal{O}(U \cap V)$ defines a class in $H^1(R, L^{-\eta})$, so we can write

$$h^{\eta} = h^{\eta}_{0} + \phi_U - g^{\eta} \phi_V$$

where $\phi_U \epsilon\, \mathcal{O}(U)$, $\phi_V \epsilon\, \mathcal{O}(V)$ and $h^{\eta}_{0} \epsilon\, \mathcal{O}(U \cap V)$ is of the desired form. Make the following coordinate changes:

$$u = u_0 - \phi_U u_0^{\eta+1}, \qquad v = v_0 - \phi_V v_0^{\eta+1},$$

and substitute into (6.8). We obtain

$$v_0 - \phi_V v_0^{\eta+1} = g\left[u_0 - \phi_U u_0^{+1} + \sum_{0<\nu<\eta+1} h_0 u_0^{+1} + \phi_U u_0^{\eta+1} - g^{\eta} \phi_V u_0^{\eta+1} + 0(u_0^{\eta-1})\right].$$

Since $\phi_V v_0^{\eta+1} = g^{\eta+1} \phi_V u_0^{\eta+1} + 0(u_0^{\eta+1})$, we end up with a transition function of the desired form.

(B). Let the transition functions (6.6) for j_ζ, j_ξ be given by

$$v_0 = g(z)\left[u_0 + \sum_{0<\nu<\nu_0} h^{\nu}_{0}(z) u_0^{\nu+1}\right], \tag{6.9}$$

$$v = g(z)\left[u + \sum_{0<\nu<\nu_0} h^{\nu}(z) u^{\nu+1}\right], \tag{6.10}$$

respectively. Suppose j_ζ and j_ξ are equivalent. Then, we can find substitutions

$$(6.11)\qquad \begin{cases} v_0 = \sum_{n>0} a_n(p)\, v^n, & p \in V \\ u_0 = \sum_{n>0} b_n(z)\, u^n, & z \in U \end{cases}$$

which takes (6.9) to (6.10). Let us compute that, using the coordinate z in $U \cap V$. Substituting (6.11) into (6.9):

$$a_1 v + a_2 v^2 + \cdots = g\left[(b_1 u + b_2 u^2 + \cdots) + \sum h_0^\nu (b_1 u + b_2 u^2 + \cdots)^{\nu+1}\right].$$

Substitute (6.10) for v in the left-hand side:

$$a_1(gu + h^1 u^2 + \cdots) + a_2(g^2 u^2 + \cdots) = g b_1 u + (g b_2 + h_0^1 b_1^2) u^2 + \cdots .$$

The coefficients of u must be equal, so $a_1 g = g b_1$, i.e., $a_1 = b_1$ in $U \cap V$. Then

$$t = \begin{cases} a_1 & \text{in } V \\ b_1 & \text{in } U \end{cases}$$

is a global holomorphic function so is constant, and nonzero. Equating coefficients of u^2, we have

$$t h^1 + a_2 g^2 = g b_2 + h_0^1 t^2 ,$$

or $h^1 - t h_0^1$ is cohomologous to zero. Since our embeddings are of the form (6.6) this means $h^1 = t h_0^1$. Using that, and computing the higher order terms inductively, we shall find $h^\nu = t^\nu h_0^\nu$ for all $\nu < \nu_0$, as required.

REMARK. The proof of this theorem shows that, no matter what the normal bundle L,

$$E\Big(\bigoplus_{\nu<0} H^1(R, L^{-\nu})^{\wedge}\Big) \tag{6.12}$$

parametrizes the space of formal HNR embeddings of R up to formal equivalence.. So long as $L > 0$ or $L < 0$, by Grauert's theorems, the elements in (6.12) represented by convergent series (in $\mathcal{O}(U \cap V)$) parametrize all HNR embeddings of R with normal bundle L. In particular, if $L > 0$ the space of embeddings is definitely infinite dimensional.

Consider, for example, $R = \mathbf{P}^1$ and suppose H is the hyperplane section. Then, for any embedding $j: R \to X$, with H as normal bundle, X is 1-concave. If the "hole" can be filled in (i.e., X becomes an open subset of a compact analytic space S) then $S - R$ is strictly pseudoconvex. By contracting the exceptional varieties in $S - R$, we may assume that $S - R$ is a (normal) Stein space.

6.13. PROPOSITION. *Let* $R \subset S$ *be as above (the normal bundle of* R *is* H, S *is compact, and* $S - R$ *is Stein). Then* $S = \mathbf{P}^2$.

Proof. We revise the argument of Theorem 4.6 of [12]. First we resolve the singularities of $S - R$ to get a smooth surface $\tilde{S}$. By [21], p. 507,

$$0 = H_3(S - R, \mathbf{Z}) \cong H_3(\tilde{S} - R, \mathbf{Z}) .$$

By duality $H^1_c(S - R, \mathbf{Z}) = 0$. The exact sequence of the pair $(R, \tilde{S})$ then implies that $H^1(\tilde{S}, \mathbf{Z}) = 0$. By Corollary 2.5 of [12], $\tilde{S}$ is algebraic and hence Kahler. Hence $0 = b_1 = 2h^{0,1}$, so $H^1(\tilde{S}, \mathcal{O}) = 0$. From the exact sequence (on $\tilde{S}$)

$$0 \longrightarrow \mathcal{O} \xrightarrow{\sigma_0} [R] \longrightarrow [R]|_R \longrightarrow 0 ,$$

we get sections $\{\sigma_0, \sigma_1, \sigma_2\} \in H^0(\tilde{S}, [R])$ with $\{\sigma_1, \sigma_2\}$ restricted to R, a basis for $H^0(R, H)$. Thus we get a map $\sigma: \tilde{S} \to \mathbf{P}^2$. Since [R] is trivial on $\tilde{S} - R$, σ maps the exceptional set of $\tilde{S} - R$ to a finite set of points. By the normality of S, σ is defined on S. Now, it is easy to see that σ maps a neighborhood of R biholomorphically to a neighborhood of $P^1 \subset P^2$. Since $P^2 - P^1$ and $S - R$ are Stein and biholomorphic outside a compact set, they are biholomorphic. Hence $S = P^2$.

This proposition shows that, apart from the linear embedding of $P^1 \subset P^2$, all embeddings of P^1 with H as normal bundle give 1-concave surfaces which cannot be compactified. The examples in [4, 18] have H^2 as normal bundle. In this case, the representation (6.6) of an embedding comes down to

$$v = \sum_{n>0} \left(\sum_{0 \leq \mu \leq n} c^n_\mu z^{-(\mu+1)} \right) u^n .$$

The specific examples of [4, 18] are given by the transition function

$$v = z^{-2}u + \varepsilon z^{-3}u^2$$

(cf. [18], p. 255).

BIBLIOGRAPHY

[1] Andreotti, A. and Tomassini, G., A remark on the vanishing of certain cohomology groups, Comp. Math., *21* (1969), 417-430.

[2] Atiyah, M., Complex analytic connections in fibre bundles, T.A.M.S. *85* (1957), 181-207.

[3] Banica, C. and Stanisla, O., Algebraic methods in the global theory of complex spaces, Wiley, New York, 1976.

[4] Burns, D., Global behavior of some tangential Cauchy-Riemann equations, Partial differential equations and differential geometry, Lecture Notes in pure and applied Mathematics, Dekker, 1979.

[5] Docquier, F. and Grauert, H., Levisches Problem und Rungescher Satz für Teilgebiete Steinscher Mannigfaltigkeiten, Math. Ann. *140* (1960), 94-123.

[6] Douady, A., Obstruction primaire à la déformation, exp. 4, Familles d'espaces complexes et fondements de la géométrie analytique, Seminaire Henri Cartan, E.N.S., 1962.

[7] Grauert, H., Uber Modifikationen und exceptionelle analytische Mengen, Math. Ann. *146* (1962), 331-368.

[8] Grauert, H. and Commichau, M., Das Formale Prinzip für Kompakte Komplexe Untermannigfaltigkeiten mit 1-Positivem Normalenbündel, appearing in these proceedings, p. 101.

[9] Griffiths, P. and Harris, J., Principles of Algebraic Geometry, J. Wiley and Sons, New York, 1978.

[10] Hironaka, H. and Rossi, H., On the equivalence of imbeddings of exceptional complex spaces, Math. Ann. *156* (1964), 313-333.

[11] Kodaira, K. and Spencer, D. C., On deformations of complex analytic structure I, Ann. of Math. *67*(1958), 328-466.

[12] Morrow, J. and Rossi, H., Some theorems of algebraicity for complex spaces, J. Math. Soc. Japan *27*(1975), 167-183.

[13] ———, Submanifolds of P^n with splitting normal bundle sequence are linear, to appear in Math. Ann.

[14] ———, Canonical Embeddings, T.A.M.S., 261 (1980), 547-565.

[15] Nirenberg, L. and Spencer, D. C., On rigidity of holomorphic imbeddings, Contributions to function theory, Tata Institute, 1960, Bombay.

[16] Noether, M., Uber die invariante Darstellung algebraischer Funktionen, *Math. Ann. 17*(1880), 263-284.

[17] Petri, K., Uber die invariante Darstellung algebraischer Funktionen einer Veränderlichen, Math. Ann. *88*(1922), 247-289.

[18] Rossi, H., Attaching analytic spaces to an analytic space along a pseudoconcave boundary, Proceedings of the Conference on Complex Analysis, Minneapolis 1964, Springer-Verlag(1965), 242-256.

[19] Saint-Donat, B., On Petri's analysis of the linear system of quadrics through a canonical curve, Math. Ann. *206*(1973), 157-175.

[20] Vo Van Tan, On the characterization of convex spaces, T.A.M.S., 256 (1979), 185-197.

[21] Narasimhan, R., A topological property of Runge pairs, Ann. of Math., *76*(1962), 499-509.

LIMITS OF BOUNDED HOLOMORPHIC FUNCTIONS ALONG CURVES

Alexander Nagel[1] and Stephen Wainger[1]

§1. *Introduction*

If $F(z,\zeta)$ is a bounded holomorphic function in the open unit ball $\{(z,\zeta)\in \mathbf{C}^2 \mid |z|^2+|\zeta|^2<1\}$, then according to a classical theorem of Fatou, $\lim_{r\to 1} F(re^{i\theta},0)$ exist for almost all $\theta \in [0,2\pi]$. Thus $F(z,\zeta)$ has radial limits along the curve $(e^{i\theta},0)$, $0\le\theta\le 2\pi$, almost everywhere with respect to one-dimensional Lebesgue measure on the curve.

The question of the existence of limits almost everywhere along more general curves for bounded holomorphic functions was studied in [NR]. To describe those results, we need some standard notation. Let D be a domain in $\mathbf{C}^n$ and assume that the boundary bD is of class C^2 –i.e. assume there is an open set W in $\mathbf{C}^n$ with $bD\subset W$ and a C^2-function $\rho: W\to \mathbf{R}$ such that

$$D\cap W = \{w\in W \mid \rho(w)<0\}$$

and such that the gradient vector

$$N(w) = \left(\frac{\partial\rho}{\partial\overline{w}_1}(w),\cdots,\frac{\partial\rho}{\partial\overline{w}_n}(w)\right) \tag{1}$$

[1]Supported in part by N.S.F. grants.

Recent Developments in Several Complex Variables
0-691-08285-5/81/000327-18 $00.90/0 (cloth)
0-691-08281-2/81/000327-18 $00.90/0 (paperback)
For copying information, see copyright page.

is never zero for $w \in bD$. Let T_w be the real $2n-1$ dimensional tangent space to bD at w. A real subspace $V \subset T_w$ is called *complex* if V, viewed as a subspace of $\mathbb{C}^n$, is closed under multiplication by i.

According to [NR], we have the following result:

THEOREM A. *Suppose* $\gamma : [a, b] \to bD$ *is a curve such that*:

(i) γ *is of class* C^1.

(ii) $\gamma'(t)$ *is not contained in any complex subspace of* $T_{\gamma(t)}$ *for* $a \le t \le b$.

(iii) γ' *is of class* Lip ε *for some* $\varepsilon > 0$.

Then any function F *which is bounded and holomorphic on* D *has nontangential limits along the curve almost everywhere* dt. *More precisely, the set of* $t \in [a, b]$ *such that* F *fails to have a nontangential limit at* $\gamma(t)$ *has Lebesgue measure zero.*

Further, according to [NR], condition (ii) is necessary in order to obtain the conclusion of Theorem A. This would lead one to think that condition (i) is rather natural, while condition (iii) is not. Our original aim was to remove condition (iii). We found however that condition (i) could be replaced by the weaker hypothesis that γ is just continuous and rectifiable.

For the sake of simplicity, we first state our result in the case that γ is absolutely continuous. For any such $\gamma : [a, b] \to bD$, $\rho(\gamma(t)) = 0$ and differentiating, we see that

$$\operatorname{Re} \langle N(\gamma(t)), \gamma'(t) \rangle = 0 \quad \text{a.e.} \tag{2}$$

(Here $\langle z, w \rangle = \sum_{j=1}^{n} z_j \bar{w}_j$). Also, condition (ii) in Theorem A is equivalent to the condition that

$$\operatorname{Im} \langle N(\gamma(t)), \gamma'(t) \rangle \neq 0 .$$

Since in Theorem A γ' is continuous, condition (ii) is equivalent to the assertion

$$\left.\begin{array}{ll} \operatorname{Im} \langle N(\gamma(t)), \gamma'(t)\rangle > 0 & \text{for all } t \\ & \text{or} \\ \operatorname{Im} \langle N(\gamma(t)), \gamma'(t)\rangle < 0 & \text{for all } t \end{array}\right\}$$

Our result is now

THEOREM 1. *Let* $\gamma: [a, b] \to bD$ *be an absolutely continuous curve. Assume*

$$\begin{array}{l} \operatorname{Im} \langle N(\gamma(t)), \gamma'(t)\rangle > 0 \ \text{a.e.} \\ \qquad\qquad\qquad\qquad \textit{or} \\ \operatorname{Im} \langle N(\gamma(t)), \gamma'(t)\rangle < 0 \ \text{a.e.} \end{array}$$

Then any function F *which is bounded and holomorphic on* D *has nontangential limits along* γ *almost everywhere* dt.

Theorem 1 can be generalized to the case where $\gamma: [a, b] \to bD$ is merely continuous and rectifiable. In that case, we define a complex measure $d\mu$ on $[a, b]$ by setting

$$\int \phi \, d\mu = \sum_{j=1}^{n} \int_a^b \phi(t) \frac{\partial \rho}{\partial \overline{w}_j} (\gamma(t)) \, \overline{d\gamma_j(t)} \tag{3}$$

for $\phi \in C[a, b]$, where $\overline{d\gamma_j}(E) = \overline{d\gamma_j(E)}$ for any set E. In analogy with equation (2), it is not hard to see that $\operatorname{Re} d\mu = 0$. We shall prove

THEOREM 2. *Let* $\gamma: [a, b] \to bD$ *be continuous and rectifiable. Assume* $i\, d\mu > 0$ *or* $i\, d\mu < 0$. *Then any function* F *which is bounded and holomorphic on* D *has nontangential limits along* γ *almost everywhere* dt.

(Note that if $d\nu$ is a real measure, the notation $d\nu > 0$ means that $\int \phi \, d\nu > 0$ if ϕ is continuous, non-negative, and not identically zero.)

At this point we note that in Theorem 1, one can ask what happens if we only assume $\operatorname{Im} \langle N(\gamma(t)), \gamma'(t)\rangle \neq 0$ for almost all t. However Theorem 2 does not seem to allow such a generalization.

Our main reason for seeking the generality of Theorem 2 rather than Theorem 1 is to establish a difference between the structure of peak sets for the ball algebra and the polydisc algebra. Recall that a set E in the distinguished boundary of a domain D is called a peak set if there exists a function F continuous on $\bar{D}$ and holomorphic on D such that $F(z)=1$ for $z \in E$, and $|f(z)|<1$ for $z \in \bar{D} \setminus E$. We consider two domains: the bidisc $\Delta = \{(z,\zeta) \in \mathbf{C}^2 \mid |z|<1 \text{ and } |\zeta|<1\}$ and the ball $B = \{(z,\zeta) \in \mathbf{C}^2 \mid |z|^2+|\zeta|^2<2\}$. The distinguished boundary of Δ is the torus $T = \{(z,\zeta) \mid |z| = |\zeta| = 1\}$. This is a subset of the distinguished boundary of B which is the topological boundary $bB = \{(z,\zeta) \mid |z|^2+|\zeta|^2 = 2\}$. It follows trivially from the definitions that if $E \subset T$ is a peak set for the ball algebra, then it is a peak set for the polydisc algebra. On the other hand, if $K = \{(e^{it}, e^{ih(t)}) \mid a \le t \le b\}$ where h is a real continuous, strictly increasing, singular function of t, Davis [D] proved that K is a peak set for the polydisc algebra. However, it is an easy consequence of Theorem 2 and Lemma 4.2 in [NR], that E is *not* a peak set for the ball algebra. In fact we have:

COROLLARY. *Let ϕ be a real, continuous increasing function defined for $a \le t \le b$, and let $K \subset bB$ be any peak set for the ball algebra. Then the set $\{t \in [a,b] \mid (e^{it}, {}^{i\phi(t)}) \in K\}$ has measure zero. In particular $K = \{(e^{it}, e^{i\phi(t)}) \in bB \mid a \le t \le b\}$ is not a peak set.*

Roughly, the strategy of the proof of Theorem A in [NR] is to embed a disc Δ into D so that $b\Delta = \gamma$ and so that Δ is as close to analytic as possible. More precisely, one finds a rectangle

$$Q = \{(x,y) \in \mathbf{R}^2 \mid a \le x \le b, 0 \le y \le \delta\}$$

and a function $\Phi: Q \to D$ such that $\Phi(x,0) = \gamma(x)$ and $\frac{\partial \Phi}{\partial \bar{z}}(x,y)$ vanishes as fast as possible as $y \to 0$. One then proves that $\lim\limits_{y \to 0} F(\Phi(x,y))$ exists for almost all x.

We too shall construct the rectangle Q and the mapping Φ. Our approach differs however in that we only make the normal component of $\frac{\partial \Phi}{\partial \bar{z}}$ go to zero; i.e. we only make $\langle N(\gamma(x)), \frac{\partial \Phi}{\partial \bar{z}}(x+iy)\rangle \to 0$ where N is defined in (1). We deal with the other components by using the fact that tangential derivatives of F grow much more slowly than the normal derivatives.

Sections 2-5 of this paper is a proof of Theorem 2. We shall assume that $i\,d\mu > 0$, and if $i\,d\mu < 0$ a similar argument works. Section 2 contains a reduction to the case that $i\,d\mu$ is Lebesgue measure. Section 3 contains the construction of Φ and a discussion of its elementary properties. Section 4 discusses a variant of Green's theorem which is required since we deal with measures rather than smooth functions. In Section 5 we complete the proof. Section 6 contains some concluding remarks and two open problems.

We would like to thank Professors Walter Rudin and E. M. Stein for useful discussions and suggestions about this paper.

§2. *A reduction*

Let $\gamma:[a,b] \to bD$ be continuous and rectifiable, and assume that the measure

$$i\,d\mu(t) = i \sum_{j=1}^{n} \frac{\partial\rho}{\partial \overline{w}_j}(\gamma(t))\,\overline{d\gamma_j}(t) = i \langle N(\gamma(t)), d\gamma(t)\rangle$$

is positive. In this section we show that it suffices to prove Theorem 2 under the additional hypothesis that $i\,d\mu(t)$ is actually Lebesgue measure dt. To this end, we note that if $\alpha:[0,A] \to [a,b]$ is a strictly monotone absolutely continuous function, then $\gamma_\alpha(s) = \gamma(\alpha(s))$ is continuous and rectifiable on $[0,A]$. Moreover, if $E \subset [0,A]$ has measure zero, then $\alpha(E) \subset [a,b]$ has measure zero. Thus it suffices to prove

LEMMA 1. *There exists* $A > 0$ *and a strictly monotone absolutely continuous function* $\alpha:[0,A] \to [a,b]$ *such that*

$$i\,d\mu_\alpha(s) = i\langle N(\gamma_\alpha(s)), d\gamma_\alpha(s)\rangle$$

is Lebesgue measure ds.

Proof. Set

$$H(t) = i \int_a^t d\mu .$$

Then $H(t)$ is continuous and strictly increasing for $a \le t \le b$. Let $A = H(b)$, and let $\alpha : [0, A] \to [a, b]$ be the inverse function of H. We claim this has the desired properties. Clearly α is strictly monotone, and since the measure $d\mu$ is positive, an argument in Saks [S], page 287, shows that α is absolutely continuous.

Note that formally

$$i\, d\mu_\alpha(s) = i < N(\gamma_\alpha(s)), \gamma'(\alpha(s)) > \alpha'(s)\, ds .$$

Also $H(\alpha(s)) = s$, so that

$$1 = H'(\alpha(s))\,\alpha'(s) = i < N(\gamma_\alpha(s)), \gamma'(\alpha(s)) > \alpha'(s) .$$

This formalism actually gives a proof if γ is absolutely continuous. To deal with the more general situation, note that since $d\mu_\alpha$ is non-atomic, it suffices to show

$$i \int_0^A \phi(s)\, d\mu_\alpha(s) = \int_0^A \phi(s)\, ds$$

for all $\phi \in C^1[0, A]$ with $\phi(0) = \phi(A) = 0$. Now since $H(\alpha(s)) = s$,

$$\int_0^A \phi(s)\, ds = - \int_0^A \phi'(s)\, H(\alpha(s))\, ds$$

$$= - \int_0^A \phi'(s) \int_a^{\alpha(s)} i < N(\gamma(t)), d\gamma(t) > ds$$

$$= \int_a^b \phi(H(t))\, i < N(\gamma(t)), d\gamma(t) >$$

by Fubini's theorem. We now make the absolutely continuous change of variables $t = \alpha(s)$, $0 \le S \le A$ to obtain the desired equality. This completes the proof.

From now on we shall assume that $\gamma: [a, b] \to bD$ is continuous and rectifiable, and the measure

$$i\, d\mu(t) = i \sum_{j=1}^{n} \frac{\partial \rho}{\partial \overline{w}_j} (\gamma(t))\, \overline{d\gamma_j}(t)$$

is actually Lebesgue measure dt.

§3. *The construction of* Φ

LEMMA 2. *There exists a function* $g: [a, b] \to \mathbf{C}^n$ *which is continuous and of bounded variation such that*

$$i\langle N(\gamma(t)), g(t)\rangle = 1 \quad \textit{for} \quad a \le t \le b\,.$$

Proof. For each $t_0 \in [a, b]$ there exists an index j_0 and an open interval U_{t_0} containing t_0 such that $\frac{\partial \rho}{\partial \overline{w}_{j_0}} (\gamma(t)) \neq 0$ for $t \in U_{t_0}$. Let V_{t_0} be an open interval containing t_0 which is compactly contained in U_{t_0}. Set

$$g_{j_0}^{(t_0)}(t) = i \left[\overline{\frac{\partial \rho}{\partial \overline{w}_{j_0}} (\gamma(t))}\right]^{-1} \quad \text{for} \quad t \in V_{t_0}$$

and

$$g_j^{(t_0)}(t) = 0 \quad \text{for} \quad j \neq j_0 \quad \text{and} \quad t \in V_{t_0}\,.$$

Clearly $g^{(t_0)}: V_{t_0} \to \mathbf{C}^n$ is continuous and of bounded variation, and $i\langle N(\gamma(t)), g(t)\rangle = 1$. We finish the proof by using a partition of unity.

Now let $\phi \in C_0^\infty(\mathbf{R})$ with $\int \phi(t)\, dt = 1$. For $y > 0$, set $\phi_y(t) = \frac{1}{y}\, \phi\left(\frac{t}{y}\right)$. Extend the function g of Lemma 2 to a function defined on all of $\mathbf{R}$ so that it is continuous, of bounded variation, and vanishes outside a compact set. Define

$$u(x,y) = \phi_y * g(x) = \int \phi_y(t)\, g(x-t)\, dt$$

and

$$\Phi(x,y) = \gamma(x) + iyu(x,y)$$

for $a \le x \le b$ and $y > 0$. It is clear that $\Phi(x,y)$ extends continuously to $y = 0$, and $\Phi(x,0) = \gamma(x)$. We shall need some elementary properties of $\Phi(x,y)$.

LEMMA 3.

a) $\lim\limits_{y \to 0} \dfrac{\partial}{\partial y}(\rho(\Phi(x,y))) = -1$ *for* $a \le x \le b$.

b) *There exists* $\delta > 0$ *and* $C > 0$ *so that if*

$$Q = \{(x,y) | a \le x \le b, 0 < y \le \delta\}$$

then $\rho(\Phi(x,y)) < -Cy$ *for* $(x,y) \in Q$.

c) $\Phi(Q) \subset D$.

d) *For each* $x \in [a,b]$, *the path* $y \to \Phi(x,y)$ *in* D *tends nontangentially to* $\gamma(x)$ *as* $y \to 0$.

e) $\dfrac{\partial \Phi}{\partial y}(x,y)$ *is bounded in* Q.

Proof. b), c), and d) follow from a). a) and e) would follow if we could show that $\dfrac{\partial \Phi}{\partial y}(x,y) \to i\, g(x)$ as $y \to 0$. Thus we only need to show that $y \dfrac{\partial u}{\partial y}(x,y) \to 0$ as $y \to 0$. But

$$y \frac{\partial u}{\partial y}(x,y) = \int_{-\infty}^{\infty} \psi_y(t)\, g(x-t)\, dt$$

where $\psi(t) = \phi(t) + t\phi'(t)$. But since $\int_{-\infty}^{\infty} \psi(t)\, dt = 0$ and g is continuous, the desired result is standard. This completes the proof.

§4. *A variant of Green's theorem*

Let F be a bounded holomorphic function on D and let Q be the rectangle constructed in Lemma 3. For $(x,y) \in Q$, define

$$f(x+iy) = F(\Phi(x,y)) .$$

We shall show that for almost every x in $[a,b]$, $\lim_{y\to 0} f(x+iy)$ exists. Then by Lemma 3d) and the extension of Lindelof's theorem due to Čirka [C], F will have nontangential limits along the curve almost everywhere.

To show that $\lim_{y\to 0} f(x+iy)$ exists almost everywhere, our plan, roughly, is to write

$$2\pi i f(z) = \int_{\partial Q} \frac{f(\zeta)}{\zeta - z} d\zeta + \iint_Q \frac{\partial f}{\partial \bar z}(\zeta) \frac{d\zeta \wedge d\bar\zeta}{\zeta - z} \tag{4}$$

(as in [NR]) where $z = x+iy$ and $(x,y) \in Q$. The first term will have limits almost everywhere since it is a Cauchy integral of a bounded function, and we shall show that the second term has a partial derivative with respect to y which is in $L^1(Q)$, and hence the second term is absolutely continuous on almost every vertical line. This will complete the proof.

However, there are difficulties with formula (4). Thus $\frac{\partial f}{\partial \bar z}$ cannot now be thought of as a function, but is only a measure. In fact $f(z) = F(\Phi(x,y))$ is of bounded variation separately in x and in y in any subregion $Q_\eta = \{(x,y) \in Q \mid y \geq \eta\}$.

Now by $\frac{\partial f}{\partial \bar z}(z)$ we shall mean the measure on Q

$$\frac{1}{2}\,[d_x f(x+iy)\,dy + i d_y f(x+iy)\,dx] .$$

Then since F is holomorphic, an application of the chain rule gives

$$\frac{\partial f}{\partial \bar z}(x+iy) = \sum \frac{\partial F}{\partial w_j}(\Phi_j(x,y)) \frac{\partial \Phi_j}{\partial \bar z} \tag{5}$$

where $\frac{\partial \Phi}{\partial \bar z}$ is the measure given by

$$\frac{\partial \Phi}{\partial \bar z}(x,y) = d\gamma(x)\,dy + [-u(x,y) + iy \frac{\partial u}{\partial x}(x,y) - y \frac{\partial u}{\partial y}(x,y)]\,dx\,dy . \tag{6}$$

Note that

$$\lim_{\sigma\to 0} \int_{|x-s|<\sigma} \int_{|y-t|<\sigma} \frac{d\gamma(s)\,dt}{(x-s)+i(y-t)} = 0$$

since

$$\left| \int_{|x-s|<\sigma} d\gamma(s) \int_{|y-t|<\sigma} \frac{dt}{(x-s)+i(y-t)} \right|$$

$$= \left| \int_{|x-s|<\sigma} d\gamma(s) \int_{|y-t|<\sigma} \frac{(x-s)\,dt}{(x-s)^2+(y-t)^2} \right|$$

$$\leq c \int_{|x-s|<\sigma} |d\gamma(s)| \to 0$$

since $d\gamma$ is non-atomic. Thus we may use Green's theorem to obtain the formula

$$2\pi i f(z) = \int_{\partial Q_\eta} \frac{f(\zeta)}{\zeta - z}\, d\zeta + \iint_{Q_\eta} \frac{1}{\zeta - z} \frac{\partial f}{\partial \bar{z}}(\zeta)\,.$$

If, as in [NR], we choose a sequence $\eta_k \to 0$ such that $f(x+i\eta_k)$ tends to a weak* limit $w(x)$, we have for $z = x+iy$ with $(x,y) \in Q$

$$2\pi i f(z) = \int_{\partial Q} \frac{f(\zeta)}{\zeta - z}\, d\zeta + \lim_{k\to\infty} \iint_{Q_{\eta_k}} \frac{1}{\zeta - z} \frac{\partial f}{\partial \bar{z}}(\zeta) \tag{7}$$

(Here $f(x+i0) = w(x)$.)

§5. *The rest of the proof*

The almost everywhere existence of limits for the first term on the right-hand side of (7) is classical, so we only need to consider the second term. According to (5), we can write $\frac{\partial f}{\partial \bar{z}}$ as a complex inner product

$$\frac{\partial f}{\partial \bar{z}}(\zeta) = \left\langle \frac{\partial \Phi}{\partial \bar{z}}(\zeta), \overline{\nabla F}(\Phi(\zeta)) \right\rangle$$

where $\overline{\nabla F}$ is the vector $\left(\frac{\overline{\partial F}}{\partial w_1}, \cdots, \frac{\overline{\partial F}}{\partial w_n}\right)$. We wish to separate this inner product into complex normal and complex tangential parts. To this end we introduce a special orthonormal basis in $\mathbf{C}^n$. If $\zeta = s+it$, we set $N_1(\zeta) = \frac{N(\gamma(s))}{|N(\gamma(s))|}$ where N is defined in (1). We then choose unit vectors $N_2(\zeta), \cdots, N_n(\zeta)$ so that $\langle N_i(\zeta), N_j(\zeta)\rangle = \delta_{ij}$ for $1 \le i, j \le n$, and each $N_j(\zeta)$ is independent of t. Then

$$\frac{\partial f}{\partial \bar{z}}(\zeta) = \sum_{\ell=1}^{n} \left\langle \frac{\partial \Phi}{\partial \bar{z}}(\zeta), N_\ell(\zeta) \right\rangle \langle N_\ell(\zeta), \overline{\nabla F}(\Phi(\zeta))\rangle$$

$$= \sum_{\ell=1}^{n} I_\ell(\zeta) .$$

We set

$$K_\ell(z) = \lim_{k\to\infty} \iint_{Q_{\eta_k}} \frac{I_\ell(\zeta)}{\zeta - z} . \tag{8}$$

It now suffices to show that each $K_\ell(x+iy)$ has a y derivative which is in $L^1(Q)$. We begin with the normal component $K_1(z)$.

We have $K_1(z) = U_1(z) + V_1(z)$ where

$$U_1(z) = \lim_{k\to\infty} \iint_{Q_{\eta_k}} \frac{1}{\zeta - z} \langle N_1(\zeta), \overline{\nabla F}(\Phi(\zeta))\rangle \cdot$$
$$\cdot \langle d\gamma(s)\,dt - g(s)\,ds\,dt, N_1(\zeta)\rangle$$

and

$$V_1(z) = \lim_{k\to\infty} \iint\limits_{Q_{\eta_k}} \frac{1}{\zeta - z} <N_1(\zeta), \overline{\nabla F}(\Phi(\zeta))> \cdot$$

$$\cdot \Big\langle (g(s) - u(s,t) + it \frac{\partial u}{\partial s}(s,t) - t \frac{\partial u}{\partial t}(s,t))\, ds\, dt, N_1(\zeta) \Big\rangle .$$

But

$$< d\gamma(s)\, dt - g(s)\, ds\, dt, N_1(\zeta)>$$

$$= \frac{1}{|N(\gamma(s))|} [< d\gamma(s), N(\gamma(s))> - < g(s), N(\gamma(s))> ds]\, dt$$

$$= 0$$

by Lemma 2, and the reduction of Section 2. Thus $U_1(z) = 0$.

To deal with $V_1(z)$, let

$$r(s,t) = <N_1(\zeta), \overline{\nabla F}(\Phi(\zeta))> \Big\langle g(s) - u(s,t) + it \frac{\partial u}{\partial s}(s,t) - t \frac{\partial u}{\partial t}(s,t), N_1(\zeta) \Big\rangle$$

with $\zeta = s + it$. Note that r is smooth in t for $t > 0$. We shall show that

$$\iint\limits_{Q} |r(s,t)|^p\, ds\, dt < \infty \tag{9}$$

for any $p < 2$. This will show first of all that Q_{η_k} can be replaced by Q in (8). Then one can use the theory of singular integrals to show that $\frac{\partial V}{\partial y}$ is in $L^p(Q)$ for $p < 2$ and hence in $L^1(Q)$. (See [NR], Lemma 2.2.)

By Lemma 3 b), $|\nabla F(\Phi(\zeta))| \le C/t$. Thus

$$|r(s,t)| \le ct^{-1}|g(s) - u(s,t)| + c\Big|\frac{\partial u}{\partial s}(s,t)\Big| + c\Big|\frac{\partial u}{\partial t}(s,t)\Big|$$

$$= II_1(s,t) + II_2(s,t) + II_3(s,t) .$$

We claim that there exists $\psi \in C_0^\infty(\mathbf{R})$ so that

$$II_j(s,t) \le \int_{-\infty}^{\infty} \frac{1}{t} \psi\Big(\frac{s-\lambda}{t}\Big) |dg(\lambda)| \quad j = 1, 2, 3 .$$

Consider first

$$\frac{1}{t}(g(s)-u(s,t)) = \frac{1}{t^2}\int \phi\left(\frac{u}{t}\right)[g(s)-g(s-u)]\,du$$

$$= \frac{1}{t^2}\int \phi\left(\frac{u}{t}\right)\int_{s-u}^{s} dg(\lambda)\,du$$

$$= \frac{1}{t^2}\int_{-\infty}^{s} dg(\lambda)\int_{s-\lambda}^{\infty}\phi\left(\frac{u}{t}\right)du - \frac{1}{t^2}\int_{s}^{\infty} dg(\lambda)\int_{-\infty}^{s-\lambda}\phi\left(\frac{u}{t}\right)du$$

$$= \int_{-\infty}^{\infty}\frac{1}{t}\,\theta\left(\frac{s-\lambda}{t}\right)dg(\lambda)$$

where

$$\theta(r) = \begin{cases} \displaystyle\int_{r}^{\infty}\phi(u)\,du & \text{if} \quad r \geq 0 \\[2ex] \displaystyle -\int_{-\infty}^{r}\phi(u)\,du & \text{if} \quad r < 0\,. \end{cases}$$

Clearly θ has compact support and is bounded, and hence can be dominated by a function of the desired type. Next

$$\frac{\partial u}{\partial s}(s,t) = -\frac{1}{t}\int_{-\infty}^{\infty}\frac{\partial}{\partial u}\left(\phi\left(\frac{s-u}{t}\right)\right)g(u)\,du$$

$$= \int_{-\infty}^{\infty}\frac{1}{t}\,\phi\left(\frac{s-\lambda}{t}\right)dg(\lambda)$$

and ϕ is a function of the desired type. Finally

$$\frac{\partial u}{\partial t}(s,t) = \int_{-\infty}^{\infty}\left[-\frac{1}{t^2}\phi\left(\frac{s-u}{t}\right) - \frac{s-u}{t^3}\phi'\left(\frac{s-u}{t}\right)\right]g(u)\,du$$

$$= -\frac{1}{t}\int_{-\infty}^{\infty}\frac{\partial}{\partial u}\left(\omega\left(\frac{s-u}{t}\right)\right)g(u)\,du$$

$$= \int_{-\infty}^{\infty}\frac{1}{t}\,\omega\left(\frac{s-\lambda}{t}\right)dg(\lambda)$$

where

$$\omega(r) = -\int_{-\infty}^{r}(\phi(u)+u\phi'(u))\,du\ .$$

Since ϕ has compact support and $\int_{-\infty}^{\infty}(\phi(u)+u\phi'(u))\,du = 0$, ω is again a function of the desired type.

Now inequality (9) follows from the following

LEMMA 4. *Let* $\psi \in C_0^\infty(R)$ *and for* $y > 0$ *set*

$$\omega(x,y) = \frac{1}{y}\int_{-\infty}^{\infty}\psi\left(\frac{x-\lambda}{y}\right)|dg(\lambda)|\ .$$

Then $\iint_Q |\omega(x,y)|^p\,dx\,dy < \infty$ *for all* $p < 2$.

Proof. Let $m(x,y) \in L^q(Q)$ with $\frac{1}{p}+\frac{1}{q} = 1$. Then

$$\left|\iint_Q \omega(x,y)\,m(x,y)\,dx\,dy\right|$$

$$\le \int_{-\infty}^{\infty}|dg(\lambda)|\int_0^{\delta}\frac{1}{y}\int_a^b |m(x,y)|\left|\psi\left(\frac{x-\lambda}{y}\right)\right|dx\,dy$$

$$\le \int_{-\infty}^{\infty}|dg(\lambda)|\int_0^{\delta}\frac{1}{y}\left(\int_a^b|m(x,y)|^q\,dx\right)^{1/q}\left(\int_a^b\left|\psi\left(\frac{x-\lambda}{y}\right)\right|^p dx\right)^{1/p}dy$$

$$\leq \int_{-\infty}^{\infty} |dg(\lambda)| \int_0^{\delta} y^{1/p-1} \left(\int_a^b (m(x,y))^q \, dx \right)^{1/q} \|\psi\|_p \, dy$$

$$\leq \|\psi\|_p \|m\|_q \int_{-\infty}^{\infty} |dg(\lambda)| \left(\int_0^{\delta} y^{1-p} \, dy \right)^{1/p}$$

$$< \infty \text{ if } p < 2 .$$

This proves Lemma 4, and finishes the discussion of $K_1(z)$. We now turn to the tangential terms $K_\ell(z)$ for $\ell = 2, \cdots, n$.

We may divide $K_\ell(z) = U_\ell(z) + V_\ell(z)$ as we divided K_1. Then the treatment of $V_\ell(z)$ for $\ell \geq 2$ is the same as for $V_1(z)$. However $U_\ell(z)$ for $\ell \geq 2$ is no longer zero.

In dealing with $U_\ell(z)$, we shall only study the term

$$\lim_{k\to\infty} \iint_{Q_{\eta_k}} \frac{1}{\zeta - z} \langle N_\ell(\zeta), \overline{\nabla F}(\Phi(\zeta)) \rangle \langle d\gamma(s)\, dt, N_\ell(\zeta) \rangle \tag{10}$$

since the term involving $g(s)\, ds\, dt$ is easier. Expression (10) is a finite sum of terms of the form

$$\Omega(z) = \Omega(x+iy) = \lim_{k\to\infty} \iint_{Q_{\eta_k}} \frac{1}{\zeta - z} h(s,t)\, dP(s)\, dt \tag{11}$$

where $P(s)$ is a continuous function of bounded variation and compact support, and $h(s,t)$ is continuous in Q and satisfies

$$\left. \begin{aligned} |h(s,t)| &\leq ct^{-1/2} \\ \left| \frac{\partial h}{\partial t}(s,t) \right| &\leq ct^{-3/2} \end{aligned} \right\} \tag{12}$$

The last estimates follow since the derivatives of F, $\langle N_\ell(\zeta), \overline{\nabla F}(\Phi(\zeta)) \rangle$ are now tangential, and $\frac{\partial \Phi}{\partial y}$ is bounded (see Lemma 3 e)). Thus we may take the limit as $k \to \infty$ to obtain

$$\Omega(x+iy) = \iint_Q \frac{h(s,t)}{(x-s)+i(y-t)}\, dP(s)\, dt\,.$$

We write $\Omega = \Omega_1 + \Omega_2$ where

$$\Omega_1(x+iy) = \int_a^b dP(s)\left(\int_0^{y/2} + \int_{3y/2}^{\delta}\right) \frac{h(s,t)}{(x-s)+i(y-t)}\, dt$$

and

$$\Omega_2(x+iy) = \int_a^b dP(s) \int_{y/2}^{3y/2} \frac{h(s,t)}{(x-s)+i(y-t)}\, dt = \int_a^b dP(s) \int_{-y/2}^{+y/2} \frac{h(s,y-t)\, dt}{(x-s)+it}.$$

Thus

$$\left|\frac{\partial\Omega_2}{\partial y}(x+iy)\right| \le Cy^{-1/2} \int_a^b \frac{1}{|x-s|+y}\, |dP(s)|$$

$$+ Cy^{-3/2} \int_a^b |dP(s)| \int_{-y/2}^{y/2} \frac{dt}{|x-s|+|t|}\,.$$

Hence

$$\iint_Q \left|\frac{\partial\Omega_2}{\partial y}(x+iy)\right| dx\, dy \le C \int_a^b |dP(s)| \iint_Q \frac{dx\, dy}{y^{1/2}(|x-s|+y)}$$

$$+ C \int_a^b |dP(s)| \iint_Q y^{-3/2} \int_{-y/2}^{y/2} \frac{dt}{|x-s|+|t|}\, dx\, dy$$

and easy estimates show both these integrals are bounded.

Finally

$$\left|\frac{\partial\Omega_1}{\partial y}(x+iy)\right| \le Cy^{-1/2} \int_a^b \frac{1}{|x-s|+y}\, |dP(s)|$$

$$+ C \int_a^b |dP(s)| \left(\int_0^{y/2} + \int_{3y/2}^{\delta}\right) \frac{dt}{t^{1/2}\,[(x-s)^2+(y-t)^2]}\,.$$

The first integral is the same as the first integral in $\frac{\partial \Omega_2}{\partial y}$. The $L^1(Q)$ norm of the second is less than

$$C\int_a^b |dP(s)| \int_0^\delta dy \left(\int_0^{y/2} + \int_{3y/2}^\delta\right) \frac{dt}{t^{1/2}} \int_a^b \frac{dx}{(x-s)^2+(y-t)^2}$$

$$\leq C\int_a^b |dP(s)| \int_0^\delta dy \left(\int_0^{y/2} + \int_{3y/2}^\delta\right) \frac{dt}{t^{1/2}|y-t|}$$

$$\leq C\int_a^b |dP(s)| \int_0^\delta \frac{\ell n(1/y)}{y^{1/2}}\, dy < \infty\,.$$

This completes the discussion of the terms $U_\ell(z)$ for $\ell \geq 2$, and thus completes the proof of Theorem 2.

§6. *Final remarks*

We wish to point out an extension of Theorem 2, and to pose two related questions. We have stated Theorems A and 1 and 2 for domains of class C^2. In fact, in [NR], Theorem A is proved under the weaker hypothesis that the defining function ρ for D is only of class C^1. An examination of the proof we have given for Theorem 2 shows that the only place where we used that $\rho \in C^2$ is in obtaining estimates (12) for the tangential derivatives of a bounded holomorphic function. If we had assumed only that $\rho \in C^1$ *and* the derivative of ρ were in lip ε for some $\varepsilon > 0$, a similar proof of Theorem 2 could be given. Estimates (12) would have to be replaced by

$$\left.\begin{aligned} &|h(s,t) \leq Ct^{-1+\varepsilon} \\ &\left|\frac{\partial h}{\partial t}(s,t)\right| \leq Ct^{-2+\varepsilon} \end{aligned}\right\}$$

and then the treatment of the function $\Omega(x+iy)$ would go as before.

However, this proof would break down if ρ were only of class C^1. Thus the first problem we ask is whether Theorem 2 is still true if D is only of class C^1. One should note, however, that the example of Davis cited in the introduction shows that Theorem 2 is false if the domain D is only Lipschitz.

The second problem we pose also has to do with peak sets on the distinguished boundary of the polydisc. Let $K = \{(e^{it}, e^{ih(t)}) | a \leq t \leq b\}$ where h is a real continuous increasing function. The theorem of Davis says that if h is strictly increasing, and if $h'(t) = 0$ almost everywhere, then K is a peak set. On the other hand, the methods of [NR] show that if h is C^1 *and* $\frac{dh}{dt}$ is in Lip ε for some $\varepsilon > 0$, then K is not a peak set. (This improves on an unpublished result of Carleson who required that h have a continuous third derivative. The proof of Carleson's result is contained in [R].) Thus we can ask whether K can be a peak set if h is only C^1.

UNIVERSITY OF WISCONSIN
MADISON, WISCONSIN

REFERENCES

[C] E. M. Čirka, "The theorems of Lindelof and Fatou in C^n," Mat. Sb. 92(134)(1973), pp. 622-644 (Math. USSR Sb. 21(1973), pp. 619-639).

[D] C. S. Davis, "A zero set for $A(U^2)$," Proc. A.M.S. 46(1974), pp. 287-288.

[NR] A. Nagel and W. Rudin, "Local boundary behavior of bounded holomorphic functions," Can. J. Math. 30(1978), pp. 583-592.

[R] W. Rudin, "Harmonic analysis in polydiscs," Actes. Congrès intem. Math., 1970, Tome 2, pp. 489-493.

[S] S. Saks, *Theory of the integral*, Dover Publications, Inc., New York, 1964.

ON APPELL'S SYSTEMS OF HYPERGEOMETRIC DIFFERENTIAL EQUATIONS

Takeshi Sasaki and Kyoichi Takano

§1. *Introduction*

In this note, we give a brief survey of investigations concerning the monodromy groups of Appell's systems of hypergeometric differential equations and related topics.

First let us review Gauss' hypergeometric equation. We denote it by (G).

$$\text{(G)} \qquad \theta(\theta+\gamma-1)z - x(\theta+\alpha)(\theta+\beta)z = 0 ,$$

where $\theta = x\frac{\partial}{\partial x}$ and α, β, γ are complex parameters. Fundamental solutions are given by the integrals

$$\int u^{\beta-\gamma}(u-1)^{\gamma-\alpha-1}(u-x)^{-\beta}\,du .$$

Among the many problems about this equation, we will single out the following.

(a) calculation of the monodromy group,

(b) characterization of (G) by its local monodromy,

(c) determination of conditions for all solutions to be algebraic,

(d) study of the mapping $(f_1, f_2): \mathbf{C}-\{0,1\}-\mathbf{P}$; where f_1, f_2 are linearly independent solutions.

Recent Developments in Several Complex Variables
0-691-08285-5/81/000345-12 $00.60/0 (cloth)
0-691-08281-2/81/000345-12 $00.60/0 (paperback)
For copying information, see copyright page.

(b) is the so-called Riemann problem. (c) was solved by Schwarz ([20]) in 1872. As for (d), taking $\alpha=\beta=1/3$, $\gamma=1$ for example, the inverse mapping of (f_1, f_2) is uniform and defined on the unit ball. This is the elliptic modular function λ.

Our aim is to consider these problems for Appell's systems.

§2. *Definition of Appell's systems*

P. Appell ([3]), generalizing Gauss' hypergeometric equation to the case of two variables, defined the following four systems:

$$(F_1)\begin{cases}\theta(\theta+\theta'+\gamma-1)z-x(\theta+\theta'+\alpha)(\theta+\beta)z=0\\ \theta'(\theta+\theta'+\gamma-1)z-y(\theta+\theta'+\alpha)(\theta'+\beta')z=0,\end{cases}\qquad (F_2)\begin{cases}\theta(\theta+\gamma-1)z-x(\theta+\theta'+\alpha)(\theta+\beta)z=0\\ \theta'(\theta'+\gamma'-1)z-y(\theta+\theta'+\alpha)(\theta'+\beta')z=0,\end{cases}$$

$$(F_3)\begin{cases}\theta(\theta+\theta'+\gamma-1)z-x(\theta+\alpha)(\theta+\beta)z=0\\ \theta'(\theta+\theta'+\gamma-1)z-y(\theta'+\alpha')(\theta'+\beta')z=0,\end{cases}\qquad (F_4)\begin{cases}\theta(\theta+\gamma-1)z-x(\theta+\theta'+\alpha)(\theta+\theta'+\beta)z=0\\ \theta'(\theta'+\gamma'-1)z-y(\theta+\theta'+\alpha)(\theta+\theta'+\beta)z=0,\end{cases}$$

where $z = z(x,y)$ is an unknown function and $\theta = x\frac{\partial}{\partial x}$, $\theta' = y\frac{\partial}{\partial y}$. These systems are defined on P^2 and known to be of Fuchsian type. G. Lauricella ([9]) generalized these systems to the case of n variables. Here we write one of his systems denoted, (F_D), which is a generalization of (F_1):

$$(F_D)\,\theta_i(\theta_1+\cdots+\theta_n+\gamma-1)z - x_i(\theta_1+\cdots+\theta_n+\alpha)(\theta_i+\beta_i)z=0 \quad (1\leqq i\leqq n),$$

where $\theta_i = x_i\frac{\partial}{\partial x_i}$, $(x_1,\cdots,x_n)\in C^n$. The order (dimension of the solution space) of system (F_2), (F_3) or (F_4) is four and that of (F_D) is $n+1$. Hereafter we denote by (F) one of (F_i), $1\leqq i\leqq 4$, or (F_D).

In the general case, we do not know the most general and natural definition of hypergeometric functions. In the case of two variables, J. Horn ([4]) gave the following definition: the double power series

$$\sum_{m,n=0}^{\infty} A_{m,n}x^m y^n$$

is a hypergeometric function if the two quotients $A_{m+1,n}/A_{m,n}$ and $A_{m,n+1}/A_{m,n}$ are rational functions of m and n. He then obtained 14 hypergeometric functions (and 20 confluent hypergeometric functions) of order two among which are Appell's hypergeometric functions. M. Sato ([19]) defined hypergeometric functions as simultaneous eigenfunctions of differential operators associated with "b-functions."

System (F_D) appears in the representation theory of Lie groups. Namely, W. Miller ([10]) proved $SL(n+1, \mathbf{C})$ is the dynamical symmetry group of (F_D). J. Sekiguchi ([21]) found that some zonal spherical functions on the symmetric space $SL(3, \mathbf{R})/SO(3)$ satisfy (F_1). It is desirable to clarify the structure of these systems by relating them to other branches of mathematics.

Now, let us return to the problems stated in §1.

§3. *Monodromy groups*

In order to calculate the monodromy group of each system (F), it is convenient to integrate (F) by single or double integrals.

Double integral representations. It is known ([3], [4]) that the integrals

$$(1)\quad \int_\Delta u^{\beta-1}v^{\beta'-1}(1-u-v)^{\gamma-\beta-\beta'-1}(1-xu-yv)^{-\alpha}\,du\wedge dv\,,$$

$$(2)\quad \int_\Delta u^{\beta-1}v^{\beta'-1}(1-u)^{\gamma-\beta-1}(1-v)^{\gamma'-\beta'-1}(1-xu-yv)^{-\alpha}\,du\wedge dv\,,$$

and

$$(3)\quad \int_\Delta u^{\beta-1}v^{\beta'-1}(1-u-v)^{\gamma-\beta-\beta'-1}(1-xu)^{-\alpha}(1-yv)^{-\alpha'}\,du\wedge dv$$

form fundamental sets of solutions of (F_1), (F_2) and (F_3) respectively. Here the Δ's are suitably chosen compact 2-cycles attached to each integrand. It is also known that (F_D) has a fundamental set of solutions expressed by n-ple integrals.

For system (F_4), K. Aomoto found the following integral representations

$$(4) \quad \int_\Delta u^{\beta-\gamma'} v^{\beta-\gamma} (u+v-uv)^{\gamma+\gamma'-\beta-2} (1-xu-yv)^{-\alpha}\, du \wedge dv\ .$$

We note that the integrands of (F_1), (F_2), (F_3) and (F_D) are power products of linear functions, but the integrand of (F_4) contains a quadratic factor.

Single integral representations. Picard pointed out that (F_1) can be integrated by the following single integrals:

$$(5) \quad \int_C u^{\alpha-1} (1-u)^{\gamma-\alpha-1} (1-xu)^{-\beta} (1-yu)^{-\beta'} du\ ,$$

where the C's are compact 1-cycles (closed curves) on the Riemann surface of the integrand. Note that (F_D) has also single integral representations.

For systems (F_2), (F_3) and (F_4), A. Erdélyi ([4]) found single integral representations whose integrands contain one or two Gauss' hypergeometric functions. We give here only those for (F_4):

$$(6) \quad \int_C u^{-\gamma}(1-u)^{-\gamma'} P \begin{Bmatrix} 0 & 1 & \infty & \\ 0 & 0 & \alpha & \frac{x}{u} + \frac{y}{1-u} \\ 2-\gamma-\gamma' & \gamma+\gamma'-\alpha-\beta-1 & \beta & \end{Bmatrix} du\ ,$$

where P{ } denotes Riemann's P-function.

For each system (F), take as a fundamental set of solutions one of the sets of solutions given above; more precisely, choose and fix cycles Δ (or C) and branches of integrands and choose generators of $\pi_1(\bar{P}^n - S)$, S being the singular set, and $n=2$ for $(F) = (F_i)$. Then, by observing the variation of the cycles and the branches of the integrands along generators of $\pi_1(P^n - S)$, we can get the monodromy group.

If we take the double integral representations as a fundamental set of solutions, we can make use of a generalized Picard-Lefschetz formula ([14]) to determine the local monodromy group at each singular set.

In the Appendix we will list generators of each monodromy group.

Let z be a solution of (F) and let us consider z as a function of one variable, say x_1. Then this function satisfies an ordinary differential equation with respect to x_1. In case $(F) = (F_D)$ this is a Jordan-Pochhammer equation (JP), and in case $(F) = (F_3)$ this is the fourth order system (O), one of equations introduced by K. Okubo ([13]). The monodromy groups of (JP) and (O) have also been calculated ([25], [17]).

REMARK. Aomoto ([2]) obtained the following general result. Let $\ell_1, \cdots, \ell_m$, $m \geqq n+1$, be linear functions on $\mathbf{C}^n$ with real coefficients and let S_j be the set $\{\ell_j = 0\}$. Then the integrals

$$\int \ell_1^{\lambda_1} \cdots \ell_m^{\lambda_m} dx_1 \cdots dx_n$$

satisfy some differential equations with respect to the coefficients of ℓ_j. In case S_j are in general position and λ_j are real positive numbers, he gave a method of computing the monodromy of these integrals which is applicable to system (F), except (F_4).

§4. *Riemann's problem*

Riemann proved that Riemann's differential equation is characterized by properties of the solution space, namely by the characteristic exponents at three singular points. It was Picard who generalized the result of Riemann to system (F_1). T. Terada ([26]) completed the work of Picard and gave a characterization of system (F_D) from Riemann's point of view.

M. Yoshida ([27]) gave another characterization of (F_1); namely, a system whose solution space is given by the following Riemann's scheme

$$P\left\{\begin{matrix} x=0 & x=1 & x=\infty & y=0 & y=1 & y=\infty & x=y \\ 0 & 0 & \alpha_\infty & 0 & 0 & \beta_\infty & 0 \\ 1 & 1 & \alpha_\infty+1 & 1 & 1 & \beta_\infty+1 & 1 \\ \alpha_0 & \alpha_1 & \alpha'_\infty & \beta_0 & \beta_1 & \beta'_\infty & \alpha_y=\beta_x \end{matrix} \quad ; x,y\right\}$$

reduces to (F_1), except trivial cases.

E. Goursat ([7]) tried to solve this problem for (F_2) and (F_3), but it seems to us that his proof is not complete.

§5. *Finiteness of monodromy groups*

Next let us recall Schwarz's investigation of (G). He asked the conditions that imply (G) has only algebraic solutions, i.e. when is its monodromy group finite; and he found that the monodromy groups are regular polyhedral groups when these conditions hold.

The monodromy groups of (JP) and (F_D) have similar characters to that of (G). These groups are generated by unitary reflections if they are of finite order. Since the finite groups generated by unitary reflections are well known ([22]), we can determine the case under which all solutions are algebraic: When $n \geqq 2$, and if the monodromy group of (F_D) is an irreducible finite group it is one of the following: (i) the Hessian group of order 648, (ii) the Hessian group of order 1296 in three variables, or (iii) the symmetry group of the complex regular polytope of order 216.6! in four variables ([18]). For the system (JP), other than the above groups, the imprimitive group $G(3, 3, 3)$ of order 54 can occur ([25]). (For notation and definition of these groups, see [22].)

On the other hand, the monodromy groups of (F_2) and (F_3) cannot be finite when they are irreducible ([31]).

It is natural to ask for the equations whose monodromy group is a given finite (unitary reflection) group. In this connection, A. Boulanger computed the equations for the Klein group and the Hessian group ([5], [6]). For regular polyhedral groups, Yoshida ([29]) found the equations of Fuchsian type of two variables.

REMARK. Contrary to the finiteness, we have the following infiniteness theorem ([31]): Picard-Vessiot group, which is the Zariski closure of the monodromy group of our system, is equal to the full general linear group when parameters take general values.

§6. *Automorphic functions*

The system (F_D) has special features in the following sense. Let $f_1, \cdots, f_{n+1}$ be linearly independent solutions of (F_D). Then $f = (f_1, \cdots, f_{n+1})$ is a multivalued locally biholomorphic mapping from $P^n - S$ into P^n. The inverse mapping is not generally uniform, but when the parameters α, β_i, γ take some special values this mapping is uniform and defined on the unit ball ([26]). It is automorphic with respect to the corresponding monodromy group. The fact that the image of f happens to be the unit ball is due to Riemann's inequality satisfied by the period matrix of the Riemann surface connected with the system (F_D). For details, see [26]. The special case when $\alpha = \beta_1 = \beta_2 = 1/3$, $\gamma = 1$ $(n=2)$ was first considered by Picard ([15]). Also, H. Shiga ([23]) reconsidered this case from a new viewpoint and gave a precise connection of the inverse mapping with a 2-parameter family of K3 surfaces.

Recently, Yoshida ([28], [30]) has studied the inverse problem; that is, he has completely solved the following problems: (i) determine all locally volume finite discrete subgroups Γ of the isotropy groups at the boundary point p of the 2-dimensional unit ball B_2, (ii) for each discrete subgroup Γ such that $B_2/\Gamma \cup \{p\}$ is nonsingular, construct the system of differential equations of Fuchsian type whose monodromy group is Γ. We note that the above procedure (ii) is carried out by making use of the generalized Schwarzian derivatives obtained by T. Oda ([12]).

Finally, we will raise a question. H. Alexander ([1]) defined a closed embedding of the one-dimensional unit ball into C^2 using the λ-function. It is interesting to find a meaningful embedding of a ball of any dimension into C^n; and we ask whether solutions of (F_D) can afford this embedding.

Appendix. *Generators of monodromy groups*

We denote by Γ_i, $1 \leq i \leq 4$, the monodromy group of (F_i). Generators of Γ_i are listed below. The fundamental set of solutions and generators of $\pi_1(P^2-S)$ used in calculation, but which we will not specify here, can be found in the references cited. The notation $e(a) = e^{2\pi i a}$ is used.

(Γ_1) (5 generators, [16], [26])

$$\begin{pmatrix} \mu_0\mu_1 & 0 & 0 \\ \mu_0(\mu_1-1) & 1 & 0 \\ \mu_0(\mu_1-1) & 0 & 1 \end{pmatrix} \quad \begin{pmatrix} 1-\mu_1(1-\mu_3) & \mu_1(1-\mu_2)(1-\mu_3) & \mu_1\mu_2(1-\mu_3) \\ 0 & 1 & 0 \\ (1-\mu_1)/\mu_2 & (1-\mu_1)(\mu_2-1)/\mu_2 & \mu_1 \end{pmatrix}$$

$$\begin{pmatrix} 1-\mu_1(1-\mu_2) & \mu_1(1-\mu_2) & 0 \\ 1-\mu_1 & \mu_1 & 0. \\ 0 & 0 & 1 \end{pmatrix} \quad \begin{pmatrix} 1 & 0 & 0 \\ 0 & 1-\mu_2(1-\mu_3) & \mu_2(1-\mu_3) \\ 0 & 1-\mu_2 & \mu_2 \end{pmatrix}$$

$$\begin{pmatrix} 1-\mu_0(1-\mu_1)(1-\mu_2) & -\mu_0\mu_1(1-\mu_2) & 0 \\ -\{(1-\mu_0)+\mu_0\mu_1(1-\mu_2)\}(1-\mu_1)/\mu_1 & \mu_0-\mu_0\mu_1(1-\mu_2) & 0 \\ -\mu_0(1-\mu_1)(1-\mu_2) & -\mu_0\mu_1(1-\mu_2) & 1 \end{pmatrix}$$

where $\mu_0 = e(\beta+\beta'-\gamma)$, $\mu_1 = e(-\beta)$, $\mu_2 = e(-\beta')$ and $\mu_3 = e(\gamma-\alpha)$.

(Γ_2) (5 generators, [11])

$$\begin{pmatrix} 1 & 0 & 0 & 0 \\ 0 & e(-\gamma) & 0 & e(-\beta)-1 \\ e(-\beta)-1 & 0 & e(-\gamma) & 0 \\ 0 & 0 & 0 & 1 \end{pmatrix}\begin{pmatrix} 1 & 0 & 0 & 0 \\ 0 & e(-\gamma') & e(-\beta')-1 & 0 \\ 0 & 0 & 1 & 0 \\ e(-\beta')-1 & 0 & 0 & e(-\beta') \end{pmatrix}$$

$$\begin{pmatrix} 1 & e(-\alpha)-1 & 0 & 0 \\ 0 & e(\delta) & 0 & 0 \\ 0 & 1-e(\delta+\beta'-\gamma') & 1 & 0 \\ 0 & 1-e(\delta+\beta-\gamma) & 0 & 1 \end{pmatrix} \begin{pmatrix} 1 & 0 & 1-e(-\alpha) & 0 \\ 0 & e(\delta+\beta') & e(\delta+\beta')(1-e(-\beta')) & 0 \\ 0 & 0 & e(\delta+\beta'-\gamma') & 0 \\ 0 & 1-e(-\alpha+\gamma') & -e(-\alpha+\gamma')(1-e(-\beta')) & 1 \end{pmatrix}$$

$$\begin{pmatrix} 1 & 0 & 0 & 1-e(-\alpha) \\ 0 & e(\delta+\beta) & 0 & e(\delta+\beta)(1-e(-\beta)) \\ 0 & 1-e(-\alpha+\gamma) & 1 & -e(-\alpha+\gamma)(1-e(-\beta)) \\ 0 & 0 & 0 & e(\delta+\beta-\gamma) \end{pmatrix}$$

where $\delta = -\alpha-\beta-\beta'+\gamma+\gamma'$.

(Γ_3) (5 generators, [11])

$$\begin{pmatrix} e(-\beta) & 0 & 0 & 0 \\ 0 & e(-\alpha) & 0 & e(-\alpha)(1-e(-\beta)) \\ e(-\alpha)(1-e(-\beta)) & 0 & e(-\alpha) & 0 \\ 0 & 0 & 0 & e(-\beta) \end{pmatrix} \begin{pmatrix} 1 & e(-\beta-\beta'+\gamma)-1 & 0 & 0 \\ 0 & e(\delta) & 0 & 0 \\ 0 & 1-e(\delta+\alpha') & 1 & 0 \\ 0 & 1-e(\delta+\alpha) & 0 & 1 \end{pmatrix}$$

$$\begin{pmatrix} e(-\beta') & 0 & 0 & 0 \\ 0 & e(-\alpha') & e(-\alpha')(1-e(-\beta')) & 0 \\ 0 & 0 & e(-\beta') & 0 \\ e(-\alpha')(1-e(-\beta') & 0 & 0 & e(-\alpha') \end{pmatrix}$$

$$\begin{pmatrix} 1 & 0 & 1-e(\delta+\alpha+\alpha') & 0 \\ 0 & e(\delta+\beta') & e(\delta+\beta')(1-e(-\beta')) & 0 \\ 0 & 0 & e(\delta+\alpha') & 0 \\ 0 & 1-e(\delta+\alpha+\beta') & -e(\delta+\alpha+\beta')(1-e(-\beta')) & 1 \end{pmatrix}$$

$$\begin{pmatrix} 1 & 0 & 0 & 1-e(\delta+\alpha+\alpha') \\ 0 & e(\delta+\beta) & 0 & e(\delta+\beta)(1-e(-\beta)) \\ 0 & 1-e(\delta+\alpha'+\beta) & 1 & -e(\delta+\alpha'+\beta)(1-e(-\beta)) \\ 0 & 0 & 0 & e(\delta+\alpha) \end{pmatrix}$$

where $\delta = -\alpha-\alpha'-\beta-\beta'+\gamma$.

(Γ_4) (3 generators, [24], [8])

$$\begin{pmatrix} 1 & 1-e(-\alpha) & 0 & 0 \\ 0 & e(-\gamma) & 0 & 0 \\ a & b & e(-\gamma) & c \\ 0 & 0 & 0 & 1 \end{pmatrix} \begin{pmatrix} 1 & 0 & 0 & 1-e(-\alpha) \\ 0 & 1 & 0 & 0 \\ a' & c' & e(-\gamma') & b \\ 0 & 0 & 0 & e(-\gamma') \end{pmatrix}$$

$$\begin{pmatrix} 1 & 0 & -(1-e(-\alpha)) & 0 \\ 0 & 1 & 1 & 0 \\ 0 & 0 & -e(\gamma+\gamma'-\alpha-\beta) & 0 \\ 0 & 0 & 1 & 1 \end{pmatrix}$$

where

$$a = -(1-e(-\beta))(1-e(\gamma))\,e(\gamma')(1-e(\gamma+\gamma'))^{-1},$$

$$b = (1-e(-\alpha))(1-e(-\beta))\,e(\gamma+\gamma')(1-e(\gamma+\gamma'))^{-1},$$

$$c = -(1+e(\gamma')-e(\gamma'-\alpha)-e(\gamma'-\beta)-e(\gamma+\gamma')+e(\gamma+2\gamma'-\alpha-\beta))(1-e(\gamma+\gamma'))^{-1},$$

$$a' = -(1-e(-\beta))(1-e(\gamma'))\,e(\gamma)(1-e(\gamma+\gamma'))^{-1},$$

$$c' = -(1+e(\gamma)-e(\gamma-\alpha)-e(\gamma-\beta)-e(\gamma+\gamma')+e(2\gamma+\gamma'-\alpha-\beta))(1-e(\gamma+\gamma'))^{-1}.$$

NAGOYA UNIVERSITY AND KOBE UNIVERSITY

REFERENCES

[1] Alexander, H.: Explicit imbedding of the (punctured) disc into $\mathbb{C}^2$, Comm. Math. Helv. 52 (1977), 539-544.

[2] Aomoto, K.: On the structure of integrals of power product of linear functions, Sci. Papers of the Coll. of Gen. Educ., Univ. of Tokyo, 27 (1977), 46-61.

[3] Appell, P. and J. Kampé de Fériet: Fonctions hypergéométriques et hypersphériques-Polynomes d'Hérmite, Gauthiers-Villars 1926.

[4] Bateman, H. (ed. by A. Erdélyi): Higher transcendental functions, Vol. 1, McGraw-Hill 1963.

[5] Boulanger, A.: Contribution à l'étude des équations différentielles linéaires homogènes intégrables algébriquement, J. Ec. Pol. 4 (1898), 1-122.

[6] Boulanger, A.: Détermination des invariants différentielles fondamentaux attachés au groupe G_{168} de M. Klein, J. Ec. Pol. 6 (1900), 121-146.

[7] Goursat, E.: Extension du problème de Riemann à des fonctions hypergéométriques de deux variables, C. R. XCV (1882), 903-906.

[8] Kaneko, J.: Monodromy group of (F_4), to appear.

[9] Lauricella, G.: Sulle funcioni ipergeometriche a piu variabili, Rend. Circ. Mat. Palermo 7 (1893), 111-158.

[10] Miller, W.: Lie theory and the Lauricella functions F_D, J. Math. Phys. 13 (1972), 1393-1399.

[11] Nakagiri, E.: Monodromy representations of hypergeometric differential equations of two variables, to appear.

[12] Oda, T.: On Schwarzian derivatives in several variables (in Japanese), Kokyoroku of RIMS, Kyoto Univ., 226 (1974).

[13] Okubo, K.: Connection problems for systems of linear differential equations, Springer lecture note 243 (1971), 238-248.

[14] Pham, F.: Introduction a l'étude topologique des singularités de Landau, Gauthiers-Villars, 1967.

[15] Picard, E.: Sur les fonctions de deux variables indépendantes analogues aux fonctions modulaires, Acta Math. 2 (1883), 114-135.

[16] ________: Sur les groupes de certaines équations différentielles linéaires, Bull. Soc. Math. Ser. II, 9 (1885), 202-209.

[17] Sasai, T.: On a monodromy group and irreducibility conditions of a fourth order Fuchsian differential system of Okubo type, J. für Math. 299/300 (1978), 38-50.

[18] Sasaki, T.: On the finiteness of the monodromy group of the system of hypergeometric differential equations (F_D), J. Fac. Sci., Univ. of Tokyo, 24 (1977), 565-573.

[19] Sato, M.: Singular orbits in prehomogeneous vector spaces (written by K. Aomoto in Japanese), Univ. of Tokyo, 1971.

[20] Schwarz, H. A.: Über diejenigen Fälle, inwelchen die Gaussische hypergeometrische Reihe eine algebraische Funktion ihres vierten Elementes darstellt, J. Reine Angew. Math. 75 (1872), 292-335.

[21] Sekiguchi, J.: Zonal spherical functions on the symmetric space SL(3, R)/SO(3) and related topics (in Japanese), Master Thesis in Nagoya Univ., 1976.

[22] Shephard, G. C. and J. A. Todd: Finite unitary reflection groups, Can. J. Math. 6 (1954), 274-304.

[23] Shiga, H.: One attempt to the K3 modular function I, Ann. Sc. Norm. Sup. Pisa 6 (1979), 609-635.

[24] Takano, K.: Monodromy group of the system for Appell's F_4, Funkcial. Ekvac. 23 (1980), 97-122.

[25] Takano, K. and E. Bannai: A global study of Jordan-Pochhammer differential equations, Funkcial. Ekvac. 19 (1976), 85-99.

[26] Terada, T.: Problèmes de Riemann et fonctions automorphes provenant des fonctions hypergeométriques de plusieures variables, J. Math. Kyoto Univ. 13 (1973), 557-578.

[27] Yoshida, M.: Fuchsian differential equations of several complex variables II – Complementary calculations concerning E. Picard's paper, Math. Japonicae 19 (1974), 325-331.

[28] ________: Local theory of Fuchsian systems with certain discrete monodromy groups I, Funkcial. Ekvac. 21 (1978), 105-137.

[29] ________: Local theory of Fuchsian systems with certain discrete monodromy groups II, Funkcial. Ekvac. 21 (1978), 203-221.

[30] Yoshida, M. and S. Hattori: Local theory of Fuchsian systems with certain discrete monodromy groups III, Funkcial. Ekvac. 22 (1979), 1-49.

[31] Sasaki, T.: Picard-Vessiot group of Appell's system of hypergeometric differential equations and infiniteness of monodromy group, Kumamoto J. of Sci. (Math.) 14 (1980), 85-100.

A CLASS OF HYPERBOLIC MANIFOLDS

Nessim Sibony

The purpose of this paper is to study the relation between the existence of bounded plurisubharmonic functions on complex manifolds and hyperbolicity in the sense of Kobayashi.

Let M be a complex manifold of dimension n. We define an infinitesimal pseudometric P_M on the tangent bundle of M. The definition of P_M is quite similar to that of the infinitesimal Carathéodory metric E_M, but instead of bounded holomorphic functions we use a class of bounded p.s.h. functions. The pseudometric P_M is decreasing under holomorphic mappings and is bigger than E_M but smaller than the infinitesimal Kobayashi metric F_M.

We use the pseudometric P to prove that if a complex manifold M possesses a p.s.h. bounded exhaustion function then M is hyperbolic (in the sense of Kobayashi) and taut. It is possible that manifolds with p.s.h. bounded exhaustion function have very few bounded holomorphic functions. In fact we give an example of a Stein domain in $\mathbb{C}^2$ diffeomorphic to $\mathbb{C}^2$ admitting a p.s.h. bounded exhaustion function but on which the bounded holomorphic functions depend only in one variable.

In the last paragraph we give an estimate of the metric P for a class of domains in $\mathbb{C}^n$ containing the strictly pseudoconvex bounded domains.

Recent Developments in Several Complex Variables
0-691-08285-5/81/000357-16 $00.80/0 (cloth)
0-691-08281-2/81/000357-16 $00.80/0 (paperback)

I. *An invariant metric on complex manifolds*

We begin first with a version of the Schwarz lemma in the unit disc in the complex plane $D = \{z/z \in \mathbb{C} \,|z| < 1\}$. We shall denote by Δ the usual Laplacian in $\mathbb{C}$.

PROPOSITION 1. *Let* u *be a function defined on* D *of class* $\mathcal{C}^2$ *in a neighborhood of the origin. Suppose that* $0 \le u \le 1$, $u(0) = 0$ *and that* $\log u$ *is subharmonic in* D. *Then*

a) $u(z) \le |z|^2$ *for* $z \in D$ *with equality at some point different from* 0 *iff* $u(z)$ *is identically equal to* $|z|^2$.

b) $\Delta u(0) \le 4$ *with equality iff* $u(z) = |z|^2$ *for every* $z \in D$.

Proof. Since $u(0) = 0$ and $u \ge 0$, grad $u(0) = 0$. Define in $D\backslash\{0\}$ the function $v(z) = u(z)|z|^{-2}$. Log v is subharmonic in $D\backslash\{0\}$, thus v is subharmonic in $D\backslash\{0\}$. Since v is bounded, there is a subharmonic extension $\tilde{v}$ of v in D. Observe that $\overline{\lim_{z \to e^{i\theta}}} v(z) \le 1$, it follows that $\tilde{v} \le 1$ and therefore $u(z) \le |z|^2$. If at some point $z_0 \ne 0$ $u(z_0) = |z_0|^2$ then $\tilde{v}(z_0) = 1$ which implies that v is identically 1.

To prove b) recall that the segment $[0, 1]$ is not thin at 0. Therefore for every $(\alpha, \beta) \in \mathbb{R}^2$ with $\alpha^2 + \beta^2 = 1$ we have

$$\tilde{v}(0) = \lim_{\substack{t \to 0 \\ t \ne 0}} v(\alpha t, \beta t) .$$

But since u is $\mathcal{C}^2$ in a neighborhood of the origin

$$\lim_{t \to 0} v(\alpha t, \beta t) = \frac{1}{2} \frac{\partial^2 u}{\partial x^2}(0)\alpha^2 + \frac{\partial^2 u}{\partial x \partial y}(0)\alpha\beta + \frac{1}{2} \frac{\partial^2 u}{\partial y^2}(0)\beta^2 .$$

As the preceding limit is independent of (α, β) we get

$$\tilde{v}(0) = \frac{1}{2} \frac{\partial^2 u}{\partial x^2}(0) = \frac{1}{2} \frac{\partial^2 u}{\partial y^2}(0) .$$

Thus we have

$$\Delta u(0) = 4\tilde{v}(0) \leq 4 .$$

If $\Delta u(0) = 4$, then $v(0) = 1$ and v is constant.

REMARK. If u is in $\mathcal{C}^2(D)$, $0 \leq u < 1$, and $\log u$ is subharmonic then the preceding proposition shows that $\Delta u(z) \leq 4(1-|z|^2)^{-2}$ in a neighborhood of $u^{-1}(0)$. But this inequality does not hold in the whole disc.

We fix some notations and definitions. If u is a function of class $\mathcal{C}^2$ in an open set of $\mathbf{C}^n$ and $w = (w_1, \cdots, w_n) \in \mathbf{C}^n$ we denote

$$<\mathcal{L}u(p)\, w, w> = \sum_{i,j=1}^{n} \frac{\partial^2 u(p)}{\partial z_i \partial \bar{z}_j} w_i \bar{w}_j$$

the Levi form of u at p. When u is a function on a neighborhood U of a point p in a complex manifold M and $\xi \in T_pM$, where T_pM is the tangent space to M at p, we define $<\mathcal{L}u(p)\xi, \xi>$ as follows. Let Φ be a chart from a neighborhood of p to an open neighborhood of the origin 0 in $\mathbf{C}^n$ with $\Phi(p) = 0$. Then

$$<\mathcal{L}u(p)\xi, \xi> = <\mathcal{L}(u \circ \Phi^{-1})(0)\, \Phi'(p)\xi, \Phi'(p)\xi> .$$

It is clear that the preceding definition is independent of the chart Φ. In fact if Ψ is another chart then

$$\begin{aligned} <\mathcal{L}(u \circ \Psi^{-1})(0)\Psi'(p)\xi, \Psi'(p)\xi> &= <\mathcal{L}(u \circ \Phi^{-1} \circ \Phi \circ \Psi^{-1})(0)\Psi'(p)\xi, \Psi'(p)\xi> \\ &= <\mathcal{L}(u \circ \Phi^{-1}(0)(\Phi \circ \Psi^{-1})'(0)\Psi'(p)\xi, (\Phi \circ \Psi^{-1})(0)\Psi'(p)\xi> \\ &= <\mathcal{L}(u \circ \Phi^{-1})(0)\Phi'(p)\xi, \Phi'(p)\xi> . \end{aligned}$$

Let $\mathcal{S}_p$ be the family of functions u defined on M satisfying $0 \leq u \leq 1$, $u(p) = 0$, u of class $\mathcal{C}^2$ in a neighborhood of p and $\log u$ p.s.h. on M. We define the pseudometric P_M on the tangent bundle as follows:

$$P_M(p,\xi) = \sup_{u\in\mathcal{S}_p} (<\mathcal{L}u(p)\xi,\xi>)^{1/2}$$

where $<p,\xi>$ is an element of the tangent bundle.

PROPOSITION 2. *The pseudometric* P_M *is locally bounded on the tangent bundle* TM. *Furthermore we have*

a) $P_M(p,\lambda\xi) = |\lambda| P_M(p,\xi)$ *for* $\lambda \in \mathbf{C}$, $<p,\xi> \in TM$.

b) $P_M(p,\xi_1+\xi_2) \le P_M(p,\xi_1) + P_M(p,\xi_2)$.

If f *is a holomorphic map from* M_1 *to* M_2 *then*

$$P_{M_2}(f(p), f'(p)\xi) \le P_{M_1}(p,\xi).$$

Proof. Let Φ be a chart from a neighborhood of p to the ball $B(0,r)$ in $\mathbf{C}^n$. For $u \in \mathcal{S}_p$ the function $\log(u\circ\Phi^{-1})$ is p.s.h. in $B(0,r)$ and $(u\circ\Phi^{-1})(0) = 0$. Therefore by Proposition 1

$$<\mathcal{L}(u\circ\Phi^{-1})(0)w,w> \le \frac{1}{r^2}|w|^2 \quad \text{for} \quad w \in \mathbf{C}^n.$$

Consequently

$$P(p,\xi) \le \frac{1}{r}|\Phi'(p)\xi| < \infty.$$

Properties a), b) are clear from the definitions.

Suppose f is a holomorphic map from a complex manifold M_1 to a complex manifold M_2. It is clear that for $p \in M_1$ $f^*(\mathcal{S}_{f(p)}(M_2)) \subset \mathcal{S}_p(M_1)$.

If $q = f(p)$ and $u \in \mathcal{S}_q(M_2)$ we have

$$<\mathcal{L}u(q)f'(p)\xi, f'(p)\xi> = <\mathcal{L}(u\circ f)(p)\xi,\xi>.$$

It follows then that

$$P_{M_2}(f(p), f'(p)\xi) \le P_{M_1}(p,\xi).$$

We recall here the definitions of the infinitesimal Carathéodory metric E and the infinitesimal Kobayashi metric F. If M, N are two complex manifolds let $H(M,N)$ denote the family of holomorphic maps from M to N. We have then

$$E_M(p,\xi) = \sup\{|f'(p)\xi|/f \in H(M,D)\, \xi \in T_pM\}$$

$$F_M(p,\xi) = \inf\{a/a>0, f \in H(D,M)\, f(0)=p,\ f'(0)=\xi/a\}.$$

We refer to [5], [6], [7] for the study of these metrics.

From Proposition 1 it follows that for the unit disc D we have $P_D(p,\xi) = E_D(p,\xi) = F_D(p,\xi)$. Therefore, since P is distance decreasing for holomorphic maps, it follows from the functorial properties of E and F, [6] Th. 2.2, that for every complex manifold M we have

$$E_M(p,\xi) \le P_M(p,\xi) \le F_M(p,\xi).$$

REMARKS. a) If Ω is a bounded symmetric domain then $P_\Omega = E_\Omega = F_\Omega$. But in general even for strictly pseudoconvex domains with smooth boundary the three pseudometrics are different.

b) The indicatrix I_p of the metric P defined by

$$I_p = \{\xi \mid \xi \in T_pM,\ P_M(p,\xi) < 1\}$$

is a convex disked domain while the indicatrix for the Kobayashi metric could be any disked pseudoconvex domain.

II. *Existence of bounded p.s.h. functions and hyperbolicity*

We first recall some results due to H. Royden. A complex manifold M is hyperbolic at a point p if there is a neighborhood U of p and a positive constant c such that $F_M(q,\xi) \ge c\|\xi\|$ for all $q \in U$, where $\|\ \|$ denotes the norm of the tangent vector ξ with respect to a fixed hermitian metric on M. The manifold M is said to be hyperbolic if the Kobayashi pseudodistance d_M is a distance; recall that d_M is the integrated form of M. In [7] Royden proved that M is hyperbolic at every point iff d_M is a distance.

THEOREM 3. *Let* M *be a complex manifold. Suppose there is a p.s.h. bounded function* u *on* M *which is strictly p.s.h. on a neighborhood of* p, *then* M *is hyperbolic at* p.

Proof. For simplicity we shall suppose that u is of class $\mathcal{C}^2$ in a neighborhood of p. The modifications needed when u is only supposed uppersemicontinuous and

$$<\mathcal{L}u(z)\xi,\xi> \ \geq c\|\xi\|^2 d\sigma$$

as distributions are given at the end of the proof (here $d\sigma$ denotes Lebesgue measure and c is a positive constant).

Let Φ be a chart from a neighborhood U of p to the ball B(0, 2r) such that $\Phi(p) = 0$. We can suppose that the function u is negative and that on B(0, 2r)

$$<\mathcal{L}(u\circ\Phi^{-1})(z)w,w> \ \geq c|w|^2 c > 0\,, \quad \text{for} \quad w \in \mathbf{C}^n\,.$$

Let θ be a smooth nondecreasing function defined on $\mathbf{R}_+$ such that $\theta(x) = x$ for $x \leq \frac{1}{2}$ and $\theta(x) = 1$ for $x \geq \frac{3}{4}$. For $\lambda > 0$ we define

$$\Psi_\lambda(z) = \theta\left(\frac{|\Phi(z)|^2}{r^2}\right)\exp(\lambda u(z)) \qquad \text{for} \quad z \in U$$

$$\Psi_\lambda(z) = \exp(\lambda u(z)) \qquad \text{for} \quad z \in M\setminus U\,.$$

It is clear that on a neighborhood of $\overline{M\setminus U}$ and on $\Phi^{-1}(B(0,r/2))$ the function $\log\Psi_\lambda$ is p.s.h. Consider in B(0, 2r) the function

$$\tilde{\Psi}_\lambda(z) = \Psi_\lambda\circ\Phi^{-1}(z) = \theta\left(\frac{|z|^2}{r^2}\right)\exp(\lambda(u_o\Phi^{-1})(z))\,.$$

Let $h = \log\theta$. An easy computation shows that there is a constant $A > 0$ such that for all $z \in B(0,1)$ and $w \in \mathbf{C}^n$

$$<\mathcal{L}h(|z|^2)w,w> \ \geq -A|w|^2\,.$$

But then

$$<\mathcal{L}\Psi_\lambda(z)w, w> \geq \left(-\frac{A}{r^2} + c\lambda\right)|w|^2 .$$

If we choose $\lambda = \frac{A}{r^2c}$ we find that $\Psi_\lambda \in \mathcal{S}_p$ and

$$<\mathcal{L}\Psi_\lambda(p)\xi, \xi> = \frac{|\Phi'(p)\xi|^2}{r^2} \exp(\lambda u(p)) .$$

Therefore

$$P(p, \xi) \geq \frac{|\Phi'(p)\xi|}{r} \exp\left(\frac{A}{2r^2c} u(p)\right) .$$

Since $F(p, \xi) \geq P(p, \xi)$ and the majorization above is independent of $p \in \Phi^{-1}(B(0, r))$ we see that M is hyperbolic at p and we have even a lower bound for P in terms of geometric objects.

When u is only upper semicontinuous (u.s.c.) we will show that the above estimate still holds for $F_M(p, \xi)$. The proof that $\operatorname{Log}\Psi_\lambda$ is p.s.h. is the same. Let f be in H(D, M) with $f(0) = p$ $f'(0) = \xi/a$. Then $\Phi(f(\zeta)) = \zeta h(\zeta)$ with $h(0) = \Phi'(p)\xi/a$. Since $(\log \Psi_\lambda \circ f)$ is subharmonic on D and $\Psi_\lambda(f(\zeta)) \cdot |\zeta|^{-2}$ is bounded on D, we find, as in Proposition 1, that

$$\frac{1}{r^2} |\Phi'(p)\xi/a|^2 \exp(\lambda u(p)) = \overline{\lim_{\zeta \to 0}} \frac{\Psi_\lambda(f(\zeta))}{|\zeta|^2} \leq 1 .$$

So

$$a \geq \left|\frac{\Phi'(p)\xi}{r}\right| \exp\left(\frac{\lambda}{2} u(p)\right)$$

and

$$F(p, \xi) \geq \frac{|\Phi'(p)\xi|}{r} \exp\left(\frac{A}{2r^2c} u(p)\right) .$$

COROLLARY 4. *Let* M *be a complex manifold with a smooth strictly p.s.h. function* v. *Suppose there exists on* M *a bounded continuous p.s.h. function* u. *Let* K *be a compact connected component of the set* $(u \leq \lambda)$ $\lambda \in \mathbf{R}$. *Then* M *is hyperbolic at every point of* K.

Proof. We can suppose that the function v, which is not necessarily bounded, is positive on M. Let V be a relatively compact neighborhood of K such that $\overline{V}$ is disjoint from the other components of $(u \leq \lambda)$. Let $\mu = \inf_{z \in \partial V} u(z)$. Since u is continuous we have $\mu > \lambda$. Let $c = \operatorname{Max}_{z \in \partial V}(u+v)$ and let h be a convex function defined on $\mathbf{R}$ such that $h(x) = x$ for $x < \lambda + \frac{\mu - \lambda}{2}$ and $h(\mu) > c$. We define the function Ψ by

$$\Psi(z) = \sup((u+v)(z), h(u(z))) \quad \text{for} \quad z \in V$$

$$\Psi(z) = h(u(z)) \quad \text{for} \quad z \in M \setminus V .$$

The function Ψ is p.s.h. bounded and strictly p.s.h. on a neighborhood of K. Thus by Theorem 3, M is hyperbolic at every point of K.

The following corollary partially answers problem 2 of [9] raised by H. Wu. The question in [9] is the following. Suppose a complex manifold M possesses a $\mathcal{C}^\infty$ strictly p.s.h. bounded exhaustion function. Is it complete hyperbolic?

COROLLARY 5. *Let* M *be a complex manifold admitting a smooth strictly p.s.h. function. If there exists on* M *a bounded p.s.h. exhaustion function, then* M *is hyperbolic. In fact* M *is a taut manifold.*

Proof. Let u be the bounded exhaustion function. We suppose that u has its values in $[0,1]$; since u is an exhaustion function, for every $a < 1$ $u^{-1}[0,a]$ is relatively compact in M. We recall that a complex manifold M is taut if for every sequence (f_n) in $H(D,M)$ either there exists a convergent subsequence for the compact open topology or (f_n) is compactly divergent. For results concerning taut manifolds we refer to [5], [6].

The proof of Corollary 4 can be easily adapted to show that M is hyperbolic at every point. Hence by Royden's result [7] M is hyperbolic.

Consider now a sequence $(f_n) \in H(D,M)$. Let (u_n) be the sequence of subharmonic functions in the unit disc defined by $u_n(z) = u(f_n(z))$.

Suppose there is a point $z_o \in D$ such that $\underline{\lim}_{n\to\infty} u_n(z_o) < 1$. Then there is a subsequence f_{n_j} such that $f_{n_j}(z_o)$ is in a compact set $L \subset M$. Since M is hyperbolic the sequence f_n is equicontinuous. Recall that the Kobayashi metric d is distance decreasing under holomorphic maps and that this metric induces the topology of M. Thus as u is an exhaustion function there exist $r > 0$ and $\varepsilon > 0$ such that $\overline{\lim}_j u_{n_j}(z) \le 1-\varepsilon$ for $|z-z_o| < r$.

Consider now the function v defined on D by

$$v(z) = \overline{\lim_j}\, u_{n_j}(z) .$$

The upper semicontinuous regularization v^* of v is subharmonic on D; moreover it is well known that $v = v^*$ almost everywhere in D. Since $v \le 1-\varepsilon$ on a disc it follows by the maximum principle that $v^* < 1$ everywhere. Consequently for every $r < 1$ there exists a constant $c(r) < 1$ such that $v^* \le c(r)$ on the disc $D(0,r)$. But then by Hartog's lemma, [4] Th. 1.6.13, for $r_1 < r$ and $\varepsilon > 0$, such that $c(r)+\varepsilon < 1$, one has $u_{n_j}(z) \le c(r)+\varepsilon$ for $|z| < r_1$ and $j > j_o$. Therefore for every $z \in D$ $f_{n_j}(z)$ is relatively compact, so Ascoli's lemma implies that a subsequence of (f_{n_j}) is convergent in the compact open topology.

The second possibility is that for every $z \in D$ $\underline{\lim}_n u_n(z) = 1$. Since $u < 1$, this implies that for every $z \in D$ $f_n(z)$ tends to infinity on M. Let K be compact in D and L a compact in M. Choose $c < 1$ such that $L \subset (u \le c)$. Let $\delta > 0$ be such that $L_1 = \{z | z \in M \ d(z,L) \le \delta\}$ is compact. Then there is $c_1 < 1$ with $L_1 \subset (u(z) \le c_1)$. Choose $\eta > 0$ so that if z, z' are in K and $|z-z'| \le \eta$ then $d(f_n(z), f_n(z')) \le \delta$ for every n. As K is compact there is a finite covering of K by discs $D(p_j, \eta)$ $1 \le j \le s$. Choose n_o so that $u_n(p_j) > c_1$ for $n \ge n_o$ and $1 \le j \le s$.

This implies that for $n \geq n_o$ and $z \in K$, $f_n(z) \notin L$. In fact if $z \in K$ there is j such that $z \in D(p_j, \eta)$ and $d(f_n(z), f_n(p_j)) \leq \delta$. Thus if $f_n(z) \in L$ then $f_n(p_j) \in L_1$. This contradicts the choice of n_o, consequently the sequence f_n is compactly divergent.

REMARK. Corollary 5 and the pseudometric P give some information about Theorem F of R. Greene and H. Wu [3]. In particular their inequality (6.3), p. 100 can be read as a lower bound estimate for the Kobayashi metric.

III. *An example*

We give an example of a Stein domain Ω in $\mathbb{C}^2$ diffeomorphic to $\mathbb{C}^2$ which is complete hyperbolic and which admits a bounded p.s.h. exhaustion function but such that the bounded holomorphic functions depend only on one variable.

This example gives a negative answer to question 3′ in [2] raised by R. E. Greene and H. Wu. The construction of Ω is a modification of an idea in [8].

To prove the complete hyperbolicity of Ω we shall use the following result due to P. Kiernan.

THEOREM (P. Kiernan). *Let* E, M *be two complex manifolds and* π *a holomorphic map from* E *onto* M. *Suppose that* M *is complete hyperbolic and admits an open cover* (U_i) *such that, for every* i, $\pi^{-1}(U_i)$ *is complete hyperbolic. Then* E *is complete hyperbolic.*

The result is still true with hyperbolic instead of complete hyperbolic. In fact Kiernan proves the theorem for holomorphic fibre bundles but his proof [5] can be easily adapted to give the preceding result.

We now construct the announced example. Let (a_p) be a discrete sequence in the unit disc D such that every point in ∂D is a nontangential limit of a subsequence of (a_p). Consider the negative subharmonic function u defined on D by

$$u(z) = \sum_{p=1}^{\infty} \frac{1}{n_p} \sup\left(\log\left|\frac{z-a_p}{2}\right|, -pm_p\right)$$

where n_p, m_p are sequences of positive integers such that

$$\sum_{p=1}^{\infty} \frac{1}{n_p} \log\left|\frac{a_p}{2}\right| > -\infty, \quad m_p \geq n_p$$

and the discs

$$D_p = \{z/z \in \mathbf{C}\,|z-a_p| \leq 3\exp(-pm_p)\}$$

are pairwise disjoint and contained in the unit disc.

The function u is continuous on D. In fact in D_p, we have

$$u = \frac{1}{n_p} \sup\left(\log\frac{|z-a_p|}{2}, -pm_p\right) + u_p \tag{1}$$

where u_p is harmonic. The function u is harmonic outside of $\bigcup_{p=1}^{\infty} D'_p$ where D'_p is defined by

$$D'_p = \{z/z \in \mathbf{C}\,|z-a_p| \leq 2\exp(-pm_p)\}.$$

Consider now the domain Ω in $\mathbf{C}^2$

$$\Omega = \{(z,w)/(z,w) \in \mathbf{C}^2\,|z| < 1, |w| < \exp(-u(z))\}.$$

Clearly Ω is a Stein domain diffeomorphic to $\mathbf{C}^2$ and the function

$$v(z,w) = \sup\,[|z|, |w|\exp(u(z))]$$

is a p.s.h. continuous exhaustion function with $0 \leq v < 1$.

Moreover let

$$f(z,w) = \sum_{n \geq 0} h_n(z)\, w^n$$

be a bounded holomorphic function on Ω with uniform norm $\|f\|_\infty \leq 1$. Cauchy's inequalities imply that for every n we have

$$|h_n(z)| \leq \exp(nu(z)) \leq 1 .$$

Therefore we get

$$|h_n(a_p)| \leq \exp(nu(a_p)) ,$$

and since we have

$$u(a_p) \leq -p \frac{m_p}{n_p} \leq -p ,$$

we conclude

$$(2) \qquad |h_n(a_p)| \leq \exp(-np) .$$

Using Fatou's theorem and (2) with the fact that every point in ∂D is the nontangential limit of a subsequence of (a_p) we see that for every $n \geq 1$ the function h_n is identically zero. Thus $f(z, w) = h_o(z)$.

We prove now that Ω is complete hyperbolic. Let π be the restriction to Ω of the projection on the first coordinate $\pi(z, w) = z$. We shall prove that for every $z \in D$ there is a disc $D(z, r) \subset D$ such that $\pi^{-1}(D(z,r))$ is Carathéodory complete. To prove this, it is enough to show that for every point (z_o, w_o) in the boundary of $\pi^{-1}(D(z, r))$ there exists a holomorphic function f continuous on the closure in $\mathbf{C}^2$ of $\pi^{-1}(D(z, r))$ such that $\|f\|_\infty \leq 1$, $f(z, 0) = 0$ and $|f(z_o, w_o)| = 1$.

If $z \in D_p$, let $D(z, r) = D_p$; using relation (1) we see that $\pi^{-1}(D_p)$ is given by $\pi^{-1}(D_p) = \{(z, w)/z \in D_p \cdot |w|^{n_p} \exp(n_p u_p(z)) \cdot \left[\sup\left(\left|\frac{z-a_p}{2}\right|, \exp(-pn_p m_p)\right)\right] < 1\}$. Let $(z_o, w_o) \in \partial\pi^{-1}(D_p)$. If $z_o \in \partial D_p$ the choice of a function f is obvious. So we can suppose z_o interior to D_p. Let $\tilde{u}_p$ be a holomorphic function in D_p with $\operatorname{Re} \tilde{u}_p = u_p$. If $z_o \in D'_p$, then the function

$$f(z, w) = w^{n_p} \exp(-pn_p m_p) \exp(n_p \tilde{u}_p(z))$$

satisfies the required conditions. If $z_o \in \overset{\circ}{D}_p \setminus D'_p$, then we consider the function

$$f(z,w) = w^{n_p}\left(\frac{z-a_p}{2}\right) \exp(n_p \tilde{u}_p(z)) .$$

Outside $\bigcup_p D'_p$ u is harmonic, so if $D(z,r) \subset D \setminus \bigcup_p D'_p$ then $u = \operatorname{Re} \tilde{u}$ with $\tilde{u}$ holomorphic in $D(z,r)$. The function

$$f(z,w) = w \exp(\tilde{u}(z))$$

satisfies the required conditions. So far we have proved that for every $z \in D$, there exists $D(z,r) \subset D$ such that $\pi^{-1}(D(z,r))$ is Carathéodory complete hence complete hyperbolic. It then suffices to apply the result of Kiernan to see that Ω is complete hyperbolic.

IV. *Estimates for the Kobayashi metric of a domain in* C^n

Let Ω be a domain in C^n. For every $(p,w) \in \Omega \times C^n$ define

$$\delta(p,w) = \sup\{r / p + rwD \subset \Omega\} .$$

Clearly we have

$$P(p,w) \le F(p,w) \le \delta^{-1}(p,w) .$$

It suffices to consider the function f in $H(D,\Omega)$ defined by $f(z) = p + rwz$. Then $f(0) = p$, $f'(0) = rw$ and consequently $F(p,w) \le 1/r$ for every $r < \delta(p,w)$.

The problem is in fact to give a lower estimate for F. We are going to estimate the pseudometric P and use the fact that $P(p,w) \le F(p,w)$.

PROPOSITION 6. *Let Ω be a domain in C^n, not necessarily bounded. Suppose there is a negative p.s.h. function u on Ω, such that*

$$<\mathcal{L}u(z)w, w> \ \ge c|w|^2 \quad \textit{for} \quad w \in C^n$$

where c is a positive constant. Then there is a constant A, independent *of* u, *such that*

$$F(p, w) \geq P(p, w) \geq \left(\frac{C}{A}\right)^{1/2} \frac{|w|}{|u(p)|^{1/2}}.$$

Proof. Consider the function Ψ defined by

$$\Psi(z) = \theta\left(\left|\frac{z-p}{r}\right|^2\right) \exp(\lambda u(z)).$$

As before θ is a smooth nondecreasing function on $\mathbf{R}_+$ such that $\theta(x) = x$ for $x \leq \frac{1}{2}$ and $\theta(x) = 1$ for $x \geq \frac{3}{4}$. Let A_1 be an upperbound for the second derivative of $\log \theta$. An easy calculation shows that

$$<\mathcal{L} \log \Psi(z) w, w> \geq -A_1 \frac{|w|^2}{r^2} + \lambda c |w|^2.$$

Let $\lambda = \frac{1}{|u(p)|}$ and $r = \left(\frac{A_1}{c}|u(p)|\right)^{1/2}$. Then $\log \Psi$ is p.s.h. and we have

$$<\mathcal{L}\Psi(p) w, w> = \frac{|w|^2}{r^2} \exp(\lambda u(p)) = |w|^2 \frac{c \exp(-1)}{|u(p)| A_1}.$$

Consequently we get

$$F(p, w) \geq P(p, w) \geq \left(\frac{c}{A_1 e}\right)^{1/2} \frac{1}{|u(p)|^{1/2}} |w|.$$

We have used implicitly that u is of class $\mathcal{C}^2$. The modifications needed in the general case are easy. The preceding proposition is of interest when u goes to zero on the boundary of Ω.

The asymptotic behavior of the infinitesimal Kobayashi metric on strictly pseudoconvex domains is well known [1]. We mention nonetheless the following global result which can be refined to give the known asymptotic behavior in the case of a bounded strictly pseudoconvex domain with smooth boundary.

PROPOSITION 7. *Let* Ω *be a domain in* $\mathbb{C}^n$ *and* r *a negative p.s.h. function on* Ω *of class* $\mathcal{C}^2$. *Suppose that the derivatives of* r *up to second order are bounded and that there is a positive constant* c *such that*

$$<\mathcal{L}r(z)w, w> \geq C|w|^2 \quad \textit{for} \quad w \in \mathbb{C}^n .$$

Then there is a constant K *depending only on the* $\mathcal{C}^2$ *norm of* r *such that*

$$F(p, w) \geq P(p, w) \geq K\left[c^2 \frac{|<\partial r(p), w>|^2}{|r(p)|^2} + c\frac{|w|^2}{|r(p)|}\right]^{1/2} .$$

Proof. It is enough to compute the Levi form at p of the function Ψ defined by

$$\Psi(z) = \theta\left[\frac{|<\partial r(p), z-p>|^2}{\alpha^2} + \frac{|z-p|^2}{\beta^2}\right] \exp(\lambda u(z) + \mu r(z))$$

where

$$u(z) = -(-r)^{1/2}(z), \quad \alpha = \frac{|r(p)|}{c}, \quad \beta = \left(\frac{|r(p)|}{c}\right)^{1/2},$$

$$\lambda = \frac{M}{|r|^{1/2}(p)}, \quad \mu = \frac{M}{|r(p)|}$$

the constant M is chosen in term of the $\mathcal{C}^2$ norm of r to have $\log \Psi$ p.s.h. The function θ is the same as the one used in Proposition 6. We have that

$$<\mathcal{L}\Psi(p)w, w> \geq \exp(-M)\left[c^2 \frac{|<\partial r(p), w>|^2}{|r(p)|^2} + c\frac{|w|^2}{|r(p)|}\right].$$

The result follows since $\Psi \in \mathcal{S}_p(\Omega)$.

Note added in proof. J. Fornaess and the author have constructed a bounded domain Ω in $\mathbb{C}^2$ which possesses a smooth strictly p.s.h. function and which is not complete hyperbolic. This answers the problem of H. Wu considered in paragraph II.

UNIVERSITÉ DE PARIS SUD

REFERENCES

[1] Graham, I. Boundary behavior of the Carathéodory and Kobayashi metrics on strictly pseudo-convex domains in C^n. Trans. Amer. Math. Soc. 207 (1975), 219-240.

[2] Greene, R. E. and Wu, H. Analysis on noncompact Kähler manifolds. (Proc. Symp. Pure Math.), vol. 30, A.M.S. Providence (1977), 69-100.

[3] ———. Function theory on manifolds which possesses a pole. Lecture Notes in Math. 699, Springer-Verlag, Berlin (1979).

[4] Hörmander, L. An introduction to complex analysis in several variables. Van Nostrand (1966).

[5] Kobayashi, S. Hyperbolic manifolds and holomorphic mappings. New York, M. Dekker Inc. (1970).

[6] ———. Intrinsic distances, measures and geometric function theory. Bull. Amer. Math. Soc. 82 (1976), 357-416.

[7] Royden, H. L. Remarks on the Kobayashi metric. Lecture Notes in Math. 185, Berlin, Springer-Verlag (1977), 125-137.

[8] Sibony, N. Prolongement des fonctions holomorphes bornées et métrique de Carathéodory. Inv. Math. 29 (1975), 205-230.

[9] Wu, H. Some open problems in the study of noncompact Kähler manifolds in geometric theory of several complex variables. Kyoto (1978).

INTERPOLATION MANIFOLDS

Edgar Lee Stout[1]

I. *Introduction*

In this paper we discuss some geometric questions concerning interpolation manifolds in the boundary of a strongly pseudoconvex domain in C^N. In Section II we begin by recalling some known results on these manifolds. We then take up the special case of interpolation manifolds in the boundary of the ball in B_N with special attention to their geometric properties related to the projection $\pi: \partial B_N \to P^{N-1}$. We obtain a characterization of the submanifolds of P^{N-1} that are projections of interpolation manifolds under π. In particular, every $\mathcal{C}^1$ curve is. In Section III of the paper we obtain a result which implies that every compact totally real manifold can be realized as an interpolation manifold. We also show that each compact totally real manifold can be realized as the intersection of the closures of two disjoint strongly pseudoconvex domains.

In the course of this work I have profited from conversations with Martin Bendersky, James King, and Vivian Klein; to each of them I express my thanks.

[1]This material is based in part upon work supported by the National Science Foundation under Grant MCS78-02139.

Recent Developments in Several Complex Variables
0-691-08285-5/81/000373-19 $00.95/0 (cloth)
0-691-08281-2/81/000373-19 $00.95/0 (paperback)
For copying information, see copyright page.

II. *Interpolation manifolds*

Given a smoothly bounded domain D in $\mathbf{C}^N$ and a point $p \in \partial D$, the tangent space $T_p(\partial D)$ contains a maximal complex subspace $T_p^{\mathbf{C}}(\partial D)$; it has complex dimension $N-1$. If D is strongly pseudoconvex and if $\Sigma \subset \partial D$ is a submanifold, we say that Σ is an *interpolation manifold* provided $T_p(\Sigma) \subset T_p^{\mathbf{C}}(\partial D)$ for each $p \in \Sigma$. Interpolation manifolds have been studied in recent years by several authors [4, 5, 6, 7, 10, 12, 17, 19]. The name *interpolation manifold* is chosen because various interpolation theorems are known on interpolation manifolds. Thus, e.g. if Σ is a $\mathcal{C}^1$ interpolation manifold contained in ∂D, D strongly pseudoconvex, then every compact subset of Σ is a peak interpolation set for the algebra $A(D)$ of functions continuous on $\bar{D}$, holomorphic on D [19]. There are also $\mathcal{C}^\infty$ and analytic versions of this result [4, 10, 12].

The geometry of interpolation manifolds can be studied with some success. The condition that Σ be an interpolation manifold may be formulated analytically as follows. Any vector tangent to Σ at p must be orthogonal in the sense of the Hermitian inner product on $\mathbf{C}^N$ to the complex line through p containing the real normal to ∂D. If D has the strongly plurisubharmonic defining function Q so that $D = \{Q<0\}$ with $dQ \neq 0$ on ∂D, then the condition that $T_p(\Sigma) \subset T_p^{\mathbf{C}}(\partial D)$ is simply the condition that the differential form $\eta = d^c Q$ vanish on Σ. More carefully put, if $\iota : \Sigma \to \partial D$ is the inclusion, then the condition is that $\iota^*\eta = 0$. [Note: $d^c = i(\bar{\partial}-\partial)$, so if we take $z_j = x_{2j-1} + ix_{2j}$, η is the form

$$-\sum_{j=1}^{N}\left(\frac{\partial Q}{\partial x_{2j}}\, dx_{2j-1} - \frac{\partial Q}{\partial x_{2j-1}}\, dx_{2j}\right).]$$

The condition that $\iota^*\eta = 0$ implies, of course, that $0 = d\iota^*\eta = \iota^* dd^c Q = \iota^*(2i\partial\bar{\partial}Q)$, *i.e.*, that the Levi form $\partial\bar{\partial}Q$ vanishes on Σ. This implies that Σ is totally real: If $X \in T_p(\Sigma)$ is such that $iX \in T_p(\Sigma)$, then $2i\partial\bar{\partial}Q(X, iX) = 0$ since $d\iota^*\eta$ is the zero form, but also, by strong plurisubharmonicity, $2i\partial\bar{\partial}Q(X, iX) > 0$ unless $X = 0$. Thus $T_p\Sigma \cap iT_p\Sigma = 0$, so Σ is totally real. We see then that

$\dim \Sigma \leq N-1$ because $T_p(\Sigma)$ is a totally real subspace of $T_p^{\mathbf{C}}(\partial D)$ and so has dimension no more than $N-1$ since $\dim_{\mathbf{C}} T_p^{\mathbf{C}}(\partial D) = N-1$.

Notice that the form η on ∂D is a *contact form* in that $\eta \wedge (d\eta)^{N-1}$ vanishes at no point of ∂D. To see this, fix $p \in \partial D$ and choose coordinates $z_1, \cdots, z_N$ with respect to which p is the origin, $T_0(\partial D) = \{x_{2N} = 0\}$. Then $T_0^{\mathbf{C}}(\partial D) = \{z_N = 0\}$. At 0, $\eta = \frac{\partial Q}{\partial x_{2N}} dx_{2N-1}$. Since the restriction of $(d\eta)^{N-1} = (2i\partial\bar{\partial}Q)^{N-1}$ to $\mathbf{C}^{N-1} = \{z_N = 0\}$ gives a volume form on $\mathbf{C}^{N-1}$ because Q is strongly plurisubharmonic, we see that $\eta \wedge (d\eta)^{N-1}$ gives a nonzero form on $T_0(\partial D)$ as asserted. Thus, ∂D is a contact manifold with contact form η. For the theory of contact manifolds, see [1, 2].

Since the form η is a contact form, a theorem of Darboux [1, p. 362; 15, Appendix] says that given a point $p \in \partial D$, there are local coordinates $s_1, \cdots, s_{N-1}, t_1, \cdots, t_{N-1}, u$, in a neighborhood U of p in ∂D so that in terms of these coordinates, we have

$$\eta = s_1 dt_1 + \cdots + s_N dt_N + du .$$

If we let Σ be the submanifold of U defined by the equations $s_1 = \cdots = s_{N-1} = u = 0$, then Σ is an integral manifold for the form η; we see that ∂D contains (not necessarily closed) interpolation manifolds of dimension $N-1$. It seems to be a difficult question as to whether in general ∂D must contain compact interpolation manifolds of dimension $N-1$. (It is evident that there are compact interpolation manifolds of dimension $N-2$: Any compact submanifold of the Σ just constructed is one.) Somewhat more is true [5]: Given an interpolation manifold $\Sigma \subset \partial D$, and given a point $p \in \Sigma$, there is an interpolation manifold Σ' in a neighborhood U of p such that $\Sigma \cap U \subset \Sigma'$ and $\dim \Sigma = N-1$; locally at least, an interpolation manifold extends to one of maximal dimension.

It is of particular interest to examine the case that D is the unit ball $B_N = \{z \in \mathbf{C}^N : |z| < 1\}$. We may take $Q(z) = |z|^2 - 1$. The form η is

$i(\bar{\partial}-\partial)(|z|^2-1) = i\left(\sum_{j=1} z_j d\bar{z}_j - \bar{z}_j dz_j\right)$ so for example, the manifold in ∂B_N on which $z = \bar{z}$, *i.e.*, the (N–1)-sphere $\Sigma = \mathbf{R}^N \cap \partial B_N$, is seen to be an interpolation manifold.

In the setting of the sphere, it is natural to consider the projection $\pi: \mathbf{C}^N \setminus \{0\} \to \mathbf{P}^{N-1}$. A submanifold Σ of ∂B_N is an interpolation manifold exactly when π acts as a local isometry of Σ into $\mathbf{P}^{N-1}$ with the Fubini-Study metric (and $\mathbf{C}^N$ with the usual Hermitian metric). This is so simply because π_* carries $T_p^{\mathbf{C}}(\partial B_N)$ isometrically onto $T_{\pi(p)}(\mathbf{P}^{N-1})$ by the definition of the Fubini-Study metric and because π_* annihilates the orthogonal complement of $T_p^{\mathbf{C}}(\partial B_N)$ in $T_p(\partial B_N)$ (or in $T_p\mathbf{C}^N$).

We ask which submanifolds of $\mathbf{P}^{N-1}$ are π-images of interpolation manifolds in ∂B_N. One necessary condition is immediate: Since π_* acts as a $\mathbf{C}$-isomorphism from $T_p^{\mathbf{C}}(\partial D)$ onto $T_{\pi(p)}(\mathbf{P}^{N-1})$, we find that if $\Sigma \subset \partial B_N$ is an interpolation manifold, then necessarily $\pi(\Sigma)$ is a totally real immersed submanifold of $\mathbf{P}^{N-1}$.

We can say more. Denote by Ω the fundamental form of the Fubini-Study metric on $\mathbf{P}^{N-1}$ so that in homogeneous coordinates $(\zeta_1 : \cdots : \zeta_N)$ on $\mathbf{P}^{N-1}$

$$\Omega = \frac{i}{2} |\zeta|^{-4} \{ |\zeta|^2 \sum_{j=1}^{N} d\zeta_j \wedge d\bar{\zeta}_j - \sum_{j,k=1}^{N} \zeta_k \bar{\zeta}_j d\zeta_j \wedge d\bar{\zeta}_k \} .$$

This form is invariant under the action of the unitary group on $\mathbf{P}^{N-1}$, and if we consider the action of $\pi^*\Omega$ on $T^{\mathbf{C}}_{(0,\cdots,0,1)}(\partial B_N)$, we see that it agrees there with the action of the form $\frac{1}{4} dd^c(|z|^2-1)$. We know that $dd^c(|z|^2-1)$ restricts to the zero form on an interpolation manifold, so we find that if $\Sigma \subset \mathbf{P}^{N-1}$ is the π-image of an interpolation manifold, then Ω restricts to Σ as the zero form. This obviously necessary condition is also sufficient.

THEOREM. *If* $\Sigma \subset \mathbf{P}^{N-1}$ is an immersed $\mathcal{C}^1$ *submanifold to which* Ω *restricts as the zero form, then there is an immersed submanifold* $\Sigma' \subset \partial B_N$ *that is an interpolation manifold and that satisfies* $\pi(\Sigma') = \Sigma$.

Precisely, the hypothesis is that for some $\mathcal{C}^1$ manifold $\tilde{\Sigma}$, there is a regular surjective $\mathcal{C}^1$ map $\mu : \tilde{\Sigma} \to \Sigma$ with $\mu^*\Omega = 0$ and $\dim \tilde{\Sigma} = \dim \Sigma$.

Proof. With μ and $\tilde{\Sigma}$ as above, there exists a regular $\mathcal{C}^1$ map ν of $\mathbf{R}^k$, $k = \dim \Sigma$, onto $\tilde{\Sigma}$.[2] Thus, if we set $\phi = \nu \circ \mu$, we have a regular map from $\mathbf{R}^k$ onto Σ with $\phi^*\Omega = 0$. We shall show that there exists a regular map $\psi : \mathbf{R}^k \to \partial B_N$ with $\pi \circ \psi = \phi$ and with $\psi(\mathbf{R}^k)$ an immersed interpolation manifold in that $\psi^* d^c(|z|^2 - 1) = 0$. Moreover, ψ is uniquely determined if we specify the value of $\psi(0) \in \pi^{-1}(\phi(0))$.

In homogeneous coordinates, ϕ is given by

$$\phi(x) = (\phi_1(x) : \cdots : \phi_N(x)) \qquad x \in \mathbf{R}^k .$$

We seek $\psi = (\psi_1, \cdots, \psi_N)$ with $\psi_j/\psi_k = \phi_j/\phi_k$, with $\sum_{j=1}^{N} \psi_j \bar{\psi}_j = 1$ and

$$\sum_{j=1}^{N} \psi_j \, d\bar{\psi}_j - \bar{\psi}_j \, d\psi_j = 0 . \tag{1}$$

To construct ψ, introduce the form θ given by

$$\theta = \frac{\langle d\phi, \phi \rangle}{\langle \phi, \phi \rangle} = \left(\sum_{j=1}^{N} \phi_j \bar{\phi}_j \right)^{-1} \left(\sum_{j=1} \bar{\phi}_j \, d\phi_j \right) ,$$

a 1-form on $\mathbf{R}^k$. The form θ is closed:

$$\begin{aligned} d\theta &= (\langle \phi,\phi \rangle)^{-2} \left\{ \langle \phi,\phi \rangle \left(\sum_{j=1}^{N} d\bar{\phi}_j \wedge d\phi_j \right) - \left(\sum_{j=1}^{N} \bar{\phi}_j d\phi_j + \phi_j d\bar{\phi}_j \right) \wedge \left(\sum_{j=1}^{N} \bar{\phi}_j d\phi_j \right) \right\} \\ &= (\langle \phi,\phi \rangle)^{-2} \left\{ \langle \phi,\phi \rangle \sum_{j=1}^{N} d\bar{\phi}_j \wedge d\phi_j - \sum_{j,k=1}^{N} \phi_j \bar{\phi}_k d\bar{\phi}_j \wedge d\phi_k \right\} \\ &= \phi^*(-2i\Omega) = 0 . \end{aligned}$$

---

[2] One way to get ν, though probably not the easiest, is this: The $\mathcal{C}^1$ manifold $\tilde{\Sigma}$ admits an analytic structure in the sense that there exists an analytic manifold $\mathfrak{M}$ and a $\mathcal{C}^1$ diffeomorphism $\sigma : \mathfrak{M} \to \tilde{\Sigma}$. See [13] and the references cited there. By [8], $\mathfrak{M}$ is the image of a Euclidean space under a finite, regular analytic map.

Thus we may define a function F on R^k by

$$F(x) = C \exp\left\{-\int_0^x \theta\right\}$$

where the constant C is chosen so that $F(0) = \left(\sum_{j=1}^N |\phi_j(0)|^2\right)^{-1}$. Define ψ_j, $1 \leq j \leq N$, by

$$\psi_j = \phi_j F .$$

The map $\psi = (\psi_1, \cdots, \psi_N)$ carries R^k into $C^N \backslash \{0\}$, and it satisfies $\pi \circ \psi = \phi$. Since ϕ is regular, so is ψ. We have that $\psi(R^k) \subset \partial B_N$, for

$$\sum_{j=1}^N \psi_j \bar{\psi}_j = \sum_{j=1}^N |C|^2 \phi_j \bar{\phi}_j \exp\left\{-\int_0^x (\theta + \bar{\theta})\right\}$$

$$= |C|^2 \left(\sum_{j=1}^N \phi_j \bar{\phi}_j\right) \exp\left\{-\int_0^x \frac{<d\phi,\phi> + <\phi,d\phi>}{<\phi,\phi>}\right\}$$

$$= |C|^2 \left(\sum_{j=1}^N \phi_j \bar{\phi}_j\right) \exp\left(-\log <\phi,\phi>\Big|_0^x\right)$$

$$= 1 .$$

We also have $\sum_{j=1}^N \bar{\psi}_j d\psi_j = 0$. This is so, for

$$d\psi_j = F d\phi_j + \phi_j dF = F d\phi_j - \phi_j F \theta$$

so

$$\sum_{j=1}^N \bar{\psi}_j d\psi_j = \bar{F} F \sum_{j=1}^N \left(\bar{\phi}_j d\phi_j - \bar{\phi}_j \phi_j \frac{<d\phi,\phi>}{<\phi,\phi>}\right) = 0 .$$

It follows that the map ψ satisfies 1), so we have proved the existence portion of the theorem.

The uniqueness assertion is immediate, for if $\gamma: \mathbf{R}^k \to \partial B_N$ satisfies $\pi \circ \gamma = \pi \circ \psi$ and $\gamma(0) = \psi(0)$, then $\gamma = g\psi$ for some $g: \mathbf{R}^k \to \mathbf{C}$ with $|g| = 1$ and $g(0) = 1$. If, in addition, $\gamma(\mathbf{R}^k)$ is an interpolation manifold, we have

$$0 = \sum_{j=1}^{N} g\psi_j d(\bar{g}\bar{\psi}_j) - \bar{g}\bar{\psi}_j d(g\psi_j)$$

$$= g\bar{g} \sum_{j=1}^{N} (\psi_j d\bar{\psi}_j - \bar{\psi}_j d\psi_j) + \left(\sum_{j=1}^{N} \psi_j \bar{\psi}_j\right) (g d\bar{g} - \bar{g} d g) .$$

Since $\psi(\mathbf{R}^k)$ is an interpolation manifold, the first term is zero, and since $|g| = 1$, we have $0 = \bar{g} dg + g d\bar{g}$ so that

$$0 = -2|\psi|^2 \bar{g} dg$$

whence $dg = 0$. Thus, g is identically 1, and we have the uniqueness assertion.

We can give an example of this proposition. Define $\phi: \mathbf{R}^{N-1} \to \mathbf{P}^{N-1}$ by

$$\phi(x) = (e^{ix_1} : \cdots : e^{ix_{N-1}} : 1) .$$

We then have $|\phi|^2 = \langle \phi, \phi \rangle = N$, we have $d\phi_j \wedge d\bar{\phi}_j = 0$ for all j, and we have $\sum_{j,k=1}^{N} \bar{\phi}_j \phi_k d\phi_j \wedge d\bar{\phi}_k = \sum_{j,k=1}^{N-1} dx_j \wedge dx_k = 0$. Thus, $\phi^* \Omega = 0$. Accordingly, ϕ lifts to a map $\psi: \mathbf{R}^{N-1} \to \partial B_N$ with $\psi(\mathbf{R}^{N-1})$ an immersed interpolation manifold. By the construction given, we have $\psi = F\phi$ with the function F determined by the condition that $d \log F = -\theta$. In this example,

$$\theta = \frac{<d\phi,\phi>}{<\phi,\phi>} = \frac{i}{N}\sum_{j=1}^{N-1} dx_j, \text{ so } F(X) = C\exp\left\{-\tfrac{i}{N}(x_1+\cdots+x_{N-1})\right\} \text{ with } C = \frac{1}{\sqrt{N}}.$$

This yields

$$\psi_j(x) = \frac{1}{\sqrt{N}}\exp\left\{-\frac{i}{N}(x_1+\cdots x_{j-1}+(1-N)x_j+x_{j+1}+\cdots+x_{n-1}\right\}$$

$j = 1, 2, \cdots, N-1$, and

$$\psi_N = \frac{1}{\sqrt{N}}\exp\left\{-\frac{i}{N}(x_1+\cdots+x_{N-1})\right\}.$$

The map $\psi = (\psi_1, \cdots, \psi_N)$ satisfies

$$\psi(x + r\varepsilon_j) = \psi(x)$$

if $\varepsilon_j = (0, \cdots, 0, 2\pi N, 0, \cdots, 0)$ with $2\pi N$ in the jth position and if r is any integer. Thus $\Sigma = \psi(R^{N-1})$ is a compact immersed interpolation manifold in ∂B_N that maps by π onto the torus $T = \phi(R^{N-1}) \subset P^{N-1}$.

If we define $g \in \mathcal{O}(C^N)$ by

$$g(z) = z_1 \cdots z_N,$$

then

$$g(F(x)) = N^{-N/2}$$

for all x, so if V is the variety on which g assumes the value $N^{-N/2}$, $\Sigma \subset V \cap \partial B_N = V \cap \bar{B}_N$. ($V \cap \partial B_N$ is an *analytic interpolation* manifold in the sense of [4].) The restriction of g to the torus

$$T^N = \{z \in C^N : |z_1| = \cdots = |z_N| = 1\}$$

is a character, so $T^N \cap g^{-1}(1)$ is a (connected) closed subgroup of T^N and so is a smooth manifold of real dimension $N-1$. Consequently, $S = g^{-1}(N^{-1/2N}) \cap \{z \in C^N : |z_1| = \cdots = |z_n| = N^{-1/2}\}$ is a smooth manifold, and as Σ is an immersed manifold of dimension $(N-1)$ contained in S, Σ is *itself* smooth. Thus Σ is an $(N-1)$-dimensional torus.

Let us note a very special case of the proposition: If Σ is one-dimensional *i.e.*, a curve, then of course Ω restricts to 0 on Σ, so every $\mathcal{C}^1$ curve in P^{N-1} is the π-projection of an interpolation curve in ∂B_n.[3]

We conclude this section with a question: Which compact (N–1)-dimensional manifolds can occur in ∂B_N as interpolation manifolds? We have tori – the Σ just constructed – and spheres – $R^N \cap \partial B_N$ – but it seems likely that there are topological restrictions. This is related, of course, to the question of which (N–1)-dimensional manifolds can be realized as to totally real submanifolds of P^{N-1}. Tori and real projective spaces can be, *e.g.*, $\pi(\Sigma)$, Σ the torus given above, and $\pi(R^N \cap \partial B_N)$ respectively. In particular, when $N = 3$, are there any others? (See [21] and [16].)

III. *Totally real manifolds*

We shall now prove a result that exhibits every compact totally real submanifold as an interpolation manifold. We regard C^N as contained in C^{N+1} in the usual way.

THEOREM. *Let Σ be a totally real submanifold of class $\mathcal{C}^k$, $k \geq 1$, of an open set in C^N, and let $E \subset \Sigma$ be compact.*

(a) *There exist strongly pseudoconvex domains D_+ and D_- in C^{N+1} and a compact neighborhood W of E in Σ such that ∂D_+ and ∂D_- are of class $\mathcal{C}^{k+1}$, such that $\bar{D}_+ \cap \bar{D}_- = W$ and such that for all $p \in E$, $T_p(\Sigma) \subset T_p^C(\partial D_+) = T_p^C(\partial D_-)$.*

(b) *If $\dim \Sigma \leq \frac{1}{3}(2N-1)$ and $k \geq 3$, then D_+ and D_- can be chosen to the lie in C^N though with ∂D_+ and ∂D_- of class $\mathcal{C}^{k-1}$.*

(c) *If $\dim \Sigma \leq N-1$ and $k \geq 3$, there exist strongly pseudoconvex domains D_+ and D_- in C^N and a compact neighborhood W of E such that ∂D_+ and ∂D_- are of class $\mathcal{C}^{k-1}$ and $\bar{D}_+ \cap \bar{D}_- = W$.*

---

[3] Walter Rudin has informed me that he too has obtained this result for curves.

REMARKS. (1) Note carefully that the conclusion of (b) includes $T_p \subset T_p^{\mathbf{C}}(\partial D_+) = T_p^{\mathbf{C}}(\partial D_-)$ for all p, but the conclusion of (c) does not.

(2) If Σ is compact, we may take $W = \Sigma$; (a) or (c) then exhibits each compact totally real submanifold as the intersection of the closures of two disjoint strongly pseudoconvex domains.

(3) Since an interpolation manifold in ∂D, $D \subset \mathbf{C}^N$, has dimension not more than $N-1$, in (a) we cannot expect to find D_+ and D_- in $\mathbf{C}^N$ when $\dim \Sigma = N$.

(4) If Σ is a curve and $N \geq 2$, we have $\dim \Sigma = 1 \leq \frac{2N-1}{3}$, and thus each closed $\mathcal{C}^3$ curve can be realized as an interpolation manifold in the boundary of a strongly pseudoconvex domain in $\mathbf{C}^N$.

IV. *Proof of the theorem*

Given a $\mathcal{C}^k$ totally real submanifold Σ of $\mathbf{C}^N$, regard Σ as a submanifold of $\mathbf{C}^{N+1}$ and invoke Corollary 1.6 of [18] to find an open set Ω in $\mathbf{C}^{N+1}$ and a strongly plurisubharmonic function ρ on Ω, ρ of class $\mathcal{C}^{k+1}$ and nonnegative, such that

$$\Sigma = \{z \in \Omega : \rho(z) = 0\}.$$

Take coordinates $z_1, \cdots, z_{N+1}$ on $\mathbf{C}^{N+1}$ with $z_j = x_j + iy_j$, and let F be defined on $\mathbf{C}^{N+1}$ by $F(z) = y_{N+1}$. Define functions σ_+ and σ_- on Ω by

$$\sigma_+(z) = \rho(z) + F(z) \quad \text{and} \quad \sigma_-(z) = \rho(z) - F(z).$$

The functions σ_+ and σ_- are strictly plurisubharmonic on Ω, for each has the same Levi form as ρ. Also, since $d\rho = 0$ along Σ, $d\sigma_+$ and $d\sigma_-$ are both nonzero along Σ and hence in a neighborhood of Σ. By shrinking Ω if necessary, we may suppose that $d\sigma_+$ and $d\sigma_-$ never vanish on Ω.

Define domains D'_+ and D'_- by

$$D'_+ = \{z \in \Omega : \sigma_+(z) < 0\}$$

and

$$D'_- = \{z \in \Omega : \sigma_-(z) < 0\}.$$

These domains are disjoint, and $\bar{D}'_+ \cap \bar{D}'_- = \Sigma$: If $\sigma_+(z) \leq 0$ and $\sigma_-(z) \leq 0$, then $0 \leq \rho(z) \leq F(z)$ and $0 \leq \rho(z) \leq -F(z)$ whence $F(z) = \rho(z) = 0$ and thus $z \in \Sigma$. Thus we have constructed in Ω two strongly pseudoconvex hypersurfaces that abut precisely along Σ.[4]

We may shrink Ω if necessary and assume that it is Stein and that every compact subset of Σ is $\mathcal{O}(\Omega)$-convex [11]. Let τ be a $\mathcal{C}^\infty$ strongly plurisubharmonic exhaustion function for Ω such that $\tau < -1$ on E. Let W_o be a relatively compact neighborhood of E in Σ, and let χ be a $\mathcal{C}^\infty$ function on Ω, $0 \leq \chi \leq 1$, with $\chi = 0$ on $(\Sigma \setminus W_o) \cup \left\{z \in \Omega : \tau(z) \geq -\frac{1}{2}\right\}$ and $\chi = 1$ on $\{z \in \Omega : \tau(z) \leq -1\}$. Choose a $\mathcal{C}^\infty$ function $\tilde{\rho}$ that approximates ρ very well in the $\mathcal{C}^2$-sense on Ω and that satisfies $\rho < \tilde{\rho}$ everywhere. Put $\tilde{\sigma}_\pm = \tilde{\rho} \pm F$, and take

$$\lambda_+ = \chi\sigma_+ + (1-\chi)\tilde{\sigma}_+ = \chi\rho + (1-\chi)\tilde{\rho} + F$$

$$\lambda_- = \chi\sigma_- + (1-\chi)\tilde{\sigma}_- = \chi\rho + (1-\chi)\tilde{\rho} - F.$$

We have $\lambda_+ \geq \sigma_+$ and $\lambda_- \geq \sigma_-$, so

$$D''_+ = \{z \in \Omega : \lambda_+(z) < 0\} \subset D'_+$$

$$D''_- = \{z \in \Omega : \lambda_-(z) < 0\} \subset D'_-.$$

If $\tilde{\rho}$ approximates ρ closely enough in the $\mathcal{C}^2$-sense, then λ_+ and λ_- will both be strongly plurisubharmonic on Ω, and neither $d\lambda_+$ nor $d\lambda_-$ will vanish there. We have

$$E \cap \bar{D}''_+ \cap \bar{D}''_- \subset \bar{D}'_+ \cap \bar{D}'_- = \Sigma,$$

and the intersection $\bar{D}''_+ \cap \bar{D}''_-$ contains a neighborhood W_1 of E in Σ.

---

[4]It has come to my attention that Henkin [6, p. 666] has used this construction as the basis for a proof of an approximation theorem.

On the set where $\tau \geq -\frac{1}{2}$, we have $\lambda_+ = \tilde{\sigma}_+$ and $\lambda_- = \tilde{\sigma}_-$, so $\partial D''_+ \cap \left\{z \in \Omega : \tau > -\frac{1}{2}\right\}$ and $\partial D'_- \cap \left\{z \in \Omega : \tau > -\frac{1}{2}\right\}$ are $\mathcal{C}^\infty$ submanifolds of $\left\{z \in \Omega : \tau > -\frac{1}{2}\right\}$. By Sard's theorem, almost every value of τ on each of these manifolds is a regular value (of the restriction of τ to the manifold).

From this point on, we deal only with D''_+; D''_- may be handled in a similar way. Suppose for convenience of notation that 0 is a regular value of τ on $\partial D''_+$. Define $\phi : \Omega \to \mathbf{R}^2$ by

$$\phi(z) = (\lambda_+(z), \tau(z)) .$$

If $\Omega(0) = \{z \in \Omega : \tau(z) < 0\}$, then under ϕ, $\bar{D}''_+ \cap \overline{\Omega(0)}$ goes into the third quadrant, and the set E goes into the part of the τ-axis defined by $\tau < -1$. The map ϕ is proper as follows from the choice of τ as an exhaustion function for Ω which implies that τ is a proper map from Ω to $\mathbf{R}$. Thus, ϕ takes closed sets to closed sets.

Since 0 is a regular value for $\tau | \partial D''_+$, the map ϕ is regular on a neighborhood of $\partial D''_+ \cap \partial\Omega(0)$. Thus $0 \in \mathbf{R}^2$ is not in the closed ϕ-image of the set in Ω on which ϕ is not regular, so there is a neighborhood U of $0 \in \mathbf{R}^2$ such that ϕ is regular on $\phi^{-1}(U)$.

Denote by h a $\mathcal{C}^\infty$ convex function on $\mathbf{R}^2$ whose first partial derivatives are everywhere nonnegative and whose differential vanishes nowhere. Require also that h be strictly convex, *i.e.*, have positive definite Hessian, at each point of the third quadrant of $\mathbf{R}^2$. If γ is the level curve $h = 0$, we want

$$\gamma \supset \{(x_1, 0) : x_1 \leq -\delta\} \cup \{(0, x_2) : x_2 \leq -\delta\}$$

with $\delta < 0$ so small that the neighborhood U of 0 chosen above contains a disc of radius δ around 0.

Let $D_+ = \phi^{-1}(\{x \in \mathbf{R}^2 : h(x) < 0\})$. Then $D_+ \subset D''_+ \cap \Omega(0)$, so D_+ is a relatively compact subset of Ω. We have $\partial D_+ = \phi^{-1}(\gamma)$, and this is a $\mathcal{C}^{k+1}$, strongly pseudoconvex manifold. The latter point follows from our construction: The part of ∂D_+ over the τ-axis is contained in the strongly pseudoconvex boundary $\partial D''_+$, that over the λ_+-axis in the strongly pseudoconvex boundary $\partial\Omega(0)$. As ϕ is regular on $\phi^{-1}(N)$, $\phi^{-1}(\gamma \cap N)$ is a manifold, and it is strongly pseudoconvex as follows from this observation:

LEMMA 1. *Let* h *be a convex* $\mathcal{C}^2$ *function defined on the convex domain* Δ *in* $\mathbb{R}^2$. *If* f *and* g *are plurisubharmonic functions on the domain* $D \subset \mathbb{C}^N$, *if* $(f(z), g(z)) \in \Delta$ *for* $z \in D$, *and if the partial derivatives* $\frac{\partial h}{\partial x_1}$ *and* $\frac{\partial h}{\partial x_2}$ *are nonnegative, then the function* G *given by* $G(z) = h(f(z), g(z))$ *is plurisubharmonic on* D. *If* h *is strictly convex and* $dh \neq 0$, *and if* f *and* g *are strongly plurisubharmonic, then* G *is strongly pseudoconvex.*

Granted this lemma, we see that D_+ is strongly pseudoconvex with $\mathcal{C}^{k+1}$ boundary and that near E, $\partial D_+ = \partial D'_+$.

We construct the corresponding strongly pseudoconvex domain D_-. The domains D_+ and D_- are exterior to each other, and they abut along a set contained in Σ that is a neighborhood of E in Σ. Near E, ∂D_+ is given by the vanishing of $\sigma_+ = \rho + F$. As $d\rho = 0$ along Σ, it follows that for $p \in \Sigma$, $T_p(\partial D_+) = T_p(\{F=0\}) = \{y_{N+1} = 0\}$. Thus, $T_p{}^{\mathbb{C}}(\partial D_+) = \mathbb{C}^N$ for $p \in \Sigma$, and we see that $T_p(\Sigma) \subset T_p{}^{\mathbb{C}}(\partial D)$. Thus, except for the lemma, we have proved the first part of the theorem.

Proof of Lemma 1. For $\lambda_j \in \mathbb{C}$, $j = 1, \cdots, N$, we have

$$\sum_{j,k} G_{z_j\bar{z}_k}\lambda_j\bar{\lambda}_k = h_1 \sum_{j,k} f_{z_j\bar{z}_k}\lambda_j\bar{\lambda}_k + h_2 \sum_{j,k} g_{z_j\bar{z}_k}\lambda_j\bar{\lambda}_k$$

$$+ h_{11}\Big|\sum_j \lambda_j f_{z_j}\Big|^2 + h_{12}\sum_j \lambda_j f_{z_j} \sum_k \bar{\lambda}_k g_{\bar{z}_k}$$

$$+ h_{21}\sum_j \bar{\lambda}_j f_{\bar{z}_j} \sum_k \lambda_k g_{z_k} + h_{22}\Big|\sum_j \lambda_j g_{z_j}\Big|^2 .$$

If we put $u = \sum \lambda_j f_{z_j}$ and $v = \sum \lambda_j g_{z_j}$, and then write $u = u' + iu''$ and $v = v' + iv''$, we find

$$\sum_{j,k} G_{\bar{z}_j z_k} \lambda_j \bar{\lambda}_k = h_j \sum_{j,k} f_{z_j \bar{z}_k} \lambda_j \bar{\lambda}_k + h_2 \sum_{j,k} g_{z_j \bar{z}_k} \lambda_j \bar{\lambda}_k$$

$$+ h_{11}(u'^2 + u''^2) + 2h_{12}(u'v' + u''v'') + h_{22}(v'^2 + v''^2).$$

The conclusions of the lemma are now clear on the basis of the positivity and convexity hypotheses.

We now take up the proof of the second part of the theorem. Suppose $\Sigma \subset \mathbf{C}^N$ is a totally real manifold, that $k \geq 3$, and that $r = \dim \Sigma \leq \frac{2N-1}{3}$. As before, we find a neighborhood Ω of Σ and a $\mathcal{C}^{k+1}$ nonnegative strongly plurisubharmonic function ρ on Ω with $\Sigma = \{\rho = 0\}$.

LEMMA 2. *There are a neighborhood* Ω' *of* Σ *and a function* F *of class* $\mathcal{C}^{k-1}$ *on* Ω' *with* dF *nowhere vanishing on* Ω', *with* $M = \{z \in \Omega' : F(z) = 0\} \supset \Sigma$ *and with* $T_p(\Sigma) \subset T_p^{\mathbf{C}}(M)$ *for all* $p \in \Sigma$.

Let us note explicitly that M is a manifold of class $\mathcal{C}^{k-1}$.

Assume the lemma for the moment. We can construct D_+ and D_-, with boundaries of class $\mathcal{C}^{k-1}$ to be sure, just as we did above. For a small $\delta > 0$, the function δF, F as in the lemma, will play the rôle taken by $F = y_{N+1}$ in the earlier case, and our construction in the present setting does not have to leave $\mathbf{C}^N$. Thus, the theorem will be proved as soon as we prove the lemma.

Proof of Lemma 2. Denote by $T\Sigma$ the tangent bundle of Σ and by $T\mathbf{C}^N$ that of $\mathbf{C}^N$. As Σ is a manifold of class $\mathcal{C}^k$, $T\Sigma$ is a manifold of class $\mathcal{C}^{k-1}$. The bundle $T\Sigma$ is a subbundle of $T\mathbf{C}^N|\Sigma$, the restriction of $T\mathbf{C}^N$ to Σ. On $T\mathbf{C}^N$ we have the almost complex structure $J : T\mathbf{C}^N \to T\mathbf{C}^N$ that corresponds to multiplication by i; it is an isomorphism of the real vector bundle $T\mathbf{C}^N$ onto itself with $J^2 = -\text{identity}$.

We have the corresponding map $J : TC^N|\Sigma \to TC^N|\Sigma$, and the total reality of Σ is simply the condition that for each $x \in \Sigma$, $JT_x(\Sigma) \cap T_x(\Sigma) = 0$. We have now a well-defined subbundle $T\Sigma \oplus JT\Sigma$ of $TC^N|\Sigma$ with fiber dimension $2r$. If $x \in \Sigma$, the fiber $(T\Sigma \oplus JT\Sigma)_x$ is the r-dimensional complex subspace of C^N spanned by $T_x\Sigma$.

Let $C^N = R^{2N}$ be endowed with the usual Riemannian metric. For each x, the space $T_x\Sigma$ is orthogonal to $JT_x\Sigma$ and thus $JT\Sigma$ is a subbundle of the normal $T\Sigma^\perp$ of Σ in C^N. Denote by $\mathcal{S}$ the orthogonal complement of $JT\Sigma$ in $T\Sigma^\perp$. Thus $\mathcal{S}$ is a $\mathcal{C}^{k-1}$ bundle of fiber dimension $2(N-r)$ over Σ. As $r = \dim \Sigma \leq \frac{2N-1}{3}$ and $\mathcal{S}$ has fiber dimension $2(N-r)$, we have $\dim \Sigma \leq$ (fiber dimension)-1, so by [14, p. 99], $\mathcal{S}$ admits a trivial one-dimensional bundle as a direct summand: For some trivial line bundle $\mathcal{B}$ and some other bundle $\mathcal{S}_1$, we have $\mathcal{S} = \mathcal{S}_1 \oplus \mathcal{B}$. The bundle $\mathcal{B}$ is trivial, so there is $\mathcal{C}^{k-1}$ bundle isomorphism $\psi : \mathcal{B} \to \Sigma \times R$. Define a function $f : \Sigma \times R \to R$ by $f(x,t) = t$; this is a regular map. Since

$$T\Sigma^\perp = JT\Sigma \oplus \mathcal{S}_1 \oplus \mathcal{B}$$

there is a natural quotient map $\eta : T\Sigma^\perp \to \mathcal{B}$. It is regular, surjective and of class $\mathcal{C}^{k-1}$. Thus $f \circ \psi \circ \eta : T\Sigma^\perp \to R$ is a regular $\mathcal{C}^{k-1}$-map so $M_o = (f \circ \psi \circ \eta)^{-1}(0)$ is a submanifold of codimension 1 in $T\Sigma^\perp$, *i.e.*, $\dim M_o = 2N-1$. Also, M_o contains the zero section of the bundle $T\Sigma^\perp$.

We now use the tubular neighborhood theorem, though, in fact, we need to look at the proof [9, p. 70] rather than the result itself. We return to C^N for a moment. The tangent bundle TC^N is trivial and may be identified with $C^N \times C^N$. There is a map $\alpha : TC^N \to C^N$ given by $\alpha(z,w) = z+w$. If we restrict this map to $TC^N|\Sigma$, we have $\alpha : \Sigma \times C^N \to C^N$ with $\alpha(x,0) = x$ for $x \in \Sigma$. Since $T\Sigma^\perp$ is a subbundle of TC^N, a further restriction gives $\alpha : T\Sigma^\perp \to C^N$, and α takes the zero section of $T\Sigma^\perp$ onto Σ. The proof of the tubular neighborhood theorem consists of showing that for some neighborhood V of the zero section of $T\Sigma^\perp$, α is a $\mathcal{C}^{k-1}$ diffeomorphism of V onto a neighborhood Ω' of Σ in C^N. By its very construction, α carries $T_x^{\perp}$ onto the $(2N-k)$-plane in $\mathbf{C}^N$ through x and normal to $T_x\Sigma$, and it carries $JT_x\Sigma$ onto $iT_x\Sigma$. Thus, if we define $F:\Omega' \to \mathbf{R}$ by $F = f\circ\psi\circ\eta\circ\alpha^{-1}$, then $F \in \mathcal{C}^{k-1}(\Omega')$, $dF \neq 0$ on Ω', $F=0$ on Σ, and the tangent space to $M = \{F=0\}$ at $x \in \Sigma$ contains $T_x\Sigma + JT_x\Sigma$. The lemma is proved as is the second part of the theorem.

For the proof of the third part of the theorem, we notice that a simplified version of the bundle argument used in Lemma 2 yields a function F of class $\mathcal{C}^{k-1}$ on a neighborhood Ω' of Σ with $\Sigma \subset M = \{z\epsilon\Omega' : F(z)=0\}$. This time we do not need $T_p(\Sigma) \subset T_p^{\mathbf{C}}(M)$, and so we do not have to require that $\dim \Sigma \leq \frac{1}{3}(2N-1)$.

The theorem is proved.

As an example of the final assertion of the theorem, consider the torus

$$T^2 = \{(z,w) \in \mathbf{C}^2 : |z|, |w| = 1\},$$

a compact, totally real manifold. Plainly $T^2 \subset \partial(B_{\sqrt{2}})$ if

$$B_{\sqrt{2}} = \{(z,w) \in \mathbf{C}^2 : |z|^2 + |w|^2 < 2\}.$$

Define $\Phi : \mathbf{C}^2 \backslash \{(z,w) : zw = 0\} \to \mathbf{C}^2$ by

$$\Phi(z,w) = \left(\frac{1}{z}, \frac{1}{w}\right).$$

The map Φ is biholomorphic from $\mathbf{C}^2 \backslash \{(z,w) : zw=0\}$ onto itself and takes T^2 onto itself. A short calculation shows that $\Phi(B_{\sqrt{2}}) \cap B_{\sqrt{2}} = \emptyset$. If we set $D'_+ = \Phi(B_{\sqrt{2}} \backslash \{(z,w) : zw = 0\})$, then D'_+ is an unbounded domain in $\mathbf{C}^2$ with real analytic strongly pseudoconvex boundary. It is disjoint from $D_- = B_{\sqrt{2}}$, but $\bar{D}_-$ and $\bar{D}'_+$ abut precisely along the torus T^2. As in the proof of Theorem 1, we can truncate D'_+ appropriately to obtain a (bounded) strongly pseudoconvex domain D_+ such that $\bar{D}_- \cap \bar{D}_+ = T^2$.

It seems reasonable to conjecture that if the manifold Σ can be exhibited as $\bar{D}_+ \cap \bar{D}_-$ with D_+ and D_- disjoint strongly pseudoconvex domains, then Σ is totally real. I do not have a proof of this, but Klas Diederich pointed out the following argument that shows Σ to be generically totally real.

First a preliminary. Let $D = \{Q<0\}$ be a strongly pseudoconvex domain, and let $\tau = J \operatorname{grad} Q$; Q is to be a strictly plurisubharmonic defining function for D. If ξ and η are vector fields on ∂D with $\xi_p, \eta_p \in T_p^C(\partial D)$ for each p so that, say $\xi = \sum_j a_j \frac{\partial}{\partial z_j} + \bar{a}_j \frac{\partial}{\partial \bar{z}_j}$ and

$\eta = \sum_j b_j \frac{\partial}{\partial z_j} + \bar{b}_j \frac{\partial}{\partial z_j}$ with $\sum_j a_j \frac{\partial Q}{\partial z_j} = 0 = \sum_j b_j \frac{\partial Q}{\partial z_j}$, then [5]

$$([J\xi, \eta]_p, \tau_p) + i([\xi, \eta]_p, \tau_p) = 4 \sum_{j,k} \frac{\partial^2 Q}{\partial z_j \partial \bar{z}_k}(p)\, a_j \bar{b}_j .$$

Applied with $\xi = \eta$, this yields

$$\begin{aligned} ([J\xi, \xi]_p, \tau_p) &= 4 \sum_{j,k} \frac{\partial^2 Q}{\partial z_j \partial z_k}(p)\, a_j \bar{a}_k \\ &\geq \text{const } \|\xi_p\|^2 . \end{aligned}$$

(Here (,) denotes the Euclidean innerproduct.) As J is a Euclidean isometry,

(2) $$(J[J\xi, \xi]_p, J\tau_p) > 0$$

unless $\xi_p = 0$.

Now let Σ be a $\mathcal{C}^1$ manifold with $\Sigma = \bar{D}_+ \cap \bar{D}_-$, D_+ and D_- disjoint strongly pseudoconvex domains with strictly plurisubharmonic defining functions Q_+ and Q_- respectively. Let ξ be a zero-free vector field along the open subset U of Σ such that $J\xi$ is also tangent to Σ along U. For $p \in U$, $\xi_p \in T_p^C(\partial D_+) \cap T_p^C(\partial D_-)$.

Let ξ^+ be a section of $T^{\mathbf{C}}(\partial D)$ over an open set U_+ in ∂D_+ with $\xi^+_p = \xi_p$ for $p \in U$. For $p \in U$ we have

$$J[J\xi^+, \xi^+]_p = X_p + s(p)\operatorname{grad} Q_+(p) .$$

with X_p tangent to ∂D_+ and $s(p)\operatorname{grad} Q_+(p)$ normal to it. We know that $s(p)\operatorname{grad} Q_+(p)/\|\operatorname{grad} Q_+(p)\|$ is the projection of $J[J\xi, \xi]_p$ into the line normal to ∂D at p. As grad $Q_+(p)$ points in the outward normal direction, we see from (2) that the normal component of $J[J\xi^+, \xi^+]_p = J[J\xi, \xi]_p$ points into the domain D_+. If we work in the same way with D_-, we find that this normal component must point into D_-, and as D_+ and D_- are mutually disjoint, we have a contradiction.

Thus, generically, Σ is totally real. In particular, it has dimension not more than n.

Note that the argument just given does not exclude the possibility that Σ may fail to be totally real at points of a small closed subset.

If a compact set $E \subset \mathbf{C}^n$ is of the form $\bar{D}_+ \cap \bar{D}_-$ as above, then it is the intersection of a sequence of domains of holomorphy and so by a theorem of Serre [20] and a result on the continuity of cohomology [3], $H^k(E, \mathbf{C}) = 0$ for $k > n$. If E is an orientable topological manifold, not necessarily smooth, it must have dimension no more than n.

THE UNIVERSITY OF WASHINGTON

REFERENCES

[1] V. I. Arnold, *Mathematical Methods of Classical Mechanics*, Springer, New York, Heidelberg, Berlin, 1978.

[2] D. E. Blair, *Contact Manifolds in Riemannian Geometry*, Springer, New York, Heidelberg, Berlin, 1976.

[3] A. Browder, Cohomology of maximal ideal spaces, Bull. Amer. Math. Soc. 67 (1961), 515-516.

[4] D. Burns and E. L. Stout, Extending functions from submanifolds of the boundary, Duke Math. J. 43 (1976), 391-404.

[5] J. Chaumat and A. M. Chollet, Ensembles pics pour $A^\infty(D)$, Ann. Inst. Fourier (Grenoble) 29(3) (1979), 171-201.

[6] E. M. Chirka and G. M. Khenkin, Boundary properties of holomorphic functions of several complex variables, J. Sov. Math. 5 (1976), 612-687.

[7] A. M. Davie and B. Øksendal, Peak interpolation sets for some algebras of analytic functions, Pacific J. Math. 41 (1972), 81-87.

[8] J. E. Fornaess and E. L. Stout, Spreading polydiscs on complex manifolds, Amer. J. Math. 99 (1977), 933-960.

[9] M. Golubitsky and V. Guillemin, *Stable Mappings and Their Singularities*, Springer, New York, Heidelberg, Berlin, 1973.

[10] M. Hakim and N. Sibony, Ensembles pics dans des domaines strictement pseudoconvexes, Duke Math. J. 45 (1978), 601-617.

[11] F. R. Harvey and R. O. Wells, Holomorphic approximation and hyperfunction theory on a C^1 totally real submanifold of a complex manifold, Math. Ann. 197 (1972), 287-318.

[12] G. M. Henkin and A. E. Tumanov, Interpolation submanifolds of pseudoconvex manifolds, AMS Translations, Ser 2, vol. 115.

[13] G. Hirsch, *Differential Topology*, Springer, New York, Heidelberg, Berlin, 1976.

[14] D. Husemoller, *Fiber Bundles*, McGraw-Hill, New York, 1966.

[15] S. Kobayashi, *Transformation Groups in Differential Geometry*, Springer, New York, Heidelberg, Berlin, 1972.

[16] H. F. Lai, Characteristic classes of real manifolds immersed in complex manifolds, Trans. Amer. Math. Soc. 172 (1972), 1-33.

[17] A. Nagel, Smooth zero sets and interpolation sets for some algebras of holomorphic functions on strictly pseudoconvex domains, Duke Math. J. 43 (1976), 323-348.

[18] R. M. Range and Y. T. Siu, $\mathcal{C}^k$ Approximation by holomorphic functions and $\bar{\partial}$-closed forms on $\mathcal{C}^k$ submanifolds of a complex manifold, Math. Ann. 210 (1974), 105-122.

[19] W. Rudin, Peak interpolation sets of class $\mathcal{C}^1$, Pacific J. Math. 75 (1978), 267-279.

[20] J. P. Serre, Applications de la théorie générale à diverse problèmes globaux, Séminare H. Cartan 1951-52, W. A. Benjamin, New York, Amsterdam, 1967.

[21] R. O. Wells, Compact real submanifolds of a complex manifold with nondegenerate holomorphic tangent bundle, Math. Ann. 179 (1969), 123-129.

A SURVEY OF SOME RECENT RESULTS IN C^∞ AND REAL ANALYTIC HYPOELLIPTICITY FOR PARTIAL DIFFERENTIAL OPERATORS WITH APPLICATIONS TO SEVERAL COMPLEX VARIABLES

David S. Tartakoff*

Introduction

Starting with the work of Garabedian and Spencer [20], Morrey [45] and Kohn [33], the field of several complex variables has given rise to problems in pure partial differential equations which have had strong independent interest and whose results have been of significance in both fields. The prime example of this phenomenon is the celebrated $\bar\partial$-Neumann problem, which we now briefly formulate. Given a domain Ω compactly contained in C^n (for simplicity) with smooth boundary Γ, and given a (0,1) form g which is $\bar\partial$-closed in Ω and smooth in $\bar\Omega$, one seeks to solve $\bar\partial u = g$ in Ω with u smooth up to the boundary, u uniquely specified by the requirement that it be orthogonal, in L^2 for a given Hermitian metric, to the kernel of $\bar\partial$. The $\bar\partial$-Neumann problem consists in solving the second order problem $\square w = (\bar\partial\bar\partial^* + \bar\partial^*\bar\partial) w = g$ subject to the boundary condition $w \in \mathfrak{D}(\bar\partial^*)$ (L^2 adjoint of $\bar\partial$), with w smooth up to the boundary, then showing that $\bar\partial g = 0$ implies $\bar\partial w = 0$, and finally setting $u = \bar\partial^* w$. The operator $\square$ is elliptic, and though a system, a determined one with the same number of unknowns as data.

*During this work the author was partly supported by an NSF grant.

Recent Developments in Several Complex Variables
0-691-08285-5/81/000393-19 $00.95/0 (cloth)
0-691-08281-2/81/000393-19 $00.95/0 (paperback)

This boundary value problem for $\square$ is not coercive, however, and the known results for elliptic boundary value problems could not be applied. Kohn [33] was able to show that if g was C^∞ up to the boundary near x_0, then the same was true of w provided a certain "basic estimate" due to Morrey [45] held; this estimate has come to be called "subelliptic", since while not coercive, it exhibits a loss of only $1/2$ derivative. (The regularity proof of Kohn was later improved in the paper of Kohn and Nirenberg [38].) The basic estimate was shown to hold if and only if a certain condition $Z(q)$ held (in the case of (p,q) forms): the Levi form should have at least $n-q$ positive or $q+1$ negative eigenvalues at each point. Later, in addition to the C^∞ results of Kohn, Derridj and Zuily [12], Tartakoff [59] and Derridj [9] have proved local Gevrey regularity up to the boundary, at least in Gevrey classes G^s with $s \geq 2$.

Shortly after Kohn's original work, Hörmander [26, 29] gave a proof which required only that Ω be pseudo-convex and proved global existence of a solution $u \in C^\infty(\Omega)$ even if the boundary of Ω is not smooth. This result leads (as does Kohn's) to a solution of the Levi problem that pseudo-convex domains (strongly pseudo-convex domains with smooth boundary) are domains of holomorphy. Using a similar technique, though with weight functions which are smooth up to the boundary, Kohn [35] showed that when the boundary is smooth and pseudo-convex, $\Omega \subset\subset C^n$ for simplicity, then for each m there exists $u_m \in C^m(\Omega)$ with $\bar{\partial} u = g$, and a remark by Hörmander showed that from the u_m one can construct a single $u \in C^\infty(\Omega)$. These results are global.

From the standpoint of several complex variables, the next results followed from the observation (independently, by Komatsu [40] and Tartakoff [58]) that when there were no zero eigenvalues of the Levi form, then one could always construct a special vector field, complementary to the holomorphic and anti-holomorphic vector fields tangent to the boundary such that T together with $(T^{1,0} \oplus T^{0,1})$ spanned the tangent space to the boundary and had the property that the commutator of T with any vector field in $(T^{1,0} \oplus T^{0,1})$ again was in this space. Thus (see the estimate (3.2) below) it was possible to use an a priori estimate which failed to be coercive and yet force commutators in directions in which it gave elliptic control.

The recent, independent, local real analytic results of F. Trèves [66] and the author [61, 62] exploited this additional hypothesis that the Levi form be non-degenerate, Trèves by employing the machinery developed by Sato which requires this condition, the author by L^2 energy methods only and the observation that the construction of a special vector field T above may be applied to a localized, high power of T with even better results.

We want to illustrate here the importance of the hypothesis that the Levi form not degenerate (i.e., that the characteristic manifold of the associated "sublaplacian" be symplectic) with an argument, due to Folland, showing that when one eigenvalue is identically zero, then even though the best subellipticity holds, the operator is exactly of the form described by Baouendi and Goulaouic in [1], i.e., is not locally analytic hypoelliptic. Let n be even and at least 6, and define $Z_j = X_j + iY_j$ by: $X_1 = \partial/\partial x_1$, $Y_1 = \partial/\partial y_1$, $X_j = \partial/\partial x_j + 2y_j\,\partial/\partial t$, $Y_j = \partial/\partial y_j - 2x_j\,\partial/\partial t$ for $j = 2, \cdots, n-1$, and $T = \partial/\partial t$. The operator $\Box_b$ is constructed out of $\bar\partial_b$ just as $\Box$ is out of above, where $\bar\partial_b$ is the "tangential Cauchy-Riemann complex" formed from the Z_j above. $\Box_b$ is subelliptic with loss of 1/2 derivative whenever condition $Y(q)$ is satisfied, for (p,q) forms, as it is here for $q = n/2$. In fact, on the $(0, n/2)$ forms $u = u\,d\overline{z_1} \wedge \cdots \wedge d\overline{z_{n/2}}$, $\Box_b$ has the form $(-1/4)\sum (X_j^2 + Y_j^2)$; under the change of coordinates

$\tilde x_j = 2x_j$, $\tilde y_j = 2y_j$, and $\tilde t = t - 2\sum_{j=1}^{n-1} x_j y_j$, this becomes

$$-\Box_b u = \sum_{j=1}^{n-1} (\partial^2 u/\partial \tilde x_j^2 + \partial^2 u/\partial \tilde y_j^2) + \sum_{j=1}^{n-2} (\tilde x_j^2)\,\partial^2 u/\partial t^2 .$$

But on coefficients u independent of the $\tilde y_j$, this is just the operator of Baouendi and Goulaouic, known to be hypoelliptic up to the level of the second Gevrey class, but not in the real analytic class.

The plan of this paper is as follows. We shall seek to put the question of local regularity for these particular operators in a larger perspective by surveying recent work on hypoelliptic operators in general. Since the work was largely inspired by the $\bar{\partial}$-Neumann problem, it is appropriate to compare results in the general theory with those, given above, known for $\Box_b$ and the $\bar{\partial}$-Neumann problem. In Section 2, we set notations and make some preliminary remarks about subellipticity and hypoellipticity in general, and mention the recent work of Kohn, Diederich and Fornaess, Catlin, and others. Section 3 will contain a summary of hypoellipticity results for complex valued operators of principal type and real operators "of Hörmander type" constructed as the sums of squares of vector fields which, together with their commutators up to some fixed order, span the tangent space at each point. Section 4 is devoted to questions of the approximation of operators constructed out of non-commuting vector fields by left invariant operators on nilpotent Lie groups of arbitrary step. The conjecture of Rockland, recently proved by Helffer and Nourrigat, is discussed. Section 5 discusses real analytic results, and in Section 6 we present a new proof of a known result, namely the analytic hypoellipticity of the operator $D_x^2 + x^2 D_t^2$ in R^2, using the methods we recently developed in [61, 62].

This survey is not complete, and we have been unable to include many topics and results which should be included. The bibliography is much more comprehensive, though, and the paper as a whole should serve to give an overview of the subject, rather than as a comprehensive exposition with proofs.

2. *Notation, definitions, and more on the* $\bar{\partial}$-*Neumann problem*

We shall, for convenience, work in R^m. Partial differential operators will be denoted by P or L, and will be assumed to have smooth coefficients.

DEFINITION 2.1. P is hypoelliptic at x_0 provided whenever a distribution satisfies $Pu = f$ near x_0 with $f \in C^\infty$ in a neighborhood of x_0, then u is C^∞ in a (possibly smaller) neighborhood of x_0. P is said to be hypoelliptic at x_0 with loss of r derivatives if it is hypoelliptic at x_0 and f in the Sobolev space H^s_{loc} implies u is in H^{s+m-r}_{loc}, where m is the order of P.

DEFINITION 2.2. P is analytic hypoelliptic at x_0 if whenever a distribution satisfies $Pu = f$ near x_0 with f real analytic in some neighborhood of x_0, then u is real analytic near x_0.

DEFINITION 2.3. P is subelliptic with loss of r derivatives in an open set Ω provided that there exists a constant C such that for all v in $C_0^\infty(\Omega)$, and some $r' > r$,

$$\|v\|_{H^{m-r}} \leq C(\|Pv\|_{L^2} + \|v\|_{H^{m-r'}}) :$$

It is easy to verify that subellipticity with loss of less than one derivative in an open set implies hypoellipticity near any point in Ω with loss of the same number of derivatives (cf. [33], [59]). When $r = 0$, P is elliptic and analytically hypoelliptic. It is not necessarily true that subellipticity with loss of one or more derivatives implies hypoellipticity, but often operators which satisfy (2.1) also satisfy somewhat stronger a priori estimates from which hypoellipticity can be deduced. For example, $\Box_b$, discussed above, on a strictly pseudo-convex domain satisfies not only (2.1) with $m = 2$, $r = 1$ but also (with $\varepsilon = 1/2$)

$$\|v\|_\varepsilon^2 \leq C(|(\Box_b v, v)| + \|v\|_0^2) \tag{2.2}$$

and in fact

$$\|v\|_{1/2}^2 + \sum_{j=1}^{n-1} \|\overset{(-)}{L}_j v\|_0^2 \leq C(|(\Box_b v, v)| + \|v\|_0^2) \tag{2.3}$$

where the L_j and $\bar{L}_j$ span the "tangential holomorphic and anti-holomorphic tangent space'' (the Z_j and $\overline{Z}_j$ in Folland's example above). Thus the estimate (2.3), while only subelliptic overall, appears coercive (no loss of derivatives) in all but one direction.

Using (2.2) and the fact that $\Box_b$ is essentially self-adjoint, it is easy, as in [38], to show hypoellipticity, and in fact any norm of positive order on the left of (2.2) would suffice.

Much recent work has gone into efforts to establish (2.2), or its analogue for the $\bar{\partial}$-Neumann problem where the right-hand side is replaced by $C(\|\bar{\partial}v\|_0^2 + \|\bar{\partial}^* v\|_0^2 + \|v\|_0^2)$. Kohn [36, 37], using results of Diederich and Fornaess [13] has shown that when the boundary is real analytic near x_0 and pseudo-convex, a sufficient condition for subellipticity (in the sense of (2.2) above with any positive norm on the left) in the $\bar{\partial}$-Neumann problem is that there exist no complex analytic variety of dimension at least q through x_0 and contained in the boundary. For a partial converse, see Egorov [15]. Catlin's example of a domain in C^3, pseudo-convex but not analytic, whose boundary contains no analytic curves yet which is not hypoelliptic shows that the C^∞ result is rather delicate.

3. *Principal type and Hörmander type operators; loss of one derivative*

The above examples, $\Box_b$ and the $\bar{\partial}$-Neumann problem, are not simple problems from the standpoint of partial differential equations; they have double characteristics. The first systematic study of the relation between subellipticity and hypoellipticity was carried out for simply characteristic operators, namely those of *principal type*.

DEFINITION 3.1. A (pseudo-) differential operator P is of principal type if its leading symbol, $p_m(x, \xi)$, satisfies: $\nabla_\xi p_m(x, \xi) \neq 0$ whenever $p_m(x, \xi) = 0$, $\xi \neq 0$.

For this class of operators, questions of local solvability and hypoellipticity have been well studied (cf. [47, 65]); to wit:

THEOREM 3.1. *Let* P *be a partial differential operator of principal type with analytic coefficients. Then* P *is hypoelliptic if and only if* P *is subelliptic (in the sense of (2.1)) with loss of less than one derivative if and only if* P *is real analytic hypoelliptic.* (F. Trèves, [65].)

A very different situation obtains in the case of real operators:

THEOREM 3.2. *Let* P *be a partial differential operator in an open set in* R^n *with smooth coefficients whose principal symbol,* $p_m(x,\xi)$ *is real. If* $\nabla_\xi p_m(x,\xi) \neq 0$ *at some* (x,ξ), $\xi \neq 0$, *where* $p_m(x,\xi) = 0$, *then* P *is not hypoelliptic near* x_0. (Hörmander, [27]; see also [48].)

This result generalizes the known result for constant coefficient equations that they must have multiple characteristics to be hypoelliptic if they are not elliptic ([28], Thm. 4.17).

Thus if an operator of order 2 with real principal part is to be hypoelliptic, then its principal symbol must be a semi-definite quadratic form and thus vanish to at least second order on its characteristic variety. Hence in any open set where the rank is constant,

$$P = \sum_1^r X_j^2 + X_0 + c \tag{3.1}$$

where the X_j are real vector fields with smooth coefficients.

Hörmander studied operators of the form (3.1) where all the X_j and c are real; observing that if in an open set, the dimension, pointwise, of the span of the X_j and all their commutators were less than n, then by the Frobenius theorem P acts in fewer than n variables and cannot be hypoelliptic, he proved the following theorem in [27]:

THEOREM 3.2 (Hörmander). *Let* P *have the form (3.1) and assume that the* X_j *and their commutators up to length at most* k *span the tangent space at each point in* Ω. *Then* P *is hypoelliptic in* Ω *with loss of* $2-\varepsilon$ *derivatives,* $\varepsilon > 0$.

Proofs of this theorem have also been given by Kohn [34] and Radkevic [49, 50] with less sharp values of ε than those obtained by

Hörmander. All proofs begin by establishing a priori inequalities whose prototype is (2.3):

$$\sum_{j=1}^{r} \|X_j v\|^2_{L^2} \leq C(|(Pv, v)| + \|v\|^2_{L^2}) . \tag{3.2}$$

It is not hard to show, then, that the right-hand side also bounds some norm $\|v\|^2_\varepsilon$ with $\varepsilon > 0$, hence is hypoelliptic. Using such estimates, or those of the form (2.1), Derridj and Zuily [12], Tartakoff [59] and Derridj [9] have proved Gevrey regularity in certain Gevrey classes, but, as the example of Baouendi and Goulaouic shows, not the real analytic class. It is interesting to note (cf. Derridj, [8]) that when the rank condition is violated at even one point, and the coefficients are analytic, then P cannot be hypoelliptic, assuming it does not have all X_j equal to zero at that point.

Somewhat more generally, if P is a pseudo-differential operator of order 2 whose leading symbol, $p_2(x, \xi)$, is nonnegative and vanishes to exactly order two on its characteristic variety, Σ, assumed to be a smooth conic submanifold of T^*R^n of codimension k, then by the Morse Lemma, $p_2(x, \xi)$ may be written as $\sum_{j=1}^{k} a_j(x, \xi)^2$, so that $P = \sum_{j=1}^{k} A_j^2 + B$ where each A_j and B has order one, and the leading symbol of A_j is a_j. The principal symbol of B, one computes, *on* Σ, is given by

$$\text{sub } \sigma(P) = b_1(x, \xi) = p_1(x, \xi) + \frac{i}{2} \sum (\partial^2 p_2(x, \xi)/\partial x_k \partial \xi_k)$$

and is invariantly defined. Using the sharp Gårding inequality, the theorem becomes that if the a_j, together with their Poisson brackets up to some finite length, generate all pseudo-differential operators of order 1, then P is hypoelliptic with loss of $2-\varepsilon$ derivatives provided Re sub $\sigma(P) \geq 0$ on Σ. When only first brackets are needed, the proof shows that one may take $\varepsilon = 1$. For example, in the case of $\Box_b$, or rather a scalar operator with the same principal part, the condition that first brackets suffice is identical with the statement that the Levi form has some non-zero entry. Taylor [57] calls this condition "not involutive", since the phrase "non-involutive" has come to mean non-degeneracy of the Levi form in this context, or symplectic characteristic variety, i.e., that the fundamental two form $\sigma = \sum d\xi_i \wedge dx_i$ is non-degenerate on Σ.

A very careful study of the hypoellipticity with loss of one derivative has been made recently [3, 4, 5, 21, 30, 42, 56, 49, 57]. For second order operators P whose principal part is non-negative and vanishes to exactly second order on Σ, one defines the fundamental matrix, F, by $\sigma(u, Fv)$ $= \mathrm{Hess}(p_2)(u, v)$. F has eigenvalues of the form $\pm i\lambda_j, \lambda_j > 0$, and 0. If V denotes the space of generalized eigenvectors corresponding to the eigenvalue zero, the theorem is that P is hypoelliptic with loss of one derivative if and only if on Σ, and for any non-negative integers a_k and $v \epsilon V$,

$$\mathrm{sub}\ \sigma(P) + \mathrm{Hess}(p_2)(v, \overline{v}) + \sum (2a_k+1)\lambda_k \neq 0 .$$

In fact, p_2 need not be non-negative; when $\mathrm{codim} > 2$ it may be allowed to take values in any proper cone, and when $\mathrm{codim} = 2$, Boutet de Monvel and Trèves [4] and Sjostrand [56] require that Σ be symplectic, that p_2 have winding number zero about Σ, and that the Hessian have "index -2" —see also Hörmander [30].

Some salient features of this result had been observed earlier. Folland [16] and Folland and Stein [19] had looked at the second order operators $\mathcal{L}_\alpha = -\frac{1}{2} \sum (Z_j \overline{Z}_j + \overline{Z}_j Z_j) + i\alpha T$, with $T = \partial/\partial t$, $Z_j = \partial/\partial z_j - i\overline{z}_j \partial/\partial t$; the characteristic variety is symplectic here, so the term with the Hessian above drops out, and the condition for hypoellipticity had been observed by Folland and Stein in [19] and of course coincides with $Y(q)$ on forms of the appropriate degree when $\square_b$ is written in the form $\mathcal{L}_\alpha$ on the Heisenberg group. The book [51] by Rockland gives another approach.

4. *Relationship with nilpotent Lie groups*

Since the observation by Folland and Stein [16, 19] that the left invariant vector fields on the Heisenberg group could be used to approximate very closely the real and imaginary parts of the holomorphic vector fields on strongly pseudo-convex manifolds (see also Rothschild and Stein [54]), and the construction of an explicit fundamental solution, by Folland, for such group theoretic "sublaplacians" from which one could read off regularity properties of solutions, much further progress has been made both in the study of homogeneous differential operators on more general nilpotent Lie groups and in the approximation of differential operators constructed as the sum of squares of real vector fields by such homogeneous operators. Rothschild and Stein [54] generalized [19] to groups of step greater than 2, and parametrices with C^∞ error (and L^p estimates) given in [55] were quite explicit; thus for second order "sublaplacians", where the vector fields involved, together with their brackets up to some length, span the tangent space, one can associate a suitable, nilpotent Lie algebra at each point, use explicit estimates and fundamental solutions, build parametrices, and read off differentiability properties of solutions in a variety of spaces.

One question of standing interest has been to determine when a homogeneous (in the sense of the Lie algebra) left invariant differential operator (other than a sublaplacian) L on a nilpotent Lie group hypoelliptic? This much one should know if there is to be any hope of answering such questions more generally. In [52], Rockland conjectured that this would be the case if and only if for every irreducible, non-trivial unitary representation π of the connected, simply connected Lie group, $\pi(L)$ was injective on the space of C^∞ vectors for the representation. He proved the conjecture in the case of the Heisenberg group, and Beals showed its necessity on the product of the Heisenberg group $\times R^k$. There followed a series of papers by Helffer and Helffer and Nourrigat culminating in [25] in which the conjecture was demonstrated in general.

Métivier, recently, studied general operators of the form $L = \sum_{|\alpha|\leq d} a_\alpha(x) X^\alpha$ where $\alpha = (\alpha_1, \alpha_2, \cdots, \alpha_{|\alpha|})$ and $X^\alpha = X_{\alpha_1} X_{\alpha_2} \cdots X_{\alpha_{|\alpha|}}$. He made two assumptions; the first is that the X_j and their brackets of some finite order span at each point, and the second that the dimension of the space spanned by the commutators of the X_j of length k be constant for each k. Then he is able to associate a nilpotent Lie group such that the given operator is well approximated by left an invariant operator $\tilde{L}_{x_0}$ on that group. Rothschild [53] is then able to analyze the hypoellipticity of such L under Métivier's conditions and prove that if $\tilde{L}_{x_0}$ is hypoelliptic at x_0 then L is also. Further, since the Rockland conjecture and its proof establish hypoellipticity by establishing a "maximal" estimate:

DEFINITION 4.1. L is maximally hypoelliptic at x_0 if one has the local estimate

$$\sum_{|\alpha|\leq d} \|X^\alpha v\|_{L^2} \leq C(\|Lv\|_{L^2} + \|v\|_{L^2})$$

Rothschild is able to prove that L is maximally hypoelliptic if and only if the representation theoretic criterion of Rockland holds.

5. *Analytic regularity*

For some years after the C^∞ regularity properties of the $\bar{\partial}$-Neumann problem and Bergman kernel function had been established, the real analyticity remained unresolved. Folland's explicit fundamental solution on the Heisenberg group (strictly pseudo-convex) led one to believe that the result should be true, at least when the Baouendi-Goulaouic example was excluded, say when the determinant of the Levi form was non-zero so that the characteristic variety of $\Box_b$ was symplectic. For a while this hypothesis led only to global analyticity results (Tartakoff [58], Derridj and Tartakoff [10], and Komatsu [40]). Trèves local result [66] applied to pseudo-differential operators, but with scalar principal part, whose characteristic variety had general codimension, and used the full weight of Sato's hyperfunction machinery plus very delicate estimates. The author's proof [61, 62] applies to general systems of the form $P = \sum a_{ij}(x) X_i X_j + \sum a_i(x) X_i + a_o(x)$, where $X_1, \cdots, X_{2n-2}$, and T span the tangent space, the characteristic variety is symplectic (i.e., if $[X_i X_j] \equiv c_{ij} T$ modulo the X_k, then $\det(c_{ij})$ should be non-zero), and the appropriate L^2 estimate be satisfied, namely that given by $Y(q)$ for $\Box_b$:

$$\sum_{i,j} \|X_i X_j v\|_{L^2} \leq C(\|Pv\|_{L^2} + \|v\|_{L^2}).$$

The proof used only L^2 estimates plus a construction of a high order differential operator of the form $gT^p + \cdots$ for general p and arbitrary $g \in C_0^\infty$ such that when commuted with an X_j, one did not get $(X_j g) T^p + \cdots$; this leading term was killed off, as were all subsequent ones in which the localizing function became differentiated with no gain in the number of T derivatives. In the next section we shall employ a related idea in a simple situation.

One can weaken the condition that the characteristic variety be symplectic, but not too much. Explicit examples by Kaplan [31] with precise fundamental solutions require in fact more than a symplectic characteristic variety, and one can construct examples which allow one to apply the example of [1] in more subtle fashion than above. The essential properties one should retain are 1) a maximal estimate and 2) the ability to construct appropriate vector fields T or gT, whose commutators with the X_j are free of the naive traps. Thus Derridj [7] is able to allow the Levi form to degenerate in some cases; in particular he obtains global results on "circular" domains in $\mathbf{C}^2$, and Tartakoff [63] has semi-global results which state that if $\det(c_{ij}) = 0$ in K but the estimate remains valid (as it well may— $Y(q)$ does not require non-degeneracy) and an analytic T

can be found, near K, which commutes appropriately with the X_j, the analytic hypoellipticity obtains in a neighborhood of K.

Very recently, Métivier [43] and Helffer [24] have examined the situation when the characteristic variety is not symplectic in more detail. One of Métivier's results, in the case of the operator (3.1) is that if $\det(c_{ij}) = 0$ in a neighborhood of x_0 than P is not analytic hypoelliptic at x_0. His main result is for general partial differential operators P with non-negative principal part such that P^* is hypoelliptic with loss of $2-\varepsilon$ derivatives, if the characteristic variety Σ, near x_0, contains an analytic variety V, isotropic for the symplectic form σ, transverse to the fibre and maximal, and if the restriction of σ to the tangent space of Σ is degenerate at one point, then P is not analytic hypoelliptic. Strikingly, Helffer has recently proved this theorem for homogeneous differential operators on a nilpotent Lie group using representation theoretic criteria, by showing that there is a particular non-unitary representation which fails to be injective. The analogue of the Rockland conjecture for step 2 groups with non-degenerate "Levi form" has very recently been proved by Métivier; perhaps the approximation of analytic vector fields by left invariant vector fields on suitable Lie groups will be made sufficiently accurate for analytic hypoellipticity results off of groups shortly.

6. *A simple example*

We will conclude by demonstrating the methods of [61, 62] to prove the analytic hypoellipticity (at zero, for convenience) of $P = D_x^2 + x^2 D_t^2$, where now $D_x = \partial/\partial x$, $D_t = \partial/\partial t$. At first glance, this operator is strikingly similar to the example of [1]; addition of D_y^2, which would seem to make it more "elliptic" in fact destroys the analytic regularity. To our knowledge, $P + t^2 D_t^2$ is not known to be, or not to be, analytically hypoelliptic. That P is was shown in [41] and [12].

We shall show that for a fixed q, we are able to do an iterative procedure to estimate up to q derivatives, in L^2 norm, with analytic growth, constants being independent of q. We shall assume that $Pu = 0$ in V_∞ and show this growth in $V_0 \subset\subset V_\infty$. We nest $2 \log_2 q$ open sets in between: $V_0 \subset\subset V_1 \subset\subset V_2 \cdots \subset\subset V_{2 \log_2 q} = V$ with the distance between V_j and the complement of V_{j+1} on the order of some constant, depending only on V_0 and V_∞, divided by j^2: This is possible since $\sum (1/j^2) < \infty$. We pick functions Q_j, identically one near $\overline{V_j}$ and supported in V_{j+1}, C^∞, and with derivatives bounded by

$$|D^s Q_k| \leq C(CN_k)^s (k^2)^s \quad \text{for} \quad s \leq N_k \tag{6.1}$$

where N_k may be freely chosen and C depends only on V_0 and V_∞. For the construction, see [62]; these types of localizing functions were introduced by Ehrenpreis and used extensively by Hörmander and others since. The reason for using a family of them will become apparent – our iterative technique drops the order of differentiation by half each time around and needs two localizing functions per cycle.

The a priori estimate, for $w \in C_0^\infty(V_\infty)$,

$$\|D_x w\|^2_{L^2} + \|xD_t w\|^2_{L^2} \leq C|(Pw, w)| \tag{6.2}$$

is evident. We shall insert $w = Q_0 D_x^r u$ (assuming that u is in $C^\infty(V_\infty)$) or $w = Q_0 (xD_t)^r u$ in (6.1), or some intermediate expression with some D_x's and some (xD_t)'s. Commuting, for example, $Q_0 D_x^r$ past P introduces constant times r terms of the form $Q_0 D_t^2 x D_x^{r-1}$ and constant times r^2 terms of the form $Q_0 D_t^2 D_x^{r-2}$ on the left of the inner product above, modulo terms in which Q_0 is differentiated. One integration by parts, again modulo such terms, and one application of the Schwarz inequality shows that, essentially,

$$\|D_x Q_0 D_x^r u\| + \|xD_t Q_0 D_x^r v\| \rightarrow$$

$$\sup r^{s_1} C^r \{\|D_x Q_0^{(s_2)} D_t^{(s_1)} u\| + \|xD_t Q_0^{(s_2)} D_t^{(s_1)} u\|\}$$

with the supremum over all (s_1, s_2) with $s_2 + 2s_1 = r$, after repeated use of (6.2) as above. That is, each time we use (6.2), either a D_x derivative lands on Q_0 or two D_x's vanish and appear as a factor of r and a new D_t. But at any rate, since it takes two D_x's to make a D_t, no more than $r/2$ D_t's will appear in the end. Likewise, we could have started with $w = Q_0(xD_t)^r$, and a similar situation would have resulted, this time again at most $r/2$ pure D_t's could appear, since it takes a D_x and an xD_t to make a D_t with no x.

Thus to estimate D_x derivatives or (xD_t) derivatives of order r, one must obtain analytic bounds for $D_t^{r/2}u$ in a slightly larger open set. Now denote D_t by T. As becomes quickly apparent, just inserting $w = Q_1 T^{r/2}$ back in (6.2) gives real trouble. But

(6.3) $$(T^{r'})_{Q_1} = Q_1 T^{r'} - Q_{1_x} x T^{r'} + Q_{1_{xx}} x^2 T^{r'}/2! - Q_{1_{xxx}} x^3 T^{r'}/3!$$

$(r'+1$ terms in all) has the property that

(6.4) $$[(T^{r'})_{Q_1}, D_x] \equiv 0 \text{ modulo } \quad Q_1{}^{(r'+1)} x^{r'+1} T^{r'}/r'!$$

and

(6.5) $$[(T^{r'})_{Q_1}, xD_t] \equiv -(T^{r'-1})_{Q_{1_t}} \circ xD_t \text{ modulo the above.}$$

This inserting $w = (T^{r'})_{Q_1} u$ in (6.2) gives errors which may either be iterated (the case of (6.5)) or give "ultimate" errors which are all $(xD_t)^{r'}$, and these we've dealt with.

A full application of what we've described so far, then, drops the order by half, picks up

$$C^r r^{s_1} Q_0^{(s_2)} Q_1^{(s_1+1)}/(s_1+1)!$$

which can be bounded by $r!$ if the N_k are well chosen. Further iterations will pick up the factors, now letting $r = q$, $1^{2q} 2^q 3^{q/2} 4^{q/4} 5^{q/8} \ldots$, but this product is bounded by C^q, with C universal. This finishes the proof.

DEPARTMENT OF MATHEMATICS
UNIV. OF ILLINOIS AT CHICAGO CIRCLE
P. O. BOX 4348
CHICAGO, ILLINOIS 60302

BIBLIOGRAPHY

[1] Baouendi, M. S., and Goulaouic, C., *Non-analytic hypoellipticity for some degenerate elliptic operators*, Bull. A.M.S. *78* (1972), 483-486.

[2] Baouendi, M. S., and Sjöstrand, J., *Régularité analytique pour des opérateurs elliptiques singuliers en un point*, Ark. för Mat. *14* (1976), 9-33.

[3] Boutet de Monvel, L., *Hypoelliptic operators with double characteristics and related pseudo-differential operators*, Comm. Pure Appl. Math. *27* (1974), 585-639.

[4] Boutet de Monvel, L. and Trèves, F., *On a class of pseudo-differential operators with double characteristics*, Inventiones Math., *24* (1974), 1-34.

[5] Boutet de Monvel, L., Grigis, A. and Helffer, B., *Parametrixes d'opérateurs pseudodifférentiels à characteristiques multiples* (preprint).

[6] Catlin, D., *Boundary behaviour of holomorphic functions on weakly pseudo-convex domains*, Thesis, Princeton Univ., 1978.

[7] Derridj, M., *Sur la régularité des solutions du problème de Neumann pour* $\bar\partial$ *dans quelques domaines faiblement pseudo-convexes*, J. Diff. Geom., to appear.

[8] ________, *Sur une classe d'opérateurs différentiels hypoelliptiques à coefficients analytiques*, Seminaire Goulaouic-Schwartz, 1970-71, exp. no. 12.

[9] ________, *Sur la régularité Gevrey jusqu'au bord des solutions du problème de Neumann pour* $\bar\partial$, Symp. in Pure Math., XXX, no. 1 (1977), 123-6.

[10] Derridj, M. and Tartakoff, D. S., *On the global real analyticity of solutions to the* $\bar\partial$*-Neumann problem*, Comm. in P.D.E.'s, I (1976), 401-35.

[11] ________, *Sur la régularité locale des solutions du problème de Neumann pour* $\bar\partial$, Journées sur les fonctions analytiques, in Springer Lecture Notes in Mathematics, vol 578 (1977), 207-16.

[12] Derridj, M. and Zuily, C., *Régularité analytique et Gevrey pour des classes d'opérateurs elliptiques paraboliques dégénérés du second ordre*, Astérisque 2 & 3, 1973.

[13] Diederich, K. and Fornaess, J. E., *Pseudoconvex domains with real-analytic boundary*, Ann. of Math., *107*(1978), 371-384.

[14] Dynin, A., *Pseudodifferential operators on the Heisenberg Group*, preprint.

[15] Egorov, Ju. V., *Subellipticity of the $\bar{\partial}$-Neumann problem*, Dokl. Akad. Nauk SSSR 235, no. 5(1977), 1009-1012.

[16] Folland, G. B., *A fundamental solution for a subelliptic operator*, Bull. A.M.S. *79*(1973), 373-376.

[17] ————, *On the Rothschild-Stein lifting theorem*, Comm. in P.D.E. *2* no. 12(1977), 165-191.

[18] Folland, G. B. and Kohn, J. J., *The Neumann problem for the Cauchy-Riemann complex*, Ann. of Math. Studies, No. 75, Princeton Univ. Press, 1972.

[19] Folland, G. B. and Stein, E. M., *Estimates for the $\bar{\partial}_b$-complex and analysis on the Heisenberg group*, Comm. Pure Appl. Math., *27*(1974), 429-522.

[20] Garabedian, P. and Spencer, D. C., *Complex boundary value problems*, Trans. A.M.S., *73*(1952), 223-242.

[21] Grigis, A., *Hypoellipticité et paramétrix pour des opérateurs pseudo-différentiels a characteristiques doubles*, Astérisque 34-5(1976), 183-205.

[22] Grushin, V. V., *On a class of hypoelliptic pseudodifferential operators degenerate on a submanifold*, Mat. Sbornik *84*(126)(1971), pp. 111-134 (Math. USSR Sbornik *13*(1971), 155-185.

[23] Helffer, B., *Sur une classe d'opérateurs hypoelliptiques à characteristiques multiples*, J. Math. Pures et Appl. *55*(1975), 207-215.

[24] ————, *Remarques sur des résultats de G. Métivier sur la non-hypoanalyticité*, preprint.

[25] Helffer, B. and Nourrigat, *Caractérisation des opérateurs hypo-elliptiques homogènes invariants à gauche sur un groupe nilpotent gradué*, preprint.

[26] Hörmander, L., *L^2 estimates and existence theorems for the operator $\bar{\partial}$*, Acta Math., *113*(1965), 89-152.

[27] ————, *Hypoelliptic second order differential equations*, Acta Math. *119*(1967), 147-171.

[28] ————, *Linear partial differential operators*, Springer-Verlag, New York, 1963.

[29] ————, *An introduction to complex analysis in several variables*, Van Nostrand, Princeton, 1966.

[30] ————, *A class of hypoelliptic pseudodifferential operators with double characteristics*, Math. Ann. *217*(1975), 165-188.

[31] Kaplan, A., *A class of nilpotent Lie groups with analytically hypoelliptic sublaplacians*, to appear, Trans., A.M.S.

[32] Kertzman, N., *The Bergman kernel function: differentiability at the boundary*, Math. Ann., *195*(1972), 149-158.

[33] Kohn, J. J., *Harmonic integrals on strongly pseudo-convex manifolds, I*, Ann. of Math., *78*(1963), pp. 112-148 and *II*, *ibid.*, *79*(1964), 450-472.

[34] ———, *Pseudo-differential operators and non-elliptic problems*, Pseudo-Differential Operators, C.I.M.E., Stresa, 1968, Edizioni Cremonese, Rome, 1969, 157-65.

[35] ———, *Global regularity for* $\bar\partial$ *on weakly pseudo-convex manifolds*, Trans. A.M.S., *181* (1973), 273-292.

[36] ———, *Subellipticity of the* $\bar\partial$*-Neumann problem on pseudo-convex domains: sufficient conditions*, Acta Math. *142*(1979), 79-122.

[37] ———, *Sufficient conditions for subellipticity on weakly pseudo-convex domains*, Proc. Nat. Acad. Sci., *74*(1977), 2214-2216.

[38] Kohn, J. J. and Nirenberg, L., *Non-coercive boundary value problems*, Comm. Pure Appl. Math., *18*(1965), 443-492.

[39] Kohn, J. J. and Spencer, D. C., *Complex Neumann problems*, Ann. of Math., *66*(1957), 89-140.

[40] Komatsu, G., *Global analytic-hypoellipticity of the* $\bar\partial$*-Neumann problem*, Töhoku Math J. Ser. 2, *28*(1976), 145-56.

[41] Matsuzawa, T., *Sur les équations* $u_{tt} + t^{\alpha}u_{xx}$, Nagoya Math. J., *42* (1971), 43-55.

[42] Menikoff, A., *Hypoelliptic operators with double characteristics*, in Seminar on Singularities of Solutions of Linear Partial Differential Equations, L. Hörmander, ed., Annals of Math. Studies, Princeton Univ. Press, 1979, 65-79.

[43] Métivier, G., *Une classe d'opérateurs non-hypoelliptiques analytiques*, preprint.

[44] ———, *Propriété des itérés et ellipticité*, Comm. P.D.E., *3*(1978), 827-876.

[45] Morrey, C. B. Jr., *The analytic embedding of abstract real-analytic manifolds*, Ann. of Math., *68*(1958), 159-201.

[46] Morrey, C. B. Jr. and Nirenberg, L., *On the analyticity of the solutions of linear elliptic systems of partial differential equations*, Comm. Pure Appl. Math., X (1957), 271-90.

[47] Nirenberg, L. and Trèves, F., *On local solvability of linear partial differential equations, I and II*, Comm. Pure Appl. Math., *23*(1970), 1-38 and 459-510.

[48] Oleinik, O. A. and Radkevic, E. V., *Second order equations with non-negative characteristic form*, Am. Math. Soc., Providence, R. I., 1973.

[49] Radkevic, E. V., *A priori estimates and hypoelliptic equations with multiple characteristics*, Dokl. Akad. Nauk SSSR *187*, pp. 274-77 (1969), Sov. Math. Dokl. *10*(1969), 849-853.

[50] ———, *Ob odnoi teoreme L. Xöpmandera*, Usp. Mat. Nauk *24* (1969), 233-234.

[51] Rockland, C., *Hypoellipticity and eigenvalue asymptotics*, Springer Lecture Notes in Mathematics, vol. 464, Springer-Verlag, New York, 1975.

[52] ———, *Hypoellipticity on the Heisenberg group-representation theoretic criteria*, to appear, Trans. A.M.S.

[53] Rothschild, L. P., *A criterion for hypoellipticity for operators constructed from vector fields*, to appear, Comm. in P.D.E.

[54] Rothschild, L. P. and Stein, E. M., *Hypoelliptic differential operators and nilpotent Lie groups*, Acta Math., *137*(1976), 247-320.

[55] Rothschild, L. P. and Tartakoff, D. S., *Parametrices with* C^∞ *error for* $\square_b$ *and operators of Hörmander type*, to appear, Proc. of Park City, Utah conference, Jan., 1977.

[56] Sjöstrand, J., *Parametrices for pseudodifferential operators with multiple characteristics*, Ark. för Mat. *12*(1974), 85-130.

[57] Taylor, M. E., *Pseudo-differential operators, part II*, to appear.

[58] Tartakoff, D. S., *On the global real analyticity of solutions to* $\square_b$, Comm. P.D.E., I(1976), 283-311.

[59] ———, *Gevrey hypoellipticity for subelliptic boundary value problems*, Comm. Pure Appl. Math., *26*(1973), 251-312.

[60] ———, *Local Gevrey and quasianalytic hypoellipticity for* $\square_b$, Bull. A.M.S., *82*(1976), 740-742.

[61] ———, *Local analytic hypoellipticity for* $\square_b$ *on non-degenerate Cauchy-Riemann manifolds*, Proc. Nat. Acad. Sci., *75*, no. 7(1978), 3027-8.

[62] ———, *On the local analyticity of solutions to* $\square_b$ *and the* $\bar\partial$*-Neumann problem*, to appear, Acta. Math.

[63] ———, *Remarks on isolated degeneracies of the Levi form and the analytic hypoellipticity of* $\square_b$ *and related partial differential operators*, to appear, proc. of conference in Williamstown, Mass., July, 1978.

[64] Trèves, F., *An invariant criterion for hypoellipticity*, Am. J. of Math., *83*(1961), 645-668.

[65] ———, *Analytic hypoelliptic partial differential equations of principal type*, Comm. Pure Appl. Math., *24*(1971).

[66] ———, *Analytic hypo-ellipticity of a class of pseudo-differential operators with double characteristics and applications to the* $\bar\partial$*-Neumann problem*, Comm. P.D.E., *3*, no. 6&7(1978), 475-642.

THE FERMAT SURFACE AND ITS PERIODS

Marvin D. Tretkoff

The Fermat surface S of degree d in P^3 is defined, using homogeneous coordinates [T:Z:W:X], by the equation $Z^d + W^d + X^d = T^d$. In this paper, we give a cellular decomposition of S, determine an explicit set of 2-cycles representing a basis for $H_2(S, \mathbf{Z})$, and calculate the periods of the holomorphic 2-forms on S along these cycles. In a sense, our results may be viewed as the extension from one to two complex dimensions of the Appendix to [1] by Rohrlich. We also note that in [1] Gross determined the periods of the Fermat surface and its n-dimensional analogues up to factors which are algebraic numbers, and that Sasakura [6] has determined the Picard number of the Fermat surface. Finally, we note a related paper by Stevenson [7].

In this paper we work exclusively with homology with integral coefficients and use the standard notation for cycles, boundaries, etc. We often use the affine coordinates $x = X/T$, $z = Z/T$, $w = W/T$, and refer to the projective line $P^1 \subset P^3$ defined by $Z = W = 0$ as the *x-axis*. Its points are labelled by $\lambda \in \mathbf{C} \cup \{\infty\}$, where $\lambda = [1:0:0:\lambda]$ if $\lambda \in \mathbf{C}$ and $\lambda = [0:0:0:1]$ if $\lambda = \infty$. Similarly, the plane $P^2_\lambda \subset P^3$ is defined by $X = \lambda T$ when $\lambda \in \mathbf{C}$ and $T = 0$ when $\lambda = \infty$. Finally, we let $\eta = e^{\pi i/d}$ and $\zeta = e^{2\pi i/d}$.

Our study of the topology of the Fermat surface in P^3 is based on an analysis of the pencil of complete curves, C_λ, cut out on it by the

Recent Developments in Several Complex Variables
0-691-08285-5/81/000413-16 \$00.80/0 (cloth)
0-691-08281-2/81/000413-16 \$00.80/0 (paperback)

planes P^2_λ, $\lambda \in \mathbf{C} \cup \{\infty\}$. Since the affine equation of C_λ is $w^d = (1-\lambda^d) - z^d$, we see that it has genus $g = \frac{(d-1)(d-2)}{2}$ when $\lambda^d \neq 1$. In case $\lambda = \zeta^j$, $1 \leq j \leq d$, C_λ is the union of d lines meeting transversally at the point $P_j = [1{:}0{:}0{:}\zeta^j]$.

From the viewpoint of the ambient projective space, it is immediate that our pencil has "base points at infinity." More precisely, the points $b_k = [0{:}1{:}\eta\zeta^k{:}0]$, $k = 1, \cdots, d$, lie on every C_λ; and these are the only points common to all curves in our pencil. Of course, the existence of this base locus prevents us from viewing our pencil as a fibre bundle with base space X, the x-axis (that is, the collection of points $[T{:}0{:}0{:}X]$) with $x = \zeta^j$, $1 \leq j \leq d$, removed. However, we shall see that S can be obtained by attaching 4-balls B_k centered at b_k to a regular 3-dimensional CW-complex W. The complex W consists of the totality of 1-skeletons, W_λ, of the classical cellular decompositions of C_λ; and the subspace of W given by $\{W_\lambda | \lambda^d \neq 1\}$ forms a fibre bundle over X. In Section 1, we shall recall some details of the cellular decomposition of C_λ for the case $\lambda = 0$; the general case is similar and should not require additional explanation. In Sections 2 and 3 we shall give a cellular decomposition of W and use it to describe a set of 2-cycles representing a basis for $H_2(W)$. Since W is the 3-skeleton of S, these 2-cycles also represent a basis for $H_2(S)$. The periods are calculated in Section 4.

1. *The Fermat curve*

The 1-skeleton, W_0, of the Fermat curve, $C = C_0$, is succinctly described as the topological join, $J = \mathbf{Z}_d * \mathbf{Z}_d$, of two copies of the cyclic group of order d. More precisely, its 0-cells are the points $0_k = [1{:}0{:}\zeta^k{:}0]$ and $R_k = [1{:}\zeta^k{:}0{:}0]$, $k = 1, \cdots, d$. Under component-wise multiplication, the points 0_k and R_k each form cyclic groups of order d. Viewing C_0 as a branched covering of the z-axis (that is, points of the form $[T{:}Z{:}0{:}0]$), the points 0_k lie above the origin, $z = 0$, and the R_k are the branch points of the Riemann surface. If e_k denotes the real oriented segment on the z-axis joining the origin to the point $z = \zeta^k$, then there is a unique oriented path $e_{j,k}$ beginning at 0_j, ending at R_k, and projecting onto e_k. The totality of $e_{j,k}$ constitutes the 1-skeleton, W_0, of C_0.

We obtain C_0 from W_0 by attaching 2-cells D_k, $k = 1, \cdots, d$, as indicated in Figure 1. In fact, the attaching maps are obtained by the classical process of analytic continuation which associates to each branch point, R_k, a permutation, π_k, of the power series solutions of $w^d = 1 - z^d$ at the origin. The power series are identified with the points 0_j; and it is easily seen that $\pi_k(0_j) = 0_{j+1}$ for each k.

It is also easy to see that $H_1(J)$ is free abelian of rank $(d-1)^2$. For example, contracting the tree formed by $e_{i,i}$, $i = 1, \cdots, d$, and $e_{i,i-1}$,

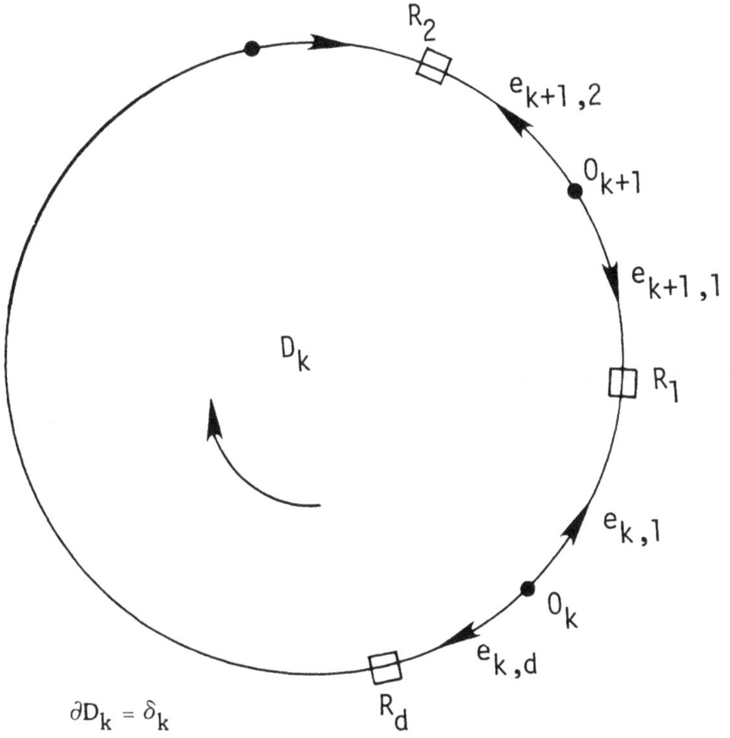

Figure 1

$i = 2, \cdots, d$, to a point yields a bouquet of $(d-1)^2$ circles. It is also easy to see that the 1-cycles

$$\gamma_{i,j} = e_{i,j} - e_{i+1,j} + e_{i+1,j+1} - e_{i,j+1}$$

with $1 \leq i,j \leq d-1$ form a basis for $H_1(W_0)$.

For later purposes it is convenient to let i and j run from 1 to d and observe that $\gamma_{d,j} = -(\gamma_{1,j} + \cdots + \gamma_{d-1,j})$ and $\gamma_{i,d} = -(\gamma_{i,1} + \cdots + \gamma_{i,d-1})$. Moreover, we have $\phi(\gamma_{i,j}) = \gamma_{i+1,j+1}$, where ϕ is the automorphism of J given by $\phi(e_{i,j}) = e_{i+1,j+1}$. Viewing J as $W_0 \subset P^3$, we see that $\phi = AB$, where $A([T:Z:W:X]) = [T:Z:\zeta W:X]$ and $B([T:Z:W:X]) = [T:\zeta Z:W:X]$ are automorphisms of P^3. Moreover, since $A^{i-1}B^{j-1}\gamma_{1,1} = \gamma_{i,j}$, we see that $H_1(W_0)$ is a cyclic $\mathbf{Z}[A,B]$ - module.

2. *A cellular decomposition of the Fermat surface*

In this section we shall exhibit a cellular decomposition of S with 3-skeleton W. Some preliminary remarks should make clear how to obtain S by attaching 4-balls B_k with centers b_k, $k = 1, \cdots, d$, to W.

First, we note that a 4-ball, B, is the join of its center, b, with its bounding 3-sphere. Next, we recall that the Hopf fibration allows us to view the 3-sphere as a family of circles $S^1(\lambda)$, $\lambda \in S^2$, parametrized by the 2-sphere. It follows that the 4-ball contains the join, $D(\lambda)$, $\lambda \in S^2$, of b with $S^1(\lambda)$, and that the union of the 2-disks $D(\lambda)$ is B. Of course, $D(\lambda_1) \cap D(\lambda_2) = b$ when $\lambda_1 \neq \lambda_2$.

Returning to the Fermat surface S, we recall that each plane section C_λ, $\lambda^d \neq 1$, is obtained from its 1-skeleton W_λ by attaching 2-disks, $D_k(\lambda)$, along 1-cycles, $\delta_k(\lambda)$, $k = 1, \cdots, d$. Of course, W_λ is just the point $P_j \in P^3$ when $\lambda = \zeta^j$; and $D_k(\lambda)$ is attached by mapping its entire boundary to P_j in this case. Now, B_k may be attached to W by mapping $S^1(\lambda)$ homeomorphically onto $\delta_k(\lambda)$ when $\lambda^d \neq 1$, and by mapping $S^1(\lambda)$ onto P_j when $\lambda = \zeta^j$. The continuity of the attaching map follows from the fact that the branches of the algebraic functions defined by $z^d = 1-x^d$ and $w^d = 1-x^d$ are continuous functions of x.

Finally, we see that the interior of B_k can be embedded in $S \subset P^3$ by mapping the interior of each $D(\lambda)$ homeomorphically onto the interior of the 2-cell $D_k(\lambda)$ belonging to C_λ. Evidently, b_k is mapped to the base point at infinity $[0:1:\eta\zeta^k:0]$.

We shall now show that W admits a decomposition as an oriented regular CW-complex. This will be accomplished by studying the loci swept out by the individual cells of W_λ as λ varies over P^1. The special nature of the Fermat curve makes this investigation possible.

It is clear from our construction that W_λ is completely determined by its 0-cells $0_j(\lambda)$ and $R_j(\lambda)$, $j=1,\cdots,d$. Now, recall that these points are given in projective coordinates by $[T:0:\zeta^j\rho:\lambda T]$ and $[T:\zeta^j\rho:0:\lambda T]$ respectively; here $\rho^d = 1-\lambda^d$ and $-\frac{\pi}{d} < \arg\rho \leq \frac{\pi}{d}$. Thus, we see that the totality of $0_j(\lambda)$, $\lambda \in P^1$, forms the Fermat curve $W^d = T^d - X^d$ in the plane defined by $Z=0$. Similarly, the $R_j(\lambda)$, $\lambda \in P^1$, form the Fermat curve $Z^d = T^d - X^d$ in the plane $W = 0$.

Now, the points $R_j(\lambda)$ may be associated with the branches $z_1(x), \cdots, z_d(x)$ of the algebraic function defined by the affine Fermat equation $z^d = 1-x^d$. The relative position of these points is known explicitly for all values of x because $z_j(x) = \zeta^j z_d(x)$. Thus, if x traverses a loop in P^1, we not only know the monodromy transformation (that is, the permutation) of the $z_j(x)$ or $R_j(\lambda)$, in effect, we even know the braid swept out by them. Of course, a similar remark applies to the $0_j(\lambda)$. It follows that we can explicitly determine the locus swept out by a 0-cell or 1-cell of W_λ as λ varies.

It will be convenient to view the x-axis (that is, the collection of points of the form $[T:0:0:X]$) as the quotient of a regular 2d-gon labelled as in Figure 2. Identification of the edges f'_j and f''_j yields the line segment f_j joining 0 to ζ^j on the x-sphere. Of course, an n-sheeted covering of the x-sphere which is branched at the ζ^j may be viewed as a quotient of the product of D and a set of n points. Similarly, we shall describe W as a quotient of $J \times D$, the product of the join $J = Z_d * Z_d$ and the 2d-gon D. It will be useful to set $J(t) = J \times \{t\} \subset J \times D$.

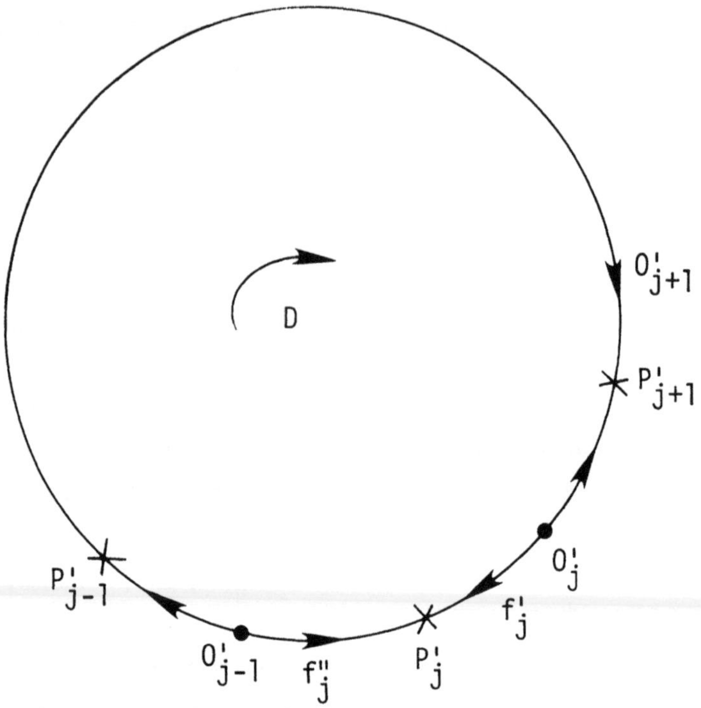

Figure 2

Our construction takes place in two stages. First, we collapse each $J(P'_j)$ to a single vertex P_j, $j = 1, \cdots, d$. Note that the boundary of the resulting complex W' consists of the 2d cones C'_j and C''_j whose vertices are at P_j and whose bases are $J(0'_j)$ and $J(0'_{j-1})$ respectively. Next, we obtain W from W' by identifying $(p, t'') \in J(t'')$ with $(\phi p, t') \in J(t')$, $p \in J$, whenever $t' \in f'_j$ and $t'' \in f''_j$ are corresponding points on the boundary of D. Recall that $\phi e_{i,j} = e_{i+1,j+1}$.

Now, the obvious cellular decomposition of W is given in Table 1. The attaching maps are determined by decomposing the boundary of each k-cell as illustrated in the accompanying figure, and identifying each of the resulting (k–1)-cells with the cell with the same label in the (k–1)-skeleton of W.

Table 1

0-cells

$0_j, R_j, P_j;\qquad 1 \le j \le d$

1-cells

$e_{ij},\qquad \partial e_{ij} = R_j - 0_i,\qquad 0_i \bullet\!\xrightarrow{\;e_{ij}\;}\!\square\, R_j$

$f_{ij},\qquad \partial f_{ij} = P_j - 0_i,\qquad 0_i \bullet\!\xrightarrow{\;f_{ij}\;}\!\times P_j \qquad 1 \le i, j \le d$

$g_{ij},\qquad \partial g_{ij} = P_j - R_i,\qquad R_i \square\!\xrightarrow{\;g_{ij}\;}\!\times P_j$

2-cells

$\Delta_{i,j,k},\qquad \partial\Delta_{i,j,k} = e_{ij} - f_{i,k} + g_{j,k}\qquad 1 \le i, j, k \le d$

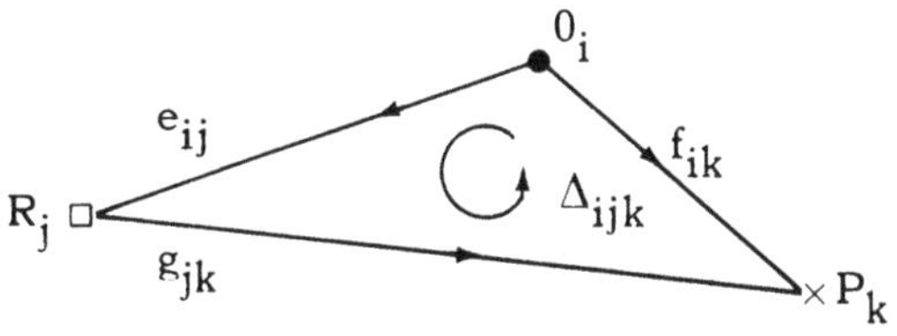

$D'_i,\qquad \partial D'_i = (-f_{i,1} + f_{i+1,1}) + (-f_{i+1,2} + f_{i+2,2})$

$\qquad\qquad + \cdots + (-f_{i+d-1,d} + f_{i+d,d})$

$D''_j,\qquad \partial D''_j = (-g_{j,1} + g_{j+1,1}) + \cdots + (-g_{j+d-1,d} + g_{j+d,d})$

$1 \le i, j \le d$

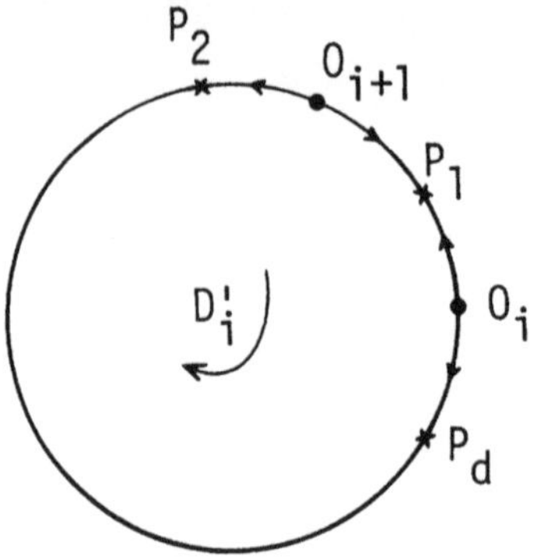

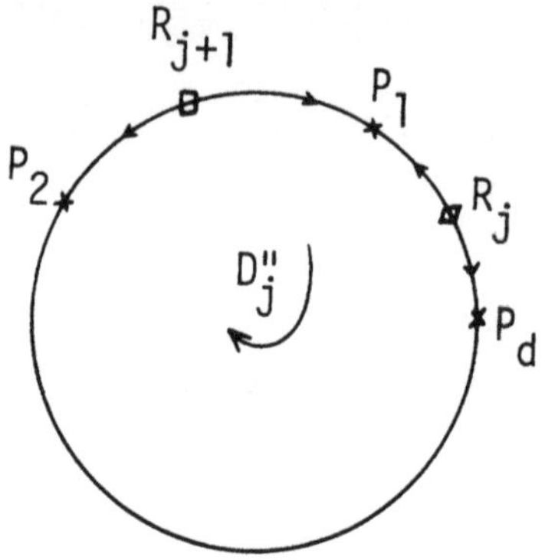

3-cells

$B_{i,j} \qquad 1 \leq i, j \leq d$

$$\partial B_{i,j} = -D'_i + D''_j + (\Delta_{i,j,1} - \Delta_{i+1,j+1,1})$$
$$+ \cdots + (\Delta_{i+d-1,j+d-1,d} - \Delta_{i+d,j+d,d}).$$

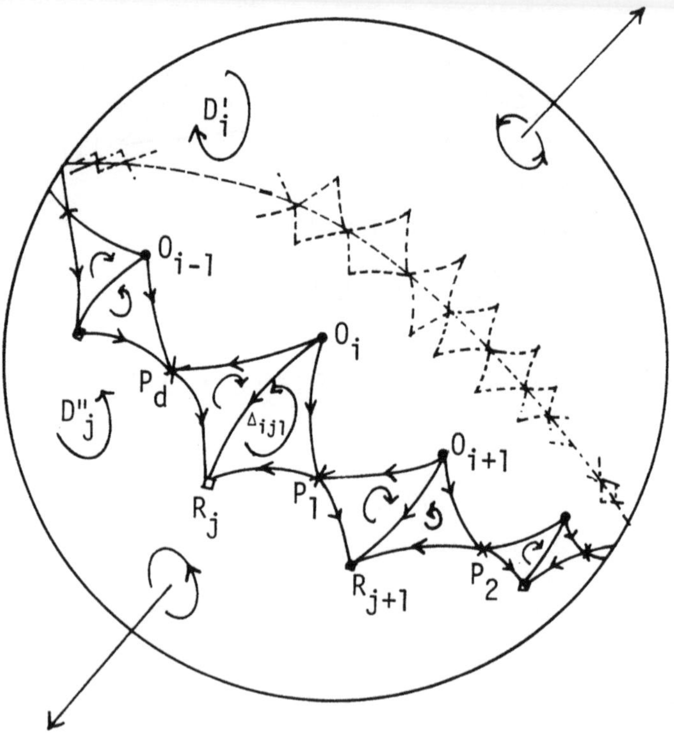

3. *A homology basis for the Fermat surface*

In this section we shall use our cell decomposition of S to explicitly determine a set of 2-cycles representing a basis for $H_2(S) = H_2(W)$. We begin by exhibiting two sets of 2-cycles which together freely generate $Z_2(W)$. First, we let $\Gamma'_{i,j,k}$ denote the locus $\gamma_{i,j}(x)$, $x \in f_k$; this locus is obviously a cone with vertex P_k and base $\gamma_{i,j} = \gamma_{i,j}(0)$, and we obtain a collection of 2-cycles by setting $\Gamma_{i,j,k} = \Gamma'_{i,j,k} - \Gamma'_{i,j,k+1}$. Observe that the union of the supports of the $\Gamma_{i,j,k}$ is the subcomplex W^* of W formed by the totality of 2-cells $\Delta_{i,j,k}$; it is immediate that W^* is homeomorphic to the topological join of three copies of Z_d. Now, it is well known ([3], [4]) and easily seen that the second homology group of this join is torsion free and has rank $(d-1)^3$. In fact, it is easily seen that $\Gamma_{i,j,k}$, $1 \leq i, j, k \leq d-1$, represents a basis for $H_2(W^*) = Z_2(W^*)$; here we only note the obvious relation $\sum_{k=1}^{d} \Gamma_{i,j,k} = 0$, $1 \leq i, j \leq d$, and that

$$\sum_{i=1}^{d} \gamma_{i,j} = 0, \quad 1 \leq j \leq d$$

and

$$\sum_{j=1}^{d} \gamma_{i,j} = 0, \quad 1 \leq i \leq d,$$

imply that

$$\sum_{i=1}^{d} \Gamma_{i,j,k} = 0, \quad 1 \leq j,k \leq d$$

and

$$\sum_{j=1}^{d} \Gamma_{i,j,k} = 0, \quad 1 \leq i,k \leq d.$$

Next, we introduce a set of 2-cycles Δ'_i and Δ''_i, $1 \leq i \leq d$, as follows:

$$\Delta'_i = D'_i + [(-\Delta_{i,1,1} + \Delta_{i+1,1,1}) + \cdots + (-\Delta_{i+d-2,d-1,d-1} + \Delta_{i+d-1,d-1,d-1})] + + [(\Delta_{i,1,d} - \Delta_{i+1,1,d}) + \cdots + (\Delta_{i+d-2,d-1,d} - \Delta_{i+d-1,d-1,d})]$$

and

$$\Delta''_j = D''_j + [(+\Delta_{1,j,1} - \Delta_{1,j+1,1}) + \cdots + (+\Delta_{d-1,j+d-2,d-1} - \Delta_{d-1,j+d-1,d-1})] + [(-\Delta_{1,j,d} + \Delta_{1,j+1,d}) + \cdots + (-\Delta_{d-1,j+d-2,d} + \Delta_{d-1,j+d-1,d})] .$$

Since the 2-cells D'_i and D''_i are members of a basis for the 2-chains on W, an integral linear combination of Δ'_i, Δ''_j, and $\Gamma_{i,j,k}$ cannot vanish if it contains Δ'_i or Δ''_i with a non-zero coefficient. Moreover, any 2-cycle containing D'_i or D''_i with a non-zero coefficient differs from a suitable integral linear combination of the Δ'_i and Δ''_i by an element of $Z_2(W^*)$. It follows that $Z_2(W)$ is freely generated by the $(d-1)^3 + 2d$ given 2-cycles Δ'_i, Δ''_i, $1 \leq i \leq d$ and $\Gamma_{i,j,k}$, $1 \leq i,j,k \leq d-1$.

Since $\partial B_{ij} + (\Delta'_i - \Delta''_j)$, $(\partial B_{i1} - \partial B_{j1}) + (\Delta'_i - \Delta'_j)$, and $(\partial B_{1i} - \partial B_{1j}) - (\Delta''_i - \Delta''_j)$ all belong to $Z_2(W^*)$, we see that

$$H_2(W) = \frac{Z_2(W)}{B_2(W)} \cong \frac{(Z_2(W^*) + Z\Delta'_1) + B_2(W)}{B_2(W)};$$

and, applying the isomorphism theorem, we obtain,

$$H_2(W) \cong \frac{(Z_2(W^*) + Z\Delta'_1)}{B_2(W) \cap (Z_2(W^*) + Z\Delta'_1)} .$$

Now, suppose that $\Gamma = \sum a_{ij} B_{ij}$ satisfies $\partial\Gamma \in B_2(W) \cap (Z_2(W^*) + Z\Delta'_1)$, but $\partial\Gamma \notin B_2(W) \cap Z_2(W^*)$. Then $\sum_j a_{1j} \neq 0$; otherwise, D'_1 would not appear in $\partial\Gamma$. Moreover, D'_i, $2 \le i \le d$ and D''_i, $1 \le i \le d$, do not belong to $\partial\Gamma$, so $\sum_j a_{ij} = 0$, $2 \le i \le d$, and $\sum_i a_{ij} = 0$, $1 \le j \le d$; and it follows that $\sum_j a_{1j} = 0$. From this contradiction, we conclude that $B_2(W) \cap (Z_2(W^*) + Z\Delta'_1) = B_2(W) \cap Z_2(W^*)$, and consequently,

$$H_2(W) \cong \frac{Z_2(W^*) + Z\Delta'_1}{B_2(W) \cap Z_2(W^*)} .$$

Now, we shall show that $B_2(W) \cap Z_2(W^*)$ is the image of an easily described homomorphism. First, we define a homomorphism, h, from $C_1(J)$ to $C_3(W)$ by setting $h(e_{ij}) = B_{ij}$. Restricting h to $Z_1(J) = H_1(J)$ and composing with the boundary operator, we obtain the desired homomorphism $d = \partial h$ from $H_1(J)$ to $B_2(W)$. Since

$$\begin{aligned} d\gamma_{ij} &= (\Gamma'_{i,j,1} - \Gamma'_{i+1,j+1,1}) + (\Gamma'_{i+1,j+1,2} - \Gamma'_{i+2,j+2,2}) \\ &\quad + \cdots + (\Gamma'_{i+d-1,j+d-1,d} - \Gamma'_{i,j,d}) \\ &= -(\Gamma_{i+1,j+1,1} + \cdots + \Gamma_{i+d,j+d,d}) , \end{aligned}$$

we see that D'_i and D''_i do not occur in the image of d and, consequently, this image belongs to $B_2(W) \cap Z_2(W^*)$. Conversely, if $\Gamma = \sum a_{ij} B_{ij}$, $a_{ij} \in Z$, satisfies $\partial\Gamma \in B_2(W) \cap Z_2(W^*)$, then $\sum_j a_{ij} = 0$, $1 \le i \le d$, and $\sum_i a_{ij} = 0$, $1 \le j \le d$. These conditions on the a_{ij} imply that the 1-chain $\gamma = \sum a_{ij}\gamma_{ij}$ is a 1-cycle; and, since $h\gamma = \Gamma$, we see the image of d is $B_2(W) \cap Z_2(W^*)$.

Thus, $H_2(W)$ is the abelian group generated by Δ'_1 and $\Gamma_{i,j,k}$, $1 \le i, j, k \le d$, with the defining relations

(1) $\sum_{i=1}^{d} \Gamma_{i,j,k} = 0, \quad 1 \leq j,k \leq d$

(2) $\sum_{j=1}^{d} \Gamma_{i,j,k} = 0, \quad 1 \leq i,k \leq d$

(3) $\sum_{k=1}^{d} \Gamma_{i,j,k} = 0, \quad 1 \leq i,j \leq d$

(4) $\Gamma_{i+1,j+1,1} + \cdots + \Gamma_{i+d,j+d,d} = 0$.

Using (1), (2), (3), and (4), it is readily seen that $H_2(W)$ is freely generated by Δ'_1, $\Gamma_{1,1,1}, \cdots, \Gamma_{d-1,1,1}$, and $\Gamma_{i,j,k}$, $1 \leq i,j \leq d-1$; $2 \leq k \leq d-1$. We note that each of these cycles is homeomorphic to the 2-sphere. Moreover, in Lefschetz's terminology, [2], the 2-cycles $\Gamma_{i,j,k}$ are "effective" because they do not meet the hyperplane section C_∞. Introducing an automorphism of P^3 by setting $C([T:Z:W:X]) = [T:Z:W:\zeta X]$, we see that $A^{i-1}B^{j-1}C^{k-1}\Gamma_{1,1,1} = \Gamma_{i,j,k}$, so the effective 2-cycles form a cyclic $Z[A, B, C]$-module. Thus, we have proved the following

THEOREM. *With the notation above,*

(1) *The 2-cycles* Δ'_1, $\Gamma_{1,1,1}, \cdots, \Gamma_{d-1,1,1}$, *and* $\Gamma_{i,j,k}$, $1 \leq i,j \leq d-1$, $2 \leq k \leq d-1$, *represent a basis for* $H_2(S) = H_2(S, Z)$.

(2) *The subgroup* $H_2(S)_0 \subset H_2(S)$ *of effective cycles is a cyclic* $Z[A, B, C]$*-module.*

This may be viewed as a generalization of Rohrlich's result [1].

Next, we note that $\Delta'_1 + \cdots + \Delta'_d$ is the sum of the plane section $W = 0$ and an effective cycle and that $\Delta'_1 - \Delta'_j$ is homologous to an effective cycle. It follows that $d\Delta'_1$ differs from the plane section $W = 0$ by an effective cycle; this difference can be determined explicitly. Now, Pham [4] determined the intersection matrix of the affine Fermat surface with respect to the basis $\Gamma_{i,j,k}$, so the intersection matrix of our homology basis can also be determined.

4. *Periods of holomorphic 2-forms*

We shall now evaluate the periods of the holomorphic 2-forms on the Fermat surface of degree d. If Γ is an effective 2-cycle and ω a holomorphic 2-form, we shall see that

$$\int_{\Gamma} \omega = K(\Gamma, \omega) \int_{\Delta_{1,1,1}} \omega ,$$

where $K(\Gamma, \omega)$ is a constant depending on Γ and ω. Now, a classical formula of M. Noether [5] asserts that a 2-form on a nonsingular surface S of degree d is holomorphic if and only if it is of the form

$$\omega = \frac{Q(x, z, w)\, dx\, dz}{f_w} ,$$

where Q is a polynomial of degree at most $d-4$ and $f(x, z, w) = 0$ is the affine equation of S. It follows that a basis for the holomorphic 2-forms on the Fermat surface is given by

$$\omega(a, b, c) = x^{a-1} z^{b-1} w^{c-d}\, dx\, dz ,$$

where $0 < a, b, c$ and $a+b+c \leq d-1$. Thus, our task is reduced to evaluating $\int_{\Delta_{1,1,1}} \omega(a, b, c)$ and determining $K(\Gamma_{i,j,k}, \omega(a, b, c))$.

We may parametrize $\Delta_{1,1,1}$ by setting *

$$x = \zeta u \qquad 0 \leq u \leq 1$$

$$z = (1-u^d)^{1/d} v \qquad 0 \leq v \leq 1$$

$$w = (1-x^d-z^d)^{1/d} = (1-u^d)^{1/d}(1-v^d)^{1/d} .$$

In terms of u and v, we now have

*See note at end of paper.

$$\omega(a, b, c) = \zeta^a u^{a-1}(1-u^d)^{((b+c)/d)-1} v^{b-1}(1-v^d)^{(c/d)-1} du\, dv ,$$

so

$$\int_{\Delta_{1,1,1}} \omega(a,b,c) = \zeta^a \iint_{0\le u,v\le 1} u^a(1-u^d)^{((b+c)/d)-1} v^{b-1}(1-v^d)^{(c/d)-1} du\, dv$$

$$= \zeta^a \int_0^1 u^a(1-u^d)^{((b+c)/d)-1} du \int_0^1 v^{b-1}(1-v^d)^{(c/d)-1} dv .$$

Setting $\tau = u^d$ and $\sigma = v^d$, we obtain

$$\int_{\Delta_{1,1,1}} \omega(a,b,c)$$

$$= (\zeta^a/d^2) \int_0^1 \tau^{(a/d)-1}(1-\tau)^{((b+c)/d)-1} d\tau \int_0^1 \sigma^{(b/d)-1}(1-\sigma)^{(c/d)-1} d\sigma$$

$$= \frac{a}{d^2} \frac{\Gamma(a/d)\Gamma((b+c)/d)}{\Gamma((a+b+c)/d)} \frac{\Gamma(b/d)\Gamma(c/d)}{\Gamma((b+c)/d)}$$

$$= \frac{a}{d^2} \cdot \frac{\Gamma(a/d)\Gamma(b/d)\Gamma(c/d)}{\Gamma((a+b+c)/d)} .$$

Since $A^{i-1}B^{j-1}C^{k-1}\Delta_{1,1,1} = \Delta_{i,j,k}$, we see that

$$\int_{\Delta_{i,j,k}} \omega = \int_{\Delta_{1,1,1}} \omega^* ,$$

where ω^* denotes the image of ω under the automorphism of 2-forms induced by $A^{i-1}B^{j-1}C^{k-1}$. Expressing $\omega(a, b, c)$, A, B, and C in terms of the affine coordinates (x, z, w), we find that $\omega(a, b, c)$ is multiplied by the constants ζ^{c-d}, ζ^b and ζ^a under application of

A, B and C respectively. Thus,

$$\int_{\Delta_{i,j,k}} \omega(a,b,c) = \zeta^{a(k-1)+b(j-1)+(c-d)(i-1)} \int_{\Delta_{1,1,1}} \omega(a,b,c)$$

and it follows that

$$\int_{\Gamma_{i,j,k}} \omega(a,b,c) = K \frac{\Gamma(a/d)\Gamma(b/d)\Gamma(c/d)}{\Gamma((a+b+c)/d)},$$

where

$$K = \frac{(1-\zeta)(1-2\zeta^b+\zeta^{b+c-d})\zeta^{a(k-1)+b(j-1)+(c-d)(i-1)+2}}{d^2}.$$

Recalling that $d\Delta'_1$ differs from the section $W=0$ by an effective cycle and that the integral of a holomorphic 2-form ω vanishes along an algebraic curve, we see that the integral of ω along Δ'_1 is determined by its values along effective cycles.

Acknowledgment

The author would like to thank the Mathematics Department of Princeton University for their kind hospitality during the academic year 1978-79 when this paper was written. He also wishes to thank Professor A. Adler for many interesting conversations about the present manuscript.

STEVENS INSTITUTE OF TECHNOLOGY
HOBOKEN, NEW JERSEY 07030

REFERENCES

[1] Gross, B. H., *On the Periods of Abelian Integrals and a Formula of Chowla and Selberg (with an appendix by D. E. Rohrlich)*, Inventiones Math. *45*(1978), 193-211.

[2] Lefschetz, S., *L'analysis situs et la geometrie algebrique*, reprinted in Selected Papers, Chelsea Publishing Co., New York, 1971.

[3] Milnor, J., *Singular Points of Complex Hypersurfaces*, Princeton University Press, 1968.

[4] Pham, F., *Formules de Picard-Lefschetz generalisees et ramification des integrales*, Bull. Soc. Math. France 93 (1965), 333-367.

[5] Picard E., and Simart, G., *Theorie des Fonctions Algebriques*, reprinted by Chelsea Publishing Co., New York 1971.

[6] Sasakura, N., *On some results on the Picard numbers of certain algebraic surfaces*, J. Math. Soc. Japan, *20*, 1-2 (1968), 297-321.

[7] Stevenson, E., *Integral representations of algebraic cohomology classes on hypersurfaces*, Pacific J. Math., LXXI (1977), 197-212.

Added in proof: The constant K is incorrect because of an error in parameterizing $\Delta_{1,1,1}$. The correct result is easily obtained by the method used in this paper. A correction is available from the author.

SHEAF COHOMOLOGY ON 1-CONVEX MANIFOLDS

Stephen S. -T. Yau*

§1. *Introduction*

Let M be a complex manifold of complex dimension n. A real-valued C^∞-function ϕ on M is said to be strongly p-plurisubharmonic if and only if the hermitian form

$$\sum_{i,j=1}^{n} \frac{\partial^2 \phi}{\partial z_i \partial \bar{z}_j} dz_i d\bar{z}_j \tag{0.1}$$

has at least n-p-1 positive eigenvalues with respect to any system of local coordinates $(z_1, \cdots, z_n)$. The complex manifold M is said to be p-pseudoconvex if there is a compact subset $B \subset M$, and a C^∞ real-valued function ϕ on M, which is strongly p-plurisubharmonic outside B, and such that for each $c \in \mathbf{R}$, the set

$$B_c = \{x \in M : \phi(x) < c\} \tag{0.2}$$

is relatively compact in M. The complex manifold M is said to be p-complete if there exists a strongly p-plurisubharmonic exhaustion function on M, i.e. the set B_c in (0.2) is relatively compact in M for all

*Research supported by NSF Grant MCS 77-15524.

Recent Developments in Several Complex Variables
0-691-08285-5/81/000429-24 \$01.20/0 (cloth)
0-691-08281-2/81/000429-24 \$01.20/0 (paperback)

$c \in \mathbf{R}$. In 1962 Andreotti-Grauert [1] generalized the finiteness theorem for compact complex spaces of Cartan-Serre. They proved the following:

THEOREM (Andreotti-Grauert). *Let* M *be a complex manifold of dimension* n. *For any coherent analytic sheaf* $\mathfrak{F}$ *on* M,

(1) *if* M *is* p-*pseudoconvex, then* $\dim H^i(M, \mathfrak{F}) < \infty$ *for all* $i \geq p$

(2) *if* M *is* p-*complete, then* $\dim H^i(M, \mathfrak{F}) = 0$ *for all* $i \geq p$.

In 1962, Narasimhan [31] made an important contribution to complex function theory. He proved that any 1-pseudoconvex manifold is actually holomorphically convex, contains a maximal compact analytic set of dimension greater than zero, and is a proper modification of a Stein space at finite numbers of isolated singular points. (The corresponding facts for p-pseudoconvex manifolds are false if $p > 1$.) On the other hand, given a Stein analytic space V with isolated singularities, by the Hironaka's beautiful theorem of resolution of singularities, there exists a 1-convex manifold M which is a proper modification of V. So the general philosophy is to relate the numerical invariants of a 1-convex manifold with the numerical invariants of the isolated singular points of the corresponding Stein analytic space. On the one hand, one tries to apply the existing theory of isolated singularities to answer concrete questions on 1-convex manifolds. On the other, the knowledge of 1-convex manifolds provides a good deal of information for isolated singularities. In recent years, considerable progress has been made in sheaf cohomology for 1-convex manifolds. In §2, we discuss the Grauert-Riemenschneider vanishing theorem and give two applications. The first application is to prove the Lefschetz theorem of Barth [3]. This is due to Schneider [35]. The second application is to give explicit formula for Hironaka numbers. This was done several years ago in [43], [44]. The purpose of this section is to supersede the preprint [44] we have circulated. It seems to us that this result cannot be found explicitly in the literature. In §4, we introduce a bunch of new invariants for isolated singularities and give a Noether type formula for general 1-convex manifold. We review some well-known theory of isolated singularities and give formulas for both Milnor number and signature of the singularity in terms of resolution data. Duality theorem for 1-convex manifold is discussed. We end our discussion with an open question.

§2. *Vanishing theorems for 1-convex manifold*

Let X be a compact complex manifold of dimension n. A vector bundle $E \to X$ is said to be positive if there exists a metric on E whose curvature form $\Theta = \{\Theta^{\rho}_{\sigma,i,\bar{j}}\}$ has the property that the hermitian quadratic form $\Theta(\xi, \eta) = \sum_{\rho,\sigma,i,j} \Theta^{\rho}_{\sigma,i,\bar{j}}\, \xi^{\sigma}\overline{\xi^{\rho}}\,\eta^{i}\overline{\eta^{j}}$ is positive definite in the two variables ξ, η. The famous Kodaira vanishing theorem says that if $L \to X$ is a positive line bundle, then $H^q(X, \Omega^p(L)) = 0$ for $p+q > n$, where Ω^p is the sheaf of germs of holomorphic p-forms on X. Later, Griffiths (cf. [11] and [12]) made a lot of important generalization of the Kodaira vanishing theorem. These are very useful tools in the theory of compact complex manifolds. On the other hand, if X is a Stein manifold, then Cartan's Theorem B asserts that $H^q(X, \mathfrak{F}) = 0$ for any coherent analytic sheaf $\mathfrak{F}$ and for any $q > 0$. Between these two extremes, there lie the p-convex manifolds. Actually, one can say something about the top dimension sheaf cohomology group by the following celebrated theorem of Siu [38].

THEOREM 2.1 (Siu). *Suppose* $\mathfrak{F}$ *is a coherent analytic sheaf on a* σ*-compact complex space* X *(not necessarily reduced). If* $\dim X = n$ *and* X *has no compact* n*-dimensional branch, then* $H^n(X, \mathfrak{F}) = 0$.

Perhaps one of the most fundamental vanishing theorems on open manifolds is the vanishing theorem of Grauert-Riemenschneider. Let us first recall two notions. Let X be an n-dimensional Kähler manifold and let $G \subset X$ be an open relatively compact set. Let $q \geq 1$ be an integer. G is called *hyper-q-convex* if for each $x_0 \in \partial G$ there exists an open neighborhood U of x_0 in X, a smooth function ϕ in U with $d\phi \neq 0$ for $x \in U$ and $U \cap G = \{x \in U : \phi(x) < 0\}$ and normal coordinates (with respect to the Kahler metric) $z_1, \cdots, z_n$ at x_0 such that:

(i) $\{z_n = 0\}$ is the complex tangent plane to ∂G at x_0.

(ii) The Levi form of ϕ on this tangent plane has diagonal form

$$\frac{\partial^2 \phi}{\partial z_i \partial \bar{z}_j}(x_0) = \delta^i_j \lambda_j \qquad 1 \leq i, j \leq n-1$$

(iii) $\lambda_{i_1} + \cdots + \lambda_{i_q} > 0$ for $1 \leq i_1 < \cdots < i_q \leq n-1$.

In particular at least $n-q$ of the eigenvalues $\lambda_1, \cdots, \lambda_{n-1}$ are positive. Hence a hyper-q-convex domain G is q-convex in the sense of Andreotti and Grauert [1]. For $q = 1$, both notions coincide. We should remark that if X is a compact Kähler manifold and if $Y \subset X$ is a closed submanifold then $X - Y$ will be called hyper-q-convex if there is a fundamental system of open neighborhoods V_n of Y in X such that $X - \bar{V}_n$ is hyper-q-convex in X. A holomorphic vector bundle E on a complex manifold X is *Nakano semi-positive* if there exists a hermitian metric h on E such that for each $x_0 \in X$, with respect to normal trivialization, the hermitian form

$$\sum_{i,j,\alpha,\beta} \frac{\partial^2 h_{\alpha\beta}}{\partial z_i \partial \bar{z}_j} \xi^{(\beta,i)} \overline{\xi^{(\alpha,j)}}$$

is negative semidefinite. The range of summation is $1 \leq i,j \leq \dim X$, $1 \leq \alpha, \beta \leq \operatorname{rank} E$. The hermitian metric h is represented via a trivialization of E by a differentiable hermitian matrix $(h_{\alpha\beta})$ in a neighborhood of x_0. Such a trivialization is called normal in x_0 if $(h_{\alpha\beta}(x_0))$ is the unit matrix and if $dh_{\alpha\beta}(x_0) = 0$ for all α, β. The matrix $-((\partial^2 h_{\alpha\beta}/\partial z_i \partial \bar{z}_j)(x_0))$ is the curvature matrix $\Theta^\alpha_{\beta i \bar{j}}$ in x_0.

THEOREM 2.2 (Grauert-Riemenschneider). *Let* $G \subset X$ *be hyper-q-convex and let* E *be a Nakano semipositive holomorphic vector bundle on* X. *Then* $H^i(G, E \otimes K_X) = 0$ *for* $i \geq q$.

In [35], Schneider proved that the following interesting theorem of Barth is a consequence of the vanishing theorem of Grauert-Riemenschneider.

THEOREM 2.3 (Barth). *If* Y *is a submanifolds of pure dimension in complex projective space, then*

$$H^i(P^n, T; \mathbb{C}) = 0 \quad \text{for} \quad i \leq 2 \dim Y - n+1 .$$

His proof roughly goes as follows. He first observed that the tangent bundle of the complex projective space is Nakano semipositive and that the complement of a k-codimensional submanifold Y of P^n is hyper-(2k−1)-convex. Then the Lefschetz theorem of Barth follows from the following proposition.

PROPOSITION 2.4. *Let* X *be an* n*-dimensional compact Kähler manifold and let* $Y \subset X$ *be a closed analytic submanifold. If* $X - Y$ *is hyper-*q*-convex and if the holomorphic tangent bundle* T_X *of* X *is Nakano semi-positive then*

$$H^i(X, Y; \mathbb{C}) = 0 \quad \text{for} \quad i \leq n-q .$$

In what follows, we shall give another application of the vanishing theorem of Grauert-Riemenschneider. This was first observed in [44] some years ago. We need to recall a few preliminaries related to the concept of depth; for more details the reader is referred, for example, to [14].

Suppose Y is a closed subspace of V defined by a sheaf of ideals I of $\mathcal{O}_V$. Let $\mathfrak{F}$ be a sheaf on V. We then have a long exact sequence for local cohomology:

$$\cdots \to H^i_Y(V, \mathfrak{F}) \to H^i(V, \mathfrak{F}) \to H^i(V-Y, \mathfrak{F}) \to \cdots .$$

If V is Stein, and $\mathfrak{F}$ is coherent, we get:

$$0 \to H^0_Y(V, \mathfrak{F}) \to H^0(V, \mathfrak{F}) \to H^0(V-Y, \mathfrak{F}) \to H^1_Y(V, \mathfrak{F}) \to 0 \tag{2.1}$$

$$H^i(V-Y, \mathfrak{F}) \xrightarrow{\approx} H^{i+1}_Y(V, \mathfrak{F}) \quad \text{for} \quad i \geq 1 . \tag{2.2}$$

Moreover, on any variety V, for coherent $\mathfrak{F}$ we have ([14], 2.8):

$$H^i_Y(V, \mathfrak{F}) = \varinjlim_{\nu} \operatorname{Ext}^i_{\mathcal{O}_V}(\mathcal{O}_V/I^\nu, \mathfrak{F}) . \tag{2.3}$$

We will be interested in the case where Y is a point p, so that $I = m_{V,p}$. We recall that $\mathrm{depth}_p(\mathfrak{F})$ is the maximum number of elements in any $\mathfrak{F}$-regular sequence contained in $m_{V,p}$ i.e., a sequence $f_1, \cdots, f_n \in m_{V,p}$ such that f_{i+1} is not a zero divisor in $\mathfrak{F}/(f_1, \cdots, f_i)\mathfrak{F}$. V is *Cohen-Macauley* if $\mathrm{depth}_p(\mathcal{O}_V) = \dim_p(\mathcal{O}_V)$ at every point p. We will need ([14], 3.7)

$$\mathrm{depth}_p(\mathfrak{F}) \geq d \Longleftrightarrow \mathrm{Ext}^i_{\mathcal{O}_V}(\mathcal{O}_V/I, \mathfrak{F}) = 0 \quad \text{for} \quad i \leq d-1 \tag{2.4}$$

where I is any ideal such that $\mathcal{O}_V/I$ is supported at p. Actually, for a coherent sheaf $\mathfrak{F}$ on V, $\mathrm{depth}_p(\mathfrak{F}) \geq d$ if and only if the local cohomology $H^i_p(V, \mathfrak{F})$ vanishes for all $i < d$.

REMARK 2.5. Every smooth variety is Cohen-Macaulay; every relative complete intersection over a Cohen-Macaulay variety is Cohen-Macaulay.

THEOREM 2.6. *Let $V \subseteq \mathbb{C}^m$ be any n-dimensional subvariety in a Stein neighborhood of the origin in $\mathbb{C}^m$. Suppose the origin is the only singularity of V and is irreducible. Let $\pi : M \to V$ be a resolution of the singularity of V. Then for $1 \leq i \leq n-1$, $H^i(M, \mathcal{O})$ is independent of resolution and*

$$\dim H^{n-1}(M, \mathcal{O}) = \dim \Gamma(V - \{0\}, \Omega^n)/L^2(V - \{0\}, \Omega^n) \tag{2.5}$$

where $L^2(V - \{0\}, \Omega^n)$ is the subspace of $\Gamma(V - \{0\}, \Omega^n)$ consisting of n-forms on $V - \{0\}$ which are square-integrable near origin i.e., $\omega \in L^2(V - \{0\}, \Omega^n)$ if $\omega \in \Gamma(V - \{0\}, \Omega^n)$ and $\int \omega \wedge \omega < \infty$ near the origin.

Suppose $n \geq 3$. Then

$$\dim H^i(M, \mathcal{O}) = \dim \mathrm{Ext}^{m-i-1}_{\mathcal{O}_{\mathbb{C}^m,0}}(\mathcal{O}_{V,0}, \mathcal{O}_{\mathbb{C}^m,0}) \quad \text{for} \quad 1 \leq i \leq n-2 . \tag{2.6}$$

If V is Cohen-Macaulay, then $H^i(M, \mathcal{O}) = 0$ for $1 \leq i \leq n-2$.

REMARK 2.7. For $n = 2$, (2.5) was proved by Laufer [26]. The general case $(n \geq 2)$ was proved by us in [43]. In case that $H^{n-1}(M, \mathcal{O})$ is zero and V is Gorenstein (i.e. canonical bundle is trivial in a delected neighborhood of origin in $V-\{0\}$), we proved in [44] that $H^i(M,\mathcal{O})=0$ for $1\le i\le n-2$. This special case was also proved independently by G. Kempf in [21]. The general theorem (Theorem 2.6) was first proved by us in [44]. More recently, Mumford shows us a proof of (2.6) by using his advanced technique. The proof given below is the simplified version of our previous proof in [44].

Proof of Theorem 2.6. By Artin's algebraization theorem [2], we may assume for a local question near $o\in V$, that V is an affine algebraic variety. As such, by [17], we may assume that we have a resolution $\pi: M\to V$ so that M is Zariski open in a smooth projective variety. In this situation, we can apply the result of Grauert-Riemenschneider to conclude that $H^i(M,\Omega^n)=0$ for $i\ge 1$. By (2.2) of [26], we have the following exact sequence

$$\begin{array}{ccccccccc} 0\to \Gamma_c(M,\Omega^n) & \to & \Gamma(M,\Omega^n) & \to & \Gamma_\infty(M,\Omega^n) & \to & H^1_c(M,\Omega^n) & \to & H^1(M,\Omega^n)\,. \\ \| & & & & & & & & \| \\ 0 & & & & & & & & 0 \end{array}$$

By Theorem 3.1 of [26], we have $\Gamma(M,\Omega^n)=L^2(V-\{0\},\Omega^n)$. On the other hand, one can prove as in [43] (Theorem A) that $\Gamma_\infty(M,\Omega^n)=\Gamma(M-A,\Omega^n)=\Gamma(V-\{0\},\Omega^n)$ by only using elementary properties of analytic cover. Since $H^{n-1}(M,\mathcal{O})$ is Serre dual to $H^1_c(M,\Omega^n)$, (2.5) follows from the exact sequence above.

Choose a non-negative C^∞ exhaustion function ϕ on M such that ϕ is strongly plurisubharmonic outside A and $A=\{x\in M:\phi(x)=0\}$. Let

$$M_r=\{x\in M:\phi(x)\le r\}\subset\subset M\,.$$

By Proposition 2.1 of [26], $H^i_\infty(M,\mathcal{O})=\text{dir lim}\ H^i(M-M_r,\mathcal{O})$. Since $H^i(M-M_r,\mathcal{O})=H^i(M-A,\mathcal{O})$ for $1\le i\le n-2$ by Theorem 15 of [1], we have

$$H^i_\infty(M,\mathcal{O})=H^i(M-A,\mathcal{O})\qquad\text{for}\qquad 1\le i\le n-2\,.$$

Consider the following exact sequence

$$H^1_c(M,\mathcal{O}) \to H^1(M,\mathcal{O}) \to H^1_\infty(M,\mathcal{O}) \to \cdots \to H^{n-2}_c(M,\mathcal{O}) \to H^{n-2}(M,\mathcal{O})$$
$$\to H^{n-2}_\infty(M,\mathcal{O}) \to H^{n-1}_c(M,\mathcal{O}) .$$

By Serre duality, we know that $H^i_c(M,\mathcal{O})$ is the strong dual of $H^{n-i}(M,\Omega^n)$ which is zero by the vanishing theorem of Grauert-Riemenschneider for $i \neq n$. So $H^i_c(M,\mathcal{O}) = 0$ for $i \neq n$. It follows that

$$H^i(M,\mathcal{O}) = H^i_\infty(M,\mathcal{O}) = H^i(V-\{0\},\mathcal{O}) \quad \text{for} \quad 1 \le i \le n-2 .$$

Look at the long exact sequence for local cohomology.

$$H^1(V,\mathcal{O}) \to H^1(V-\{0\},\mathcal{O}) \to H^2_{\{0\}}(V,\mathcal{O}) \to H^2(V,\mathcal{O}) \to H^2(V-\{0\},\mathcal{O})$$
$$\to \cdots \to H^{n-2}(V,\mathcal{O}) \to H^{n-2}(V-\{0\},\mathcal{O}) \to H^{n-1}_{\{0\}}(V,\mathcal{O}) \to H^{n-1}(V,\mathcal{O}) .$$

By Cartan Theorem B, $H^i(V,\mathcal{O}) = 0$ for $i \ge 1$. Therefore

$$\begin{aligned} H^i(M,\mathcal{O}) &= H^i(V-\{0\},\mathcal{O}) \\ &= H^{i+1}_{\{0\}}(V,\mathcal{O}) \qquad 1 \le i \le n-2 . \end{aligned} \tag{2.7}$$

As a consequence, $H^i(M,\mathcal{O}) = 0$ for $1 \le i \le n-2$ if V is Cohen-Macaulay. (2.6) follows from (2.7) by local duality (cf. [16]). Q.E.D.

DEFINITION 2.1. Let q be a point in V, a complex analytic space. Call q a rational singular point if, given a resolution of singularities $\pi: M \to V$ in the sense of [17], then $(R^{n-1}\pi_* \mathcal{O}_M)_q = 0$. q is said to be a strongly rational singular if $(R^i \pi_* \mathcal{O}_M)_q = 0$ for $i > 0$.

COROLLARY 2.8. *Let* $V \subseteq \mathbb{C}^m$ *be any* n*-dimensional subvariety in a Stein neighborhood of the origin in* $\mathbb{C}^m$. *Suppose the origin is a normal isolated rational singularity of* V. *Then the origin is strongly rational if and only if it is Cohen-Macaulay.*

Proof. Suppose the origin is strongly rational. By (2.7), we have

$$H^i_{\{0\}}(V,\mathcal{O}) = 0 \qquad 0 \le i \le n-1 .$$

Consider the following exact sequence

$$0 \longrightarrow H^0_{\{0\}}(V,\mathcal{O}) \longrightarrow H^0(V,\mathcal{O}) \xrightarrow{r} H^0(V-\{0\},\mathcal{O}) \longrightarrow H^1_{\{0\}}(V,\mathcal{O}) \longrightarrow H^1(V,\mathcal{O}) = 0 .$$

$H^0_{\{0\}}(V,\mathcal{O}) = 0$ because the map r is injective. By normality, r is also surjective. Hence $H^1_{\{0\}}(V,\mathcal{O}) = 0$. Since $H^i_{\{0\}}(V,\mathcal{O}) = 0$ for $0 \le i \le n-1$, it follows that the depth of V at the origin is greater than or equal to n. Theorem 27 of [28] says that the depth of V at the origin is at most n. Therefore V is Cohen-Macaulay at the origin. Q.E.D.

§3. *Numerical invariants for 1-convex manifolds*

DEFINITION 3.1. Let M and M' be two 1-convex manifolds with A and A' as their maximal compact analytic sets respectively. M is local holomorphically equivalent to M' if there exist open neighborhoods U and U' of A and A' respectively and a biholomorphic map $\phi: U \to U'$ such that $\phi(A) = A'$.

The natural question one can ask is to classify all 1-convex manifolds up to local holomorphically equivalent classes. This question is very interesting because it corresponds to classifying isolated singularities up to biholomorphic classes. We suggest to use numerical invariants approach. Hopefully with enough invariants invented, one can get full information of a neighborhood of the maximal compact analytic set in a 1-convex manifold. From now on we shall assume that M is a 1-convex manifold of dimension n with A as a connected maximal compact analytic set. Let $\pi: M \to V$ be a resolution of singularity of V, a subvariety in a Stein neighborhood of the origin in $\mathbb{C}^m$ of which the origin is the only singularity. We call $\dim H^i(M,\mathcal{O})$, $1 \le i \le n-1$ the i-th Hironaka number of the singularity, and will denote it by $h^{(i)}$. $h^{p,q}(M) = \dim H^q(M,\Omega^p)$ for $1 \le p \le n$, $1 \le q \le n-1$ will be called the Hironaka number of type (p,q) of the 1-convex manifold M. We remark that $h^{(i)}$, $1 \le i \le n-1$ are independent of resolutions of the singularity of V.

In [45], we introduced a bunch of invariants which are naturally attached to isolated singularities. Let $(X,\mathcal{O}_X)$ be an analytic space over C. We denote by Ω_X^p the sheaf of germs of Grauert-Grothendieck analytic differential forms on X (cf. [9], [15]) Ω_X^p is an $\mathcal{O}_X$-module, and has a C-linear differential $d:\Omega_X^p \to \Omega_X^{p+1}$ together with a canonical monomorphism $\varepsilon: \mathbb{C} \to \Omega_X^0 = \mathcal{O}_X$ such that $d\circ\varepsilon = 0$. For each morphism $f: X \to Y$ of analytic spaces, there exists a natural morphism $f^*: \Omega_Y^* \to \Omega_X^*$ which is compatible with differentials. Let U be open in $\mathbb{C}^m$, $\mathcal{O}$ the sheaf of germs of holomorphic functions on U, $\mathcal{I} \subset \mathcal{O}$ a coherent sheaf of ideals with set of zeros Y, and $(Y,\mathcal{O}_Y)$ the closed analytic subspace of U with $\mathcal{O}_Y = \mathcal{O}/\mathcal{I}$. Denote by

$$\mathcal{K}^p := \{f\alpha + dg\wedge\beta : f, g\in\mathcal{I}, \alpha\in\Omega^p, \beta\in\Omega^{p-1}\}.$$

Then

$$\Omega_Y^p := \Omega^p/\mathcal{K}^p$$

where Ω^p is the sheaf of germs of holomorphic p-forms on U and the differential on Ω_Y^* is induced by the one of Ω^* (cf. [15]; §1). Let

$$\mathcal{H}^p := \{\omega\in\Omega^p : \omega/Y^* = 0\}$$

where Y^* is the regular part of Y. Then Ferrari (cf. [8] and [5]) introduced the sheaf of germs of holomorphic p-forms on Y.

$$\tilde{\Omega}_Y^p := \Omega^p/\mathcal{H}^p$$

to be obtained by taking the sheaf of germs of holomorphic p-forms in the ambient space modulo those germs of holomorphic p-forms whose restriction on the regular part of Y is zero. He showed that this is a coherent sheaf and moreover that every holomorphic mapping $\phi: A_1 \to A_2$ induces a sheaf homomorphism $\phi^*: \Omega^p_{A_2} \to \Omega^p_{A_1}$ such that the assignment $A \mapsto \Omega^p_A$ is a contravariant functor. Recently Griffiths [13] introduced another very important concept of holomorphic forms based on the finiteness of certain L^2-norms in much the same spirit as M. Noether's original definition of differentials of the first kind on an algebraic variety. Suppose that Y is an analytic variety of dimension n with the singular locus Y_s. A holomorphic n-form ψ on Y shall be given by a holomorphic form in the usual sense on the complex manifold $Y - W = Y_1$, where W is an analytic subvariety of Y containing Y_s but not containing any irreducible component of Y, and ψ satisfies the local L^2-estimate

$$(\sqrt{-1})^{n^2} \int_{U \cap Y_1} \psi \vee \psi < \infty$$

in a neighborhood U of any point $p \in W$. For $q < n$, a holomorphic q-form ψ is given by a holomorphic form in the usual sense on $Y_1 = Y - W$ as above, and where for any local piece of q-dimensional analytic subvariety $Z \subset Y$ but $Z \subset W$, the restriction ψ/Z is holomorphic in the previous sense. Denote the sheaf of germs of holomorphic p-forms on Y in the sense of Griffiths by $\bar{\Omega}^p_Y$. Then he proved that the usual properties of forms such as admitting exterior products, exterior differentiation, and pulling back under holomorphic mappings hold. Moreover, he showed that $\bar{\Omega}^p_Y$ is nothing but $R^0\pi_*\Omega^q_{\tilde{Y}}$ the 0-th direct image sheaf of $\Omega^q_{\tilde{Y}}$ where $\pi: \tilde{Y} \to Y$ is a resolution of singularities of Y. (See also [26] and [43].)

DEFINITION 3.2. Let V be a n-dimensional Stein analytic space with $x \in V$ as its only isolated singularity. Suppose $x \in V$ is locally embedded in C^N. Let Ω^p_V, $\tilde{\Omega}^p_V$ and $\bar{\Omega}^p_V$ be sheaves of holomorphic p-forms on V in the sense of Grauert-Grothendieck, Ferrari and Griffiths respectively. Let K^p and H^p be defined by the exact sequences

$$0 \to \tilde{\Omega}^p_V \to \bar{\Omega}^p_V \to H^p \to 0$$

$$0 \to K^p \to \Omega^p_V \to \tilde{\Omega}^p_V \to 0 .$$

Then H^p, K^p are coherent sheaves with supports on x. We denote by $g^{(p)}$ the p-th Griffiths number, equal to $\dim H^p$. We call the p-th Mumford number $m^{(p)}$ to be equal to $\dim K^{(p)}$.

In response to a problem raised by Serre [36, p. 373-374], Siu [40] had the following beautiful solution. Before stating it, let us recall some notations: If $\mathfrak{F}$ is a coherent analytic sheaf on a complex analytic space X, then $S_k(\mathfrak{F})$ denotes the analytic subvariety $\{x \in X : \operatorname{cod} h\, \mathfrak{F}_x \le k\}$. If D is an open subset of X, $\bar{S}_k(\mathfrak{F}/D)$ denotes the topological closure of $S_k(\mathfrak{F}/D)$ in X. If V is an analytic subvariety of X, then $\mathcal{H}^k_V(\mathfrak{F})$ denotes the sheaf defined by the presheaf $U \mapsto H^k_V(U, \mathfrak{F})$, where $H^k_V(U,\mathfrak{F})$ is the k-dimensional cohomology group of U with coefficients in $\mathfrak{F}$ and supports in V.

THEOREM 3.1 (Siu). *Suppose V is an analytic subvariety of a complex analytic space $(X, \mathcal{O})$, q is nonnegative integer, and $\mathfrak{F}$ is a coherent analytic sheaf on X. Let $\theta : X - V \to X$ be the inclusion map. Then the following three statements are equivalent:*

(i) $\theta_0(\mathfrak{F}/X{-}V), \cdots, \theta_q(\mathfrak{F}/X{-}V)$ *(or equivalently* $\mathcal{H}^0_V(\mathfrak{F}), \cdots, \mathcal{H}^{q+1}_V(\mathfrak{F})$*) are coherent on* X.

(ii) *For every* $x \in V$, $\theta_0(\mathfrak{F}/X{-}V)_x, \cdots, \theta_q(\mathfrak{F}/X{-}V)_x$ *(or equivalently* $\mathcal{H}^0_V(\mathfrak{F})_x, \cdots, \mathcal{H}^{q+1}_V(\mathfrak{F})_x$*) are finitely generated over* $\mathcal{O}_x$.

(iii) $\dim V \cap \bar{S}_{k+q+1}(\mathfrak{F}/X{-}V) < k$ *for every* $k \ge 0$ *where* $\theta_q(\mathfrak{F})$ *is the* q*-th direct image of* $\mathfrak{F}$ *under* θ.

Let us come back to our situation. Suppose V is a complex analytic space of dimension n with x as its only singularity. Let $\theta : V - \{x\} \hookrightarrow V$ be the inclusion map. Then the 0-th direct image sheaf $\theta_* \Omega^i_{V-\{x\}} = \bar{\bar{\Omega}}_V$ is coherent by Siu's theorem. It is clear that we have an inclusion $\bar{\Omega}^i_V \hookrightarrow \bar{\bar{\Omega}}^i_V$.

DEFINITION 3.3. With the above notations, the i-th Siu number $s^{(i)}$ is defined to be

$$s^{(i)} = \dim(\bar{\bar{\Omega}}_{V,x}/\bar{\Omega}_{V,x}) .$$

DEFINITION 3.4. Let X be a complex analytic space and $x_0 \in X$ an isolated singular point. Let

$$0 \longrightarrow C \xrightarrow{d^{-1}} \mathcal{O}_{X,x_0} \xrightarrow{d^0} \Omega^1_{X,x_0} \xrightarrow{d^1} \Omega^2_{X,x_0} \longrightarrow \cdots$$

be the Poincaré complex at x_0 where d^{-1} is the inclusion map. Then the Poincaré numbers are defined as follows

$$p^{(i)} = \dim \operatorname{Ker} d^i/\operatorname{Im} d^{i-1} \quad i \geq 0 .$$

We remark that $p^{(i)}$'s are finite numbers and $p^{(i)} = 0$ for $i > N$ where N is the dimension of the Zariski tangent space of x_0. We have the following Noether's formula for arbitrary 1-convex manifolds (cf. [45]).

THEOREM 3.2. *Let* M *be a* n*-dimensional* 1*-convex manifold of dimension* n. *Suppose* M *can be blown down to a Stein analytic space* V *with* x *as its only singularity. Let* $g^{(i)}$, $m^{(i)}$, $p^{(i)}$ *and* $s^{(i)}$ *be the Griffiths numbers, Mumford numbers, Poincaré numbers and Siu numbers at* x *respectively. Let* N *be the dimension of the Zariski tangent space of* V *at* x. *Then*

(a) *If* x *is a surface singularity then*

$$\chi(M,\mathcal{O}) - \chi(M,\Omega^1) = \sum_{i=0}^{N} (-1)^i p^{(i)} - \sum_{i=0}^{N} (-1)^i m^{(i)} + \sum_{i=0}^{n} (-1)^i g^{(i)} - s^{(1)} - \chi_T(A) + 1 .$$

(b) *If* x *is a higher dimensional singularity i.e.,* $n \geq 3$, *then*

$$\sum_{i=0}^{n-1} (-1)^i \chi(M,\Omega^i) = \sum_{i=0}^{N} (-1)^i p^{(i)} - \sum_{i=0}^{N} (-1)^i m^{(i)} + \sum_{i=0}^{n} (-1)^i g^{(i)} + \sum_{i=1}^{n-1} (-1)^i s^{(i)} - \chi_T(A) + 1 .$$

where $\chi(M,\mathcal{F}) := \dim \Gamma(M-A,\mathcal{F})/P(M,\mathcal{F}) - \sum_{q=1}^{n} (-1)^q \dim H^q(M,\mathcal{F})$ *for any coherent sheaf* $\mathcal{F}$ *over* M.

It is a natural question to ask the relationship of these four sheaves mentioned above. We shall restrict ourself to the case where $V = \{f=0\}$ is a hypersurface in $\mathbb{C}^{n+1}$ with $x \in V$ as an isolated singularity. It turns out that we have (cf. [45]).

THEOREM 3.3. *Let* $f(z_0, z_1, \cdots, z_n)$ *be holomorphic in* N, *a Stein neighborhood of the origin in* $\mathbb{C}^{n+1}$ *with* $f(0,0,\cdots,0) = 0$. *Let* $V = \{(z_0, z_1, \cdots, z_n) \in N : f(z_0, z_1, \cdots, z_n) = 0\}$ *has the origin as its only singular point. Let* $\pi : M \to V$ *be a resolution of the singularity of* V. *Let* $\tau = \dim \mathbb{C}[[z_0, z_1, \cdots, z_n]]/\left(f, \frac{\partial f}{\partial z_0}, \frac{\partial f}{\partial z_1}, \cdots, \frac{\partial f}{\partial z_n}\right)$ *and* $h^{(n-1)} = \dim H^{n-1}(M, \mathcal{O})$ *be the* (n–1)-th *Hironaka number. Then*

(a) $g^{(i)} = 0$ *for* $0 \le i \le n-2$, $s^{(i)} = 0$ *for* $0 \le i \le n-2$, *and* $m^{(i)} = 0$ *for* $0 \le i \le n-1$, *i.e.*

$$\Omega^i_V = \tilde{\Omega}^i_V = \bar{\Omega}^i_V = \bar{\bar{\Omega}}^i_V \quad \text{for} \quad 0 \le i \le n-2$$

and

$$\Omega^{n-1}_V = \tilde{\Omega}^{n-1}_V \subset \bar{\Omega}^{n-1}_V \subseteq \bar{\bar{\Omega}}^{n-1}_V .$$

(b) $g^{(n)} = \tau - h^{(n-1)}$, $s^{(n)} = h^{(n-1)}$ *and* $m^{(n)} = \tau$.

(c) $\tilde{\Omega}^{n+1}_V = 0 = \bar{\Omega}^{n+1}_V = \bar{\bar{\Omega}}^{n+1}_V$, Ω^{n+1}_V *is supported on the origin and* $m^{(n+1)} = \tau$.

Let us recall Milnor's results on the topology of hypersurface singularities (cf. [29], [27]). Let $f : U \subseteq \mathbb{C}^{n+1} \to \mathbb{C}$ be an analytic function on an open neighborhood U of 0 in $\mathbb{C}^{n+1}$. We denote

$$B_\varepsilon = \{z \in \mathbb{C}^{n+1} : \|z\| \le \varepsilon\}$$

$$S_\varepsilon = \partial B_\varepsilon = \{z \in \mathbb{C}^{n+1} : \|z\| = \varepsilon\} .$$

Then:

THEOREM 3.4. *For* ε *small enough the mapping* $\phi_\varepsilon : S_\varepsilon - \{f=0\} \to S^1$ *defined by* $\phi_\varepsilon = f(z)/|f(z)|$ *is a smooth fibration.*

If 0 *is an isolated critical point of* f, *for* $\varepsilon > 0$ *small enough, the fibers of* ϕ_ε *have the homotopy type of a bouquet of* μ *spheres of dimension* n *with*

$$\mu = \dim_{\mathbf{C}} \mathbf{C}[[z_0, z_1, \cdots, z_n]] / \left(\frac{\partial f}{\partial z_0}, \cdots, \frac{\partial f}{\partial z_n} \right).$$

(A bouquet of spheres is the topological space which is a union of spheres having a single point in common.)

In [4], Bennett and the author obtained the following interesting formula for μ.

THEOREM 3.5. *Let* M *be a 1-convex manifold of dimension* $n \geq 3$. *Suppose the maximal compact analytic subset in* M *can be blown down to isolated hypersurface singularities* $q_1, \cdots, q_m$. *Let* Ω^p *be the sheaf of germs of holomorphic* p*-forms on* M. *Let* $\chi^p(M) = \sum_{i=1}^{n-1} (-1)^i \dim H^i(M, \Omega^p)$. *Then*

$$m + (-1)^n \sum_{i=1}^{m} \mu_i = \chi_T(A) + \sum_{p=2}^{n-2} (-1)^p \chi^p(M) - 2\chi^0(M) + 2\chi^1(M).$$

Here μ_i *is the Milnor number of* q_i *and* $\sum_{p=2}^{n-2} (-1)^{p+1} \chi^p(M) = 0$ *if* $n = 3$ *by convention.*

Recall that the signature σ of an arbitrary oriented $2n$-manifold W with or without boundary is defined as follows: There is a symmetric bilinear intersection pairing $(\ , \ )$ on $H_n(W; \mathbf{R})$ (this form is symmetric only if n is even) defined by setting

$$(x, y) = (x' \cup y')[W]$$

where x' and y' in $H^n(W, \partial W; \mathbf{R})$ are Lefschetz duals to x and y in $H_n(W; \mathbf{R})$ and $[W] \in H_{2n}(W, \partial W; \mathbf{R})$ is the orientation class. This bilinear form may be diagonalized, with diagonal entries $+1$, 0 and -1. The signature σ of W is the signature of the bilinear form, namely, the number of positive minus the number of negative diagonal entries.

Let $f:(\mathbb{C}^{2n+1},0)\to(\mathbb{C},0)$ be the germ of a complex analytic function with an isolated critical point at the origin. For $\varepsilon>0$ suitably small and δ yet smaller, the space $V'=f^{-1}(\delta)\cap D_\varepsilon$ (where D_ε denotes the closed disk of radius ε about 0) is a real oriented $4n$-manifold with boundary. Recently, Durfee [7] has given an interesting formula for the signature σ of V' in terms of topological invariants of resolution of the singularity at 0 of the complex surface $f^{-1}(0)$. It is a natural question to ask for a formula for the signature σ of V' for higher dimensional singularities. In [42], the signature of even dimensional singularities is given in terms of topological and analytic invariants of a resolution of the singularity.

DEFINITION 3.5. Let $\pi: M\to V=\{x\in\mathbb{C}^{2n+1}: f(x)=0\}$ be a resolution of the singularity. The signature of M, denoted by $\sigma(M)$, is defined to be the signature of $\pi^{-1}(D_\varepsilon\cap V)$ where D_ε denotes the closed ball of radius ε about 0, i.e. the signature of a closed tubular neighborhood of the exceptional set $A=\pi^{-1}(0)$.

THEOREM 3.6. *Let* $f(z_0,z_1,\cdots,z_{2n})$ *be holomorphic in* $N\subseteq\mathbb{C}^{2n+1}$, $n>1$, *a Stein neighborhood of* $(0,0,\cdots,0)$ *with* $f(0,0,\cdots,0)=0$. *Let* $V=N\cap f^{-1}(0)$ *have* $(0,0,\cdots,0)$ *as its only singular point. Let* σ *be the signature of* V'. *Let* $\pi:M\to V$ *be a resolution of* V. *Then*

$$\sigma=-\sum_{p=q}^{2n-q}\chi^p(M)-2\sum_{p=0}^{q-1}\chi^p(M)+\sigma(M)$$

for $2\le q\le n$.

For surface singularities, we have the following interesting formulas.

THEOREM 3.7. *Let* $f(x,y,z)$ *be holomorphic in* N, *a Stein neighborhood of* $(0,0,0)$ *with* $f(0,0,0)=0$. *Let* $V=\{(x,y,z)\in N: f(x,y,z)=0\}$ *have* $(0,0,0)$ *as its only singular point. Let* μ *and* $g^{(1)}$ *be the Milnor number and the Griffiths number of* $(0,0,0)$ *respectively and* σ *be the signature of* V'. *Let* $\pi: M \to V$ *be a resolution of* V *and* $A = \pi^{-1}(0,0,0)$. *Let* $\chi_T(A)$ *be the topological Euler characteristic of* A *and* n *be the number of irreducible components of* A. *Then*

$$\sigma = 2 \dim H^1(M, \mathcal{O}) + \dim H^1(M, \Omega^1) - g^{(1)} + n \tag{3.1}$$

$$1 + \mu = 2 \dim H^1(M, \mathcal{O}) - \dim H^1(M, \Omega^1) + g^{(1)} + \chi_T(A) \tag{3.2}$$

$$\sigma = -K^2 - n - 8 \dim H^1(M, \mathcal{O}) . \tag{3.3}$$

Sketch of the proof. Any holomorphic function which agrees with f to sufficiently high order defines a holomorphically equivalent singularity at $(0,0,\cdots,0)$, [2]. So we may take f to be a polynomial. Compactify $\mathbf{C}^3$ to $\mathbf{P}^3$. Let $\overline{V}_t$ be the closure in $\mathbf{P}^3$ of

$$V_t = \{(x,y,z) \in \mathbf{C}^3 : f(x,y,z) = t\} .$$

By adding a suitably general high order homogeneous term of degree e to the polynomial f, we may additionally assume that $\overline{V}_0$ has $(0,0,0) \in \mathbf{C}^3$ as its only singularity and that $\overline{V}_t$ is non-singular for small $t \neq 0$. We may also assume that the highest order term of f define, in homogeneous coordinates, a nonsingular hypersurface of order e in $\mathbf{P}^2 = \mathbf{P}^3 - \mathbf{C}^3$. $\overline{V}_t$ is then necessarily irreducible for all small t. Without loss of generality, we take $N = \mathbf{C}^3$. Then $V = V_0$.

For any 4k-dimensional topological manifold S, the signature of S is denoted by σ_S. Let B_ε be an open Milnor ball of radius ε. Let $\overline{M}$ be the resolution $\overline{V}$ which has M as an open subset. Then for small t

$$\sigma = \sigma_{\overline{V}_t} - \sigma_{\overline{V}_t - B_\varepsilon \cap \overline{V}_t}$$

$$\sigma = \sigma_{\overline{V}_t} - \sigma_{\overline{V}_0 - B_\varepsilon \cap \overline{V}_0}$$

since the family $\{\overline{V}_t\}$ is differentiably trivial away from B_ε. Hence

$$\begin{aligned}\sigma &= \sigma_{\overline{V}_t} - \sigma_{\overline{M}-\pi^{-1}(B_\varepsilon \cup V_0)} \\ &= \sigma_{\overline{V}_t} - (\sigma_{\overline{M}} - \sigma_{\pi^{-1}(\overline{B}_\varepsilon \cup V_0)}) \\ &= \sigma_{\overline{V}_t} - \sigma_{\overline{M}} + \sigma(M) \\ &= \sigma_{\overline{V}_t} - \sigma_{\overline{M}} + n\end{aligned} \tag{3.4}$$

because $\sigma(M) = n$ in view of the fact that the intersection matrix for the exceptional set is negative definite. By Hodge index theorem and Serre duality, we have

$$\begin{aligned}\sigma &= \sum_{p=0}^{2} \chi^p(\overline{V}_t) - \sum_{p=0}^{2} \chi^p(\overline{M}) + \sigma(M) \\ &= 2(\chi^0(\overline{V}_t) - \chi^0(\overline{M})) + (\chi^1(\overline{V}_t) - \chi^1(\overline{M})) + \sigma(M) .\end{aligned}$$

Recall that in [42] and [45], we have proved the following

$$\chi^p(\overline{V}_t) = \chi^p(\overline{V}_0) \quad t \neq 0 \quad 0 \leq p \leq 2$$

$$\chi^p(\overline{V}_0) - \chi^p(\overline{M}) = m^{(p)} - g^{(p)} + \dim H^1(M, \Omega^p) . \tag{3.5}$$

Therefore

$$\sigma = 2 \dim H^1(M, \mathcal{O}) + 2m^{(0)} - 2g^{(0)} + \dim H^1(M, \Omega^1) + m^{(1)} - g^{(1)} + \sigma(M). \tag{3.6}$$

(3.1) follows from (3.6) and Theorem 3.3.

Recall that the intersection of V_t with the open ε-ball is diffeomorphic with the Milnor fiber F_0. So the manifold with boundary $V_t \cap \overline{B}_\varepsilon$ is connected, with 2nd Betti number equal to μ, and with Euler number

$$\chi_T(V_t \cap \overline{B}_\varepsilon) = 1 + \mu .$$

Since the two manifolds $V_t \cap \overline{B}_\varepsilon$ and $\overline{V}_t - B_\varepsilon$ have union $\overline{V}_t$ and intersection K_t, we have the Euler number of $\overline{V}_t$

$$\begin{aligned}\chi_T(\bar{V}_t) &= \chi(V_t \cap \bar{B}_\varepsilon) + \chi(\bar{V}_t - B_\varepsilon) - \chi(K_t)\\ &= 1 + \mu + \chi(\bar{V}_0 - B_\varepsilon) - \chi(V_0 \cap S_\varepsilon)\end{aligned}$$

by the differentiable triviality of the family $\{V_t\}$ away from $(0,0,0) \in \mathbb{C}^3$. Hence

$$\begin{aligned}\chi_T(\bar{V}_t) &= 1 + \mu + \chi(\pi^{-1}(\bar{V}_0 - B_\varepsilon)) + \chi(\pi^{-1}(V_0 \cap \bar{B}_\varepsilon))\\ &\quad - \chi(V_0 \cap S_\varepsilon) - \chi(\pi^{-1}(V_0 \cap \bar{B}_\varepsilon))\\ &= 1 + \mu + \chi_T(\bar{M}) - \chi_T(A)\end{aligned}$$

since $\pi^{-1}(V_0 \cap \bar{B}_\varepsilon)$ contracts to A. Thus we have

$$\begin{aligned}1+\mu &= \chi_T(\bar{V}_t) - \chi_T(\bar{M}) + \chi_T(A)\\ &= \sum_{p=0}^{n} (-1)^p \chi^p(\bar{V}_t) - \sum_{p=0}^{2} (-1)^p \chi^p(\bar{M}) + \chi_T(A) \qquad \text{(by Hodge decomposition)}\\ &= 2(\chi^0(\bar{V}_0) - \chi^0(\bar{M})) - (\chi^1(\bar{V}_0) - \chi^1(\bar{M})) + \chi_T(A)\\ &= 2 \dim H^1(M, \mathcal{O}) - \dim H^1(M, \Omega^1) + g^{(1)} + \chi_T(A)\,.\end{aligned}$$

This proves (3.2).

$\omega = \dfrac{dx \wedge dy}{f_z} = \dfrac{dy \wedge dz}{f_x} = \dfrac{dz \wedge dx}{f_y}$ is a nonzero holomorphic 2-form on V_t, $t \neq 0$, and on $V - \{0\}$. Let $K_{\infty,t}$ be the part of the divisor of ω on $\bar{V}_t$ which is supported on $\bar{V}_t - V_t$ for t small. $K_{\infty,t} \cdot K_{\infty,t}$ is independent of t since the family $\{\bar{V}_t\}$ is differentiably trivial away from $(0,0,0) \in \mathbb{C}^3$. Let $K_\infty \cdot K_\infty$ denote this constant value for $K_{\infty,t} \cdot K_{\infty,t}$. Noether's formula says

$$\chi^0(\bar{M}) = \frac{1}{12}(K^2 + K_\infty^2 + \chi_T(\bar{M})) \tag{3.7}$$

$$\chi^0(\bar{V}_t) = \frac{1}{12}(K_\infty^2 + \chi_T(\bar{V}_t)) \qquad t \neq 0\,. \tag{3.8}$$

Hirzebruch index theorem says

$$\sigma_{\overline{V}_t} = \frac{1}{3}(K_\infty^2 - 2\chi_T(\overline{V}_t)) \qquad t \neq 0 \tag{3.9}$$

$$\sigma_{\overline{M}} = \frac{1}{3}(K^2 + K_\infty^2 - 2\chi_T(\overline{M})) . \tag{3.10}$$

(3.3) follows from (3.4), (3.9), (3.10), (3.7), (3.8) and (3.5). Q.E.D.

COROLLARY 3.12. *With the same hypothesis in Theorem 3.11,*

$$\sigma \leq -n+2 \dim H^1(M, \mathcal{O}) + \dim H^1(M, \Omega^1)$$

$$1+\mu \geq \chi_T(A) + 2 \dim H^1(M, \mathcal{O}) - \dim H^1(M, \Omega^1)$$

$$1+\mu+\sigma = \chi_T(A) - n + 4 \dim H^1(M, \mathcal{O}) .$$

Duality theorems for compact complex manifolds (such as Serre duality) are well known. Serre duality is still true for open manifolds but one has to use the cohomology with compact supports. It is a natural question for a duality theorem for 1-convex manifolds without using cohomology with compact support. In [42], we proved a first theorem in this direction. By using the same idea as in [42], we have the following refined result (cf. [47]).

THEOREM 3.13. *Let* M *be a 1-convex manifold* M *of dimension* n *which is a modification of a Stein space* V *at the isolated hypersurface singularities* $x_1, \cdots, x_m$. *Then*

(a) $h^{p,q} = h^{n-p,n-q}$ *for* $p+q \leq n-2$ $q \geq 1$ *and* $n \geq 3$
or $p+q \geq n+2$ $q \leq n-1$ *and* $n \geq 3$

(b) $h^{p,n-1-p} - h^{p,n-p} + h^{p,n-p+1} = -(h^{n-p,p+1} - h^{n-p,p} + h^{n-p,p-1})$
for $2 \leq p \leq n-2$ *and* $n \geq 4$

(c) $h^{1,n-2} - h^{1,n-1} + h^{n-1,1} - h^{n-1,2} = \tau_1 + \cdots + \tau_m - s^{(n-1)}$ *for* $n \geq 4$

where τ_i *is the number of moduli of* V *at* x_i, $h^{p,q} = \dim H^q(M, \Omega^p)$ *and* $s^{(n-1)} = \dim H^0(M-A, \Omega^{n-1})/H^0(M, \Omega^{n-1})$.

DEFINITION 3.6. Let p be a point in the complex analytic space V. Let m be the maximal ideal of $\mathcal{O}_{V,p}$. Then the Hilber function $H_{V,p}: \mathbb{N} \to \mathbb{N}$ is defined by

$$H_{V,p}(n) = \dim m^n/m^{n+1} .$$

Suppose M is a 1-convex manifold which is a proper modification of Stein analytic space V at the only isolated singularity p. The Hilber function for M is by definition the Hilber function $H_{V,p}$ of V at p.

QUESTION. Let M and M′ be 1-convex manifolds of dimension $n \geq 2$. Suppose then the maximal compact analytic sets of M and M′ are connected. If the Hilbert functions, Hironaka numbers, Griffiths numbers, Mumford numbers, Poincaré numbers and Siu numbers for M and M′ are the same, is it true that M and M′ are local biholomorphically equivalent?

DEPARTMENT OF MATHEMATICS
HARVARD UNIVERSITY
CAMBRIDGE, MASS. 02138

REFERENCES

[1] Andreotti, A., and Grauert, H., Théorémes de finitude pour la cohomologie des espaces complexes, Bull. Soc. Math. France 90 (1962), 193-259.

[2] Artin, M., Algebraic approximation of structures over complete local rings, Publ. Math. I.H.E.S. 36 (1969).

[3] Barth, W., Transplating cohomology classes in complex projective space, Amer. J. Math. 92 (1970), 951-967.

[4] Bennett, B., and Yau, Stephen S.-T., Milnor number and invariants of strongly pseudoconvex manifold. (Preprint)

[5] Bloom, T., and Herrera, M., DeRham cohomology of an analytic space, Invent. Math. 7 (1969), 275-296.

[6] Burns, D., Schneider, S., and Wells, R., Deformations of strongly pseudoconvex domains, Invent. Math. 46 (1978), 237-253.

[7] Durfee, A., The Signature of smoothings of complex surface singularities, Math. Ann., (1978), 85-98.

[8] Ferrari, A., Cohomology and holomorphic differential forms on complex analytic spaces, Annali della Scuola Norm. Sup. Pisa 24 (1970), 65-77.

[9] Grauert, H., and Kerner, H., Deformationen von Singularitäten komplex Räume, Math. Ann. 153 (1964), 236-260.

[10] Greene, R., and Wu, H., Whitney's imbedding theorem by solutions of elliptic equations and geometric consequences, Proc. Sympos. Pure Math., Vol. 27, Part II, Amer. Math. Soc. 1973, 287-296.

[11] Griffiths, Phillip A., Hermitian differential geometry, Chern classes and positive vector bundles, Global Analysis, Papers in Honor of Kodaira. Edited by D. C. Spencer and S. Iyanaga, Univ. of Tokyo Press, Princeton Univ. Press.

[12] ———, Hermitian differential geometry and the theory of positive and ample holomorphic vector bundles, J. Math. Mech. 14 (1965), 117-140.

[13] ———, Variation on a theorem of Abel, Invent. Math. 35 (1976), 321-390.

[14] Grothendieck, A., *Local cohomology*, Lecture Notes in Mathematics, Vol. 41, Berlin-Heidelberg-New York: Springer-Verlag, 1971.

[15] ———, Elements de Calcul infinitesimal Séminaire H. Cartan 1960-1961, Exp. no. 14, Secrétariat Mathematique, Paris, 1959.

[16] Harvey, R., The theory of hyperfunctions on totally real subsets of a complex manifold with application to extension problems, Amer. J. Math. *91* (1969), 853-873.

[17] Hironaka, H., Resolution of singularities of an algebraic variety over a field of characteristic 0, Ann. Math. 79 (1964), 109-326.

[18] Hirzebruch, F., and Mayer, K., O(n)-*Mannigfaltigkeiten, exotische Sparen und Singularitäten,* Lecture Notes in Math. 57, Berlin, Heidelberg, New York, Springer, 1968.

[19] Hirzebruch, F., Neumann, W., and Koh, S., *Differentiable manifolds and quaratic forms.* Lecture Notes in Pure and Applied Mathematics, Vol. 4, Marcel Dekker, New York, 1971.

[20] Kato, M., Riemann-Roch theorem for strongly pseudoconvex manifolds of dimension 2, Math. Ann. 222 (1976), 243-250.

[21] Kempf et al., *Toroidal Embeddings I.* Lecture Notes in Mathematics, Vol. 339, Berlin-Heidelberg-New York: Springer-Verlag, 1967.

[22] Kodaira, K., and Spencer, D., On deformations of complex analytic structures I, II, Ann. Math. *67* (1958), 328-466.

[23] Laufer, H., *Normal two dimensional singularities*, Annals of Mathematics Studies, No. 71, Princeton Univ. Press.

[24] ———, Ambient deformations for one dimensional exceptional sets, (preprint).

[25] ———, On μ for surface singularities, Proc. Sym. Pure Math. Vol. 30, Amer. Math. Soc. (1976), 45-49.

[26] ———, On rational singularities, Amer. J. Math. 94 (1972), 597-608.

[27] Le Dung Trang and Ramanujan, C. P., The invariance of Milnor's number implies the invariance of the topological type, Amer. J. Math. Vol. 98, No. 1 (1976), 67-78.

[28] Matsumura, H., *Commutative algebra*, W. A. Benjamin, Inc., New York, 1970.

[29] Milnor, J., *Singular points of complex hypersurfaces*, Ann. of Math. Studies No. 61, Princeton Univ. Press, Princeton, 1968.

[30] Murasugi, K., On a certain numerical invariant of link types, Trans. A.M.S., 117 (1965), 387-422.

[31] Narasimhan, R., The Levi problem for complex spaces II, Math. Ann. 146 (1962), 195-216.

[32] ———, On the homology groups of Stein spaces, Invent. Math. 2 (1967), 377-385.

[33] Palamodov, V. I., Sur la multiplicite des applications holomorphes, Founk-Anal., iievo prilojenia, tome 1, fasc. 3 (1967), 54-65 (en russe).

[34] Poincaré, H., Les fonctions analytiques de deux variables et la representation conforme. Rend. Circ. Mat. Palermo, 23 (1907), 185-220.

[35] Schneider, M., Lefschetz Theorems and a Vanishing Theorem of Grauert-Riemenschneider, Proc. Symp. Pure Math. Vol. 30, Pat 2, Amer. Math. Soc. 1977, 35-39.

[36] Serre, J.-P., Prolongement de Faisceaux Analytiques Coherents, Ann. Inst. Fourier Grenoble 16 (1966), Fasc. 1, 363-374.

[37] Shinohara, Y., On the signature of knots and links, Trans. Amer. Math. Soc. 156 (1971), 273-285.

[38] Siu, Y.-T., Analytic sheaf cohomology groups of dimension n of n-dimensional complex spaces, Trans. Amer. Math. Soc. *143* (1969), 77-94.

[39] Siu, Y.-T., and Trautmann, G., *Gap-Sheaves and the Extension of Coherent Analytic Subsheaves*, Lecture Notes in Math. Vol. 172, Berlin-Heidelberg-New York: Springer-Verlag, 1971.

[40] Siu, Y.-T., Analytic sheaves of local cohomology, Trans. A.M.S. Vol. 148 (1970), 347-366.

[41] Steenbrink, J., Intersection form for quasi-homogeneous singularities. Report 75-09, University of Amsterdam, 1975.

[42] Yau, Stephen S.-T., The signature of Milnor fibres and duality theorem for strongly pseudoconvex manifolds. Invent. Math. 46 (1978), 81-97.

[43] ———, Two theorems on higher dimensional singularities, Math. Ann. 231 (1977), 55-59.

[44] ———, Vanishing theorems for resolutions of higher dimensional singularities, Hironaka numbers and elliptic singularities in dimension > 2. (preprint)

[45] ———, Noether formula for strongly pseudoconvex manifolds and various numerical invariants for isolated singularities, (preprint).

[46] Yau, Stephen, S.-T., Deformations and equitopological deformations of strongly pseudoconvex manifolds, to appear in Nagoya Mathematical Journal (1981).

[47] ———, On Kohn-Rossi's $\bar{\partial}$b-cohomology groups and its applications to complex Plateau problem I, to appear in Annals of Math.

Library of Congress Cataloging in Publication Data
Main entry under title:

Recent developments in several complex variables.

(Annals of mathematics studies; 100)
Proceedings of a conference sponsored by the National Science Foundation and held at Princeton University, April 16-20, 1979.
1. Functions of several complex variables—Addresses, essays, lectures. I. Fornaess, John Erik.
II. National Science Foundation (U.S.) III. Series.
QA331.R38 1981 515.9′4 80-8548
ISBN 0-691-08285-5 AACR2
ISBN 0-691-08281-2 (pbk.)

www.ingramcontent.com/pod-product-compliance
Ingram Content Group UK Ltd.
Pitfield, Milton Keynes, MK11 3LW, UK
UKHW031028161224
452563UK00001BA/65

9 780691 082813